Modern Completion Technology for Oil and Gas Wells

Modern Completion Technology for Oil and Gas Wells

Ding Zhu

Kenji Furui

McGraw Hill Education

New York Chicago San Francisco
Athens London Madrid
Mexico City Milan New Delhi
Singapore Sydney Toronto

Library of Congress Control Number: 2018933148

Modern Completion Technology for Oil and Gas Wells

1 2 3 4 5 6 7 8 9 LPI 22 21 20 19 18

ISBN 978-1-259-64202-9
MHID 1-259-64202-X

Sponsoring Editor
Summers Scholl

Editing Supervisor
Donna M. Martone

Acquisitions Coordinator
Elizabeth Houde

Project Manager
Sonam Arora,
Cenveo® Publisher Services

Copy Editor
Surendra N. Shivam,
Cenveo Publisher Services

Proofreader
R. Kannan,
Data World Publishing Services

Production Supervisor
Lynn M. Messina

Composition
Cenveo Publisher Services

Art Director, Cover
Jeff Weeks

Contents

Foreword

The well completion is the critical interface that connects an oil or gas reservoir with the wellbore that provides the flow path for hydrocarbons to reach the surface. The completion can greatly enhance or restrict the productivity of the well; it can aid or hinder the well operator's ability to control reservoir flow conditions; it can facilitate or prevent the applications of advanced diagnostic measurements in wells. The well completion is also a major part of the cost of an oil or gas well. In unconventional reservoirs completed with multistage fracturing, the completion comprises about half of the total well costs, while in deepwater developments, the completion alone can cost tens of millions of dollars. For these reservoirs, engineers need the best available information about well completion practice, well completion performance, and well completion design. This knowledge is provided comprehensively in this book, covering modern well completion practices.

The book begins with a thorough treatment of well completion hardware, laying the foundation for the theories and models of how this hardware affects well behavior to follow. In subsequent chapters, the book emphasizes two key aspects of well completions—the role of the well completion in well stimulation, both hydraulic fracturing and acidizing, and intelligent well completion practices. No other book has such complete coverage of these critical aspects of well completions.

This is a book that every completion engineer should have. It will be a valuable reference material for any petroleum engineer or geoscientist who needs to better understand, and most importantly, to quantify, the effect of the well completion on well performance. It will serve as a complete textbook for undergraduate or graduate level university courses and for industry training courses on well completions.

I congratulate the authors on the timeliness, thoroughness, and clarity of this new book on well completions. I expect it to be the standard guide to well completion engineering for years to come.

Dan Hill
Texas A&M University
January 2018

Acknowledgments

During the course of developing this book, the authors had tremendous support and help from our colleagues, this includes our mentors, our peers, and our students.

We would like to thank Dr A. D. Hill, Noble Chair Professor at Texas A&M University for serving as the technical editor of the book. His encouragement, technical guidance, and invaluable critiques paved the foundation for us to write this book. Special thanks to Dr Junjing Zhang at ConocoPhillips for in-depth discussion based on field operations regarding hydraulic fracturing, to Brian Wu at Texas A&M University for editing and providing comments on improving the book to better serve as a textbook, and to Meredith Born for managing and coordinating the book development. We would like to express our appreciation to Andrew Wu for his contribution to the graphic preparations for the book.

Last, we also appreciate the great effort provided by the staff at McGraw-Hill Education and Cenveo Publisher Services in editing and handling the manuscript.

CHAPTER 1

Introduction

Well completion is a vital component for any well drilled to produce oil and gas in petroleum engineering. The functions of well completions are to provide mechanical integrity of the borehole drilled, establish the communication between the target reservoirs and the wellbore, regulate flow from the reservoir along the well, and build the paths for injection during well stimulation and enhancing oil and gas recovery. As the petroleum industry moves toward developing more unconventional reserves, today we are producing oil and gas from reservoirs that have lower transmissibility (lower permeability, higher viscosity, and thinner pay thickness) and higher heterogeneity (large contact area, multiple payzones, and targets). At the same time, it is always a goal for petroleum engineers to maximize well productivity while keeping the minimum cost per barrel in any field developments. Completion technology has advanced dramatically with many new techniques being developed in last of couple decades. These new technologies have changed the way of completing conventional wells, and are a key part of the unconventional developments being so successful. This book aims at reviewing recent advances in well completion technologies for both conventional and unconventional wells.

1.1 Brief Introduction to Petroleum Engineering

Drilling, reservoir, and production are three major components of petroleum engineering. To develop oil and gas fields, we commonly start from building a field development plan based on the geological investigation by seismic surveys during exploration. The information obtained while drilling (measurement while drilling, MWD, and/or logging while drilling, LWD), openhole logging, and testing from exploration and appraisal wells before development wells are constructed for production. Reservoir engineers then study characteristics of target formations and reservoir fluids. Well types and well spacing are selected from the reservoir study based on the characteristics of the target formations, and expected reservoir and individual well performance is predicted by reservoir simulation.

Drilling and completion engineers construct the wells according to a well plan based on the reservoir study. Once the wells are completed, production engineers take over the wells to monitor and evaluate production and efficiency of the wells. It is common for wells to perform differently than expected due to the complexity of the formation. Further testing and evaluation will be conducted to identify production problems, and if necessary, well workover and stimulation can be performed to improve well performance. Well completion is considered the bridge between drilling and production, playing a crucial role in oil and gas production. In the recent years, completion is thought to be as important as drilling, reservoir and production in petroleum engineering, and becomes the fourth individual discipline.

1.2 Roles of Well Completion and Its Progression

In the course of oil and gas production, completion plays important roles. The basic roles of completion are to provide borehole mechanical support and fluid flow conduit from a reservoir to the surface; these are minimum requirements for well completion during the early stages of oil and gas production. The well completion technology started from openhole (the simplest completion) and progressed to cementing cased hole with perforating as an alternative completion method in early days. The needs for well completion eventually exceeded solely providing borehole stability and flow path as production conditions change over time. Many wells produce unwanted fluids (i.e., gas and water) and solids (i.e., formation sands and fines), making water and sand production a leading problem in the petroleum industry. Water production control and sand control in unconsolidated formations are two of the most common production problems that must be addressed by well completion designs. Slotted liners, screens, and gravel packs were developed to solve these problems. In addition, sliding sleeves and isolation packers were invented to selectively produce reservoir fluids from multiple reservoir zones. The well completion process involves running casing, cementing, perforating, running tubing, and installing a production tree. When completing the well in an unconsolidated formation, it may require gravel-packing the well.

As reservoir pressure is depleted, water injection is often employed to maintain the reservoir pressure and well production. Enhanced oil recovery (EOR) methods require injecting steam or surfactant to improve flow performance in the reservoir, thus enhancing the recovery of hydrocarbons. The completion of injection wells controls flow conformance of the sweep front of the injected fluids recovering as much as possible oil with high efficiency.

Hydraulic fracturing, matrix acidizing, and acid fracturing are well stimulation methods to improve well productivity. Well completions for stimulation help place fractures, divert chemicals, and grow wormholes in the formation for better stimulation efficiency.

Oil and gas production had a long history in the United States since the first oil well was discovered in the 1860s. As the industry started developing challenging reservoirs such as low-permeability formations, thin pay layers, heavy oil reservoirs, and deep-water fields, well completion technology has advanced dramatically overcoming the new difficulties. Horizontal and multilateral well technologies were introduced to the industry to increase the contact area between a wellbore and the formation in order to maximize well productivity. On the other hand, the increased contact area results in unforeseen production problems associated with reservoir heterogeneity. Permeability distribution on large reservoir contact and commingled multiple target zones often require further evaluation of flow distribution and selective flow control for production optimization. Intelligent well system (IWS) was developed as multilaterals and extended-reach drilling (ERD), and commingled production in multiple reservoirs became a popular method for high-performance wells in Middle East, North Sea, and many other regions around the world. The IWS starts from downhole permanent monitoring with electric and fiber optic sensors that provide continuous measurements in time and space; this technology allows us to evaluate flow conditions miles away from the surface. Meanwhile, downhole flow control was developed to shut off and/or choke back zonal production from these high-performance wells. The equipment used in intelligent wells has advanced and been accommodated in many completion types including sand control completions and wells with downhole artificial lift equipment.

Unconventional shale development has opened a new chapter in the oil and gas industry. Multistage fracturing stimulation in a long horizontal well is the only method capable of economically producing these nano-Darcy reservoirs. Recent advances in the completion technology such as sliding sleeves, isolation plugs, perforating, and zonal isolation equipment enable multistage hydraulic fracturing to achieve its goals of efficiently stimulating shale formations for economical production performance. Today the completion technology allows the engineers to place hundreds of fractures along a horizontal well with a spacing between the fractures on an order of 20 ft. Downhole equipment for microseismic monitoring and fiber optic sensing are all a part of the well completion design to improve the success rate of fracturing stimulation.

Conventional well completions considered in this book refer to the technologies that are developed from openhole or cased/cemented/perforated completions for borehole stability and flow conduit.

In contrast, modern well completion involves the technologies beyond these basic tasks. The wells are designed to meet the challenges in the reservoirs, and the completion is the interface between the reservoirs and the wells. In today's environment of oil and gas production, we cannot emphasize enough how important is the role that completions are playing. On the other hand, if completions are designed in an incorrect manner, they can jeopardize production detrimentally. To design, install, and operate the well with modern completion, it requires thorough understandings of inflow performance and well stimulation in conjunction with complex well architecture.

1.3 Objectives of the Book

The book presents recent advances in well completions, explains the various mechanisms and functions of the modern completion technologies, helps to select the appropriate completion types, and provides guidelines for correct completion designs. The book will also discuss the new technologies and how to correctly apply them in the field. Because the completion technologies have been advancing so fast, some of the new technologies have not been discussed in any textbooks, and it would not be surprised that new technologies are developed during the time the book is published. This book covers the fundamentals of completion design, while introducing recently developed technologies. The knowledge in completion fundamentals will help the readers to understand any future completion development.

1.4 Organization of the Book

The book has eight chapters and will start from the introduction of well completion technology in Chapter 1, which summarizes the history and position of well completion as a part of the petroleum engineering. Chapter 2 carries on the discussion of type of well completions and their functions. The chapter presents the hardware structures and applications of each type of completions. We begin with conventional types of completions followed by the modern completion technologies for unconventional development.

The function of completions and their impact to well productivity and injectivity are modeled through the concept of skin factor. Completions add restriction, or additional pressure drop to the well system, and skin factor can be the representation of this additional pressure drop. In Chapter 3, we discuss the skin factor calculations for near-well formation damage, and for each completion type. We also discuss the interference between formation damage skin and completion skin for most completion types. The completion and formation damage skin factors are implemented to the well performance models to count for their impact to well flow.

Chapters 4 and 5 discuss well completion for stimulation. Chapter 4 presents completion for hydraulic fracturing, and Chapter 5 is about completion for acidizing. In both stimulation techniques, advances in completion methods are dramatic in recent years. This includes multistage hydraulic fracturing of horizontal wells in unconventional developments, and acid placement techniques in matrix acidizing for horizontal wells. These two chapters illustrate the importance of completion and stimulation in the petroleum industry.

One of the new technologies merged into the petroleum industry in recent years is intelligent well completions. An intelligent well is a well equipped with downhole monitoring sensors and downhole flow control valves. The intelligent wells allow the engineers to "see" better of what happens underground, thousands feet away from the surface. Intelligent well technology also allows us to control the downhole flow without intervention. This revolution changes the way the wells were operated before. Today, we are still in the development stage of this technology. Chapter 6 of this book explains the functions of downhole monitoring components of the intelligent well, and Chapter 7 discusses the flow control component of the technology. Examples are presented in these chapters to show how the technology should be used and their functions in oil and gas production, especially for unconventional reservoirs and conventional reservoirs with complex geological structures.

In Chapter 8, we use the examples of field applications to illustrate the benefits of modern completion technology to produce oil and gas in a more effective, efficient, and cost-effective way. The examples cover a wide range of applications, from high-permeability multilateral well with intelligent completions for production optimization in the Middle East, to offshore horizontal well matrix acidizing in the North Sea, to multistage fracture stimulation in unconventional Bakken Shale wells. These examples show how important completion design is in modern production, operation, and reservoir management. This closes the purpose of this book—to understand the functions of available modern completion technologies and utilize them correctly to meet the challenges of today's oil and gas production.

1.5 List of Reference Books as Supplement Materials

This book focuses on well completion technologies in petroleum engineering. Well completion is only one part of the well systems that is constructed to produce oil and gas. To fully master the completion technology, an engineer needs to have the fundamental knowledge about other disciplines in petroleum engineering. This includes petroleum geology, reservoir engineering, drilling engineering, production engineering, and well stimulation. In addition to the foundation, there are several books that we encourage the readers to use as companions for this book.

Petroleum Production Systems (Economides et al., 2012). Among the other disciplines, production engineering is the most-closely related subject to well completion. The goal of completions design is always to provide the means to best produce oil and gas. Many theories and equations used in this book are explained in *Petroleum Production Systems* in a more detailed manner. The examples are inflow equations for different reservoirs under different boundary conditions, the detailed hydraulic fracture designs, the detailed theory of matrix acidizing, both in sandstone and carbonate formations, and sand production control. This book should be used after *Petroleum Production Systems*.

Recent Advances in Hydraulic Fracturing (Gidley et al., 1989). Chapter 4 of this book discusses completion for fracturing. Almost all fracture-related theories for conventional reservoirs are presented in depth in *Recent Advances in Hydraulic Fracturing*. Many of these equations are directly used in this book without derivation. To design completions for efficient fracture stimulation, understanding fracture theory is a must. The readers should refer to *Recent Advances in Hydraulic Fracturing* when studying Chapter 4 of this book.

Essentials of Hydraulic fracturing: Vertical and Horizontal Wellbores (Veatch et al., 2016). Updated from *Recent Advances in Hydraulic Fracturing*, this book added modern fracture technology, especially for horizontal wells. Examples in the book for each subject also bring a great value for the readers to be able to apply the technology.

Hydraulic Fracturing (Smith and Montgomery, 2016). There are many books on in the subject of hydraulic fracturing. *Hydraulic Fracturing* presents fracturing theory, from fundamental rock mechanics to fracturing deign and execution, in a broad vision. The detailed calculation examples in each section explain how to use the theories presented. With the understanding of the objectives of fracture stimulation, one can set the goals of completion design for more efficient fracturing.

Production Enhancement with Acid Stimulation (Kalfayan, 2008) and *Acid Stimulation* (Ali et al., 2016). Acidizing is another commonly used well stimulation method to improve well performance. Acidizing is, especially in carbonate formations, a complicated statistic procedure, and completion for acid stimulation enables the successes of the stimulation. These two books present acidizing theories and field operation in detail in all aspects, and therefore build a foundation to understand how to design a completion for acid stimulation.

References

Ali, S. A., Falfayan, L., and Montgomery, C. (2016). *Acid Stimulation*. SPE Monograph Series Vol. 26, *Society of Petroleum Engineers*, Richardson ISBN: 978-1-61399-426-9.

Economides, M. J., Hill, A. D., Ehlig-Economides, C., and Zhu D. (2012). *Petroleum Production Systems*, 2nd ed., Prentice Hall, New Jersey. ISBN: 0-13-703158-0.

Gidley, J. L., Holditch, S. A., Nierode, D. E., and Veatch Jr., R. W. (1989). *Recent Advances in Hydraulic Fracturing*. SPE Monograph Series Vol. 12, *Society of Petroleum Engineers*, Richardson. ISBN: 978-1-55563-021-1.

Kalfayan, L. (2008). *Production Enhancement with Acid Stimulation*, 2nd ed., PennWell Books, Tulsa. ISBN: 1-59-370139-X, 978-159370-139-0.

Smith, B. M., and Montgomery, C. (2015). *Hydraulic Fracturing*, CRC Press, Boca Raton. Print ISBN: 978-1-4665-6685-9, eBook ISBN: 978-1-4665-6692-7.

Veatch, R. W., King, G. E., and Holditch, S. A. (2017). *Essentials of Hydraulic Fracturing: Vertical and Horizontal Wellbores*. PennWell Books, Tulsa. ISBN-13: 978-1-59370-357-8, ISBN-10: 1-59-370357-0.

CHAPTER 2

Fundamentals of Well Completion

From conventional completions such as an openhole or a cased/cemented/perforated wellbore to modern completion techniques such as an expandable screen completion and intelligent wells, this chapter presents the hardware of completions, their functions, and how they work.

2.1 Functions of the Well Completion

The objective of drilling wells is to create a conduit from the surface to the target formations for oil and gas production. The conduit, called the borehole, may not be competent enough to hold the wellbore against the earth's stresses. Thus, certain specialized hardware must be placed in the borehole to provide mechanical support and regulate the fluid flow. This equipment is collectively referred to as well completion.

Completions are essentially the interface between the hydrocarbon bearing formations and the surface production system. Completions must be able to confine reservoir fluids, preventing uncontrolled release of formation fluids to the environment throughout the life cycle of a well. Oil and gas production, or well productivity, are strongly affected by the choices of well completions. At some stage, wells may start producing formation solids, gas, and water even though these fluids are unwanted. When this happens, the well may need to be able to filter such solids and prevent or minimize water and/or gas production to the surface. The role of the completion engineers is to take over a well that has been drilled and convert it into a safe and efficient production or injection system. As we explore into more hostile areas such as deepwater or the arctic, functional requirements associated with well completions become more challenging.

Basic functions of well completions can be divided into five categories: (1) Well integrity and fluids containment; (2) well/completion flow efficiency; (3) borehole and casing stability; (4) sand/solids control; and (5) selective/zonal flow control. The following sections review each functionality of well completion.

2.1.1 Well Integrity and Fluids Containment

The most important issue in the oil and gas industry is safe operation. Keeping wells under control with sufficient barriers is crucial in well completion design. Completions are a part of the well barrier elements and remain so throughout the life of a well. Specifications for equipment used in completions are important for the overall well integrity since the equipment is mostly exposed to the well fluids over the life of the well. According to Norsok D-010 (NORSOK standard D010, 2004), well integrity is defined as "application of technical, operational and organizational solutions to reduce risk of uncontrolled release of formation fluids throughout the life cycle of a well." Inappropriate design of well completion equipment can lead to serious safety and environmental problems, which must be avoided. Key design parameters considered in the well completion include well location, lifetime of the well, fluid composition, depth, pressure, and temperature. All these parameters influence the well integrity and make the life span of a well shorter or longer. Needless to say, the working life of wells can vary from field to field and with the purpose of the well due to different operational environments. A good job in pre-engineering and design of wells will prevent well integrity problems. During the production phase, the fluid flows through a well with different temperatures, pressures, and viscosity than what was used for initial well construction. All of these factors must be taken into account when designing well completions.

2.1.2 Well/Completion Flow Efficiency

When completing wells, well deliverability is one of the most important design criteria in any field development. The role of completion engineers is to maximize the well productivity in a cost-effective manner. The overall well flow performance consists of two parts: (1) reservoir inflow performance and (2) wellbore flow performance; well completion plays an important role in both parts. The wellbore flow performance accounts for various pressure losses occurring across completion equipment including tubing, casing/liner, and flow restrictions installed on the wellbore.

The reservoir inflow performance of a well refers to as the ability of passing fluids from the reservoir to the interior of the well. The productivity index, J, is often used to measure the reservoir inflow performance. For steady-state condition and single-phase liquid wells, the productivity index is defined by (Economides et al., 2012)

$$J = \frac{q}{p_e - p_{wf}} = \frac{kh}{141.2B\mu\left(\ln\frac{r_e}{r_w} + s\right)} \tag{2.1}$$

where p_e and p_{wf} are reservoir pressure and wellbore flowing pressure, both in psi, q is flow rate in STB/d, B is formation volume factor in res bbl/STB, μ is fluid viscosity in cp, k is reservoir permeability in md, h is payzone thickness in ft, and r_e and r_w are reservoir drainage radius and wellbore radius, both in ft. The parameter in the denominator of Eq. (2.1), s, is skin factor originally introduced by Van Everdingen (1953), measuring the condition changes caused by well construction, completion, and operation in the near-well regions (s is dimensionless). The other inflow equations for different boundary conditions and different fluid types are listed in Table 2.1 and the detailed discussion is presented by Economides et al. (2012). The well skin factor, s, can be positive or negative. Because well productivity decreases with positive skin factor and increases with negative skin factor, it is important to investigate the reasons that cause skin, and to reduce skin as much as possible by optimizing the completion design.

Oil and gas wells are completed in a variety of ways, with the most common types of completions being openhole completions; slotted or perforated liner completions; cased and perforated completion; and gravel-pack completions (Fig. 2.1). Any well completion using hardware in the wellbore that alters the flow path from strictly radial flow in the near well vicinity, or through the completion itself, can lead to additional pressure drops that may have a significant impact on well performance. On the other hand, a cased and perforated completion may exhibit higher productivity than an openhole completion with the same wellbore radius. Different types of well completions should be judged using different control parameters in the reservoir inflow performance. The following sections briefly discuss the key control parameters suggested for evaluation of common completion types.

Openhole Completions and Openhole with Slotted/Perforated Liners

Pressure losses associated with this type of completion are quite simple. In this type of completion, flow from a reservoir moves to a non-horizontal (vertical or deviated), fully completed producing interval. Ideally, flow through the reservoir is radial with undisturbed parallel streamlines. As flow approaches to the wellbore, pressure losses may be increased due to reduced permeability in the mud-filtrate invaded region. If slotted or perforated liners are installed inside the borehole, flow may accelerate into the slots or perforations. Occasionally slotted or perforated liners are used in the formations where formation strengths are not sufficient. Thus when formation rock collapses, failed rock particles may plug the slots and/or perforations, causing severe pressure drops during production. Mud filtrate invasion and slot plugging are the two leading factors for increased pressure drop in openhole and openhole liner completions (Furui et al., 2005). Figure 2.2 illustrates the flow phenomenon in these completions.

	Transient Flow (Infinite acting radial flow, no boundary)	Steady-State Flow (Constant pressure at the boundary)	Pseudo-Steady-State Flow (No flow boundary)
Oil	$q = \dfrac{kh(p_i - p_{wf})}{162.6B\mu\left(\log(t) + \log\dfrac{k}{\phi\mu c_t r_w^2} - 3.23 + 0.87s\right)}$	$q = \dfrac{kh(p_e - p_{wf})}{141.2B\mu\left(\ln\dfrac{r_e}{r_w} + s\right)}$	$q = \dfrac{kh(\bar{p} - p_{wf})}{141.2B\mu\left(\ln\dfrac{0.472r_e}{r_w} + s\right)}$
Gas	$q = \dfrac{kh(p_i^2 - p_{wf}^2)}{1638ZT\mu\left(\log(t) + \log\dfrac{k}{\phi\mu c_t r_w^2} - 3.23 + Dq + 0.87s\right)}$ $q = \dfrac{kh[m(p_i) - m(p_{wf})]}{1638T\left(\log(t) + \log\dfrac{k}{\phi\mu c_t r_w^2} - 3.23 + Dq + 0.87s\right)}$	$q = \dfrac{kh(p_e^2 - p_{wf}^2)}{1424ZT\mu\left(\ln\dfrac{r_e}{r_w} + Dq + s\right)}$ $q = \dfrac{kh[m(p_i) - m(p_{wf})]}{1424T\left(\ln\dfrac{r_e}{r_w} + Dq + s\right)}$	$q = \dfrac{kh(\bar{p}^2 - p_{wf}^2)}{1424ZT\mu\left(\ln\dfrac{0.472r_e}{r_w} + Dq + s\right)}$ $q = \dfrac{kh[m(p_i) - m(p_{wf})]}{1424T\left(\ln\dfrac{0.472r_e}{r_w} + Dq + s\right)}$
2-ϕ			$\dfrac{q_o}{q_{o,\max}} = 1 - 0.2\dfrac{p_{wf}}{\bar{p}} - 0.8\left(\dfrac{p_{wf}}{\bar{p}}\right)^2$ $q_{o,\max} = \dfrac{1}{1.8}\dfrac{k_o h\bar{p}}{141.2B_o\mu_o\left(\ln\dfrac{0.472r_e}{r_w} + s\right)}$
	Horizontal Well		
Oil		$q = \dfrac{kL(p_e - p_{wf})}{141.2\mu B_o\left(\ln\left[\dfrac{hI_{ani}}{r_w(I_{ani}+1)}\right] + \dfrac{\pi y_b}{hI_{ani}} - 1.224 + s\right)}$	$q = \dfrac{\sqrt{k_y k_z}\,b(\bar{p} - p_{wf})}{141.2B_o\mu\left[\ln\left(\dfrac{A^{0.5}}{r_w}\right) + \ln C_H - 0.75 + s_R + s\right]}$

Table 2.1 Inflow Performance Relationship (IPR)

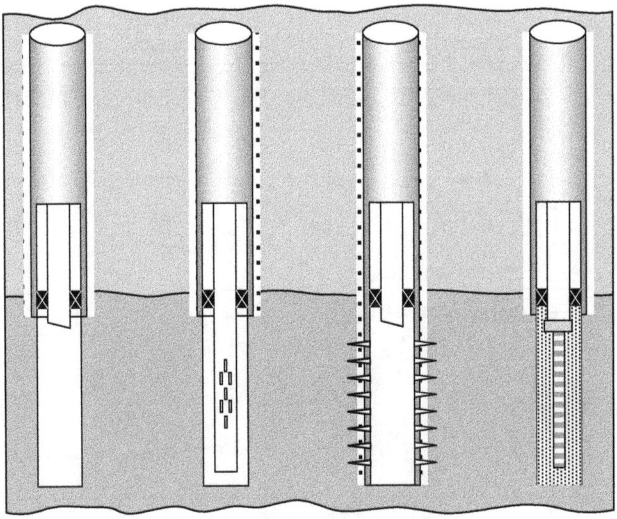

Figure 2.1 Basic option for well completions.

Perforated Cased Hole Completions

In a perforated cased hole completion, flow from a homogeneous reservoir moves to a completed producing interval. Flow through the reservoir is radial with undisturbed parallel streamlines. As flow approaches the wellbore, it deviates from the radial flow pattern in

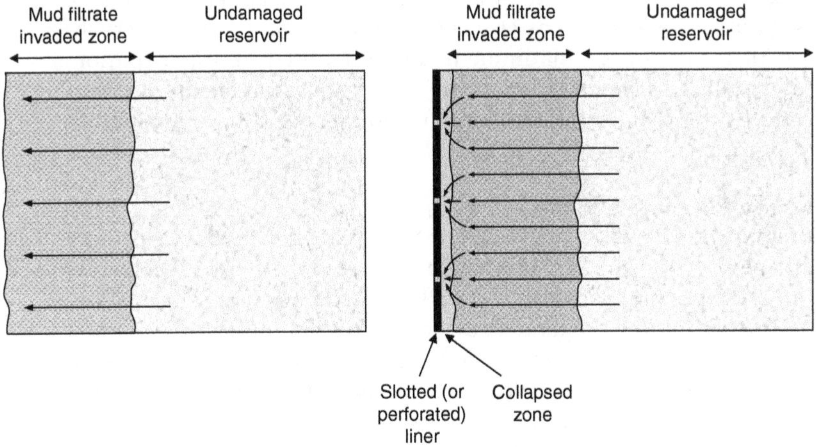

Figure 2.2 Flow into a barefoot (*a*) and openhole liner (*b*) completions.

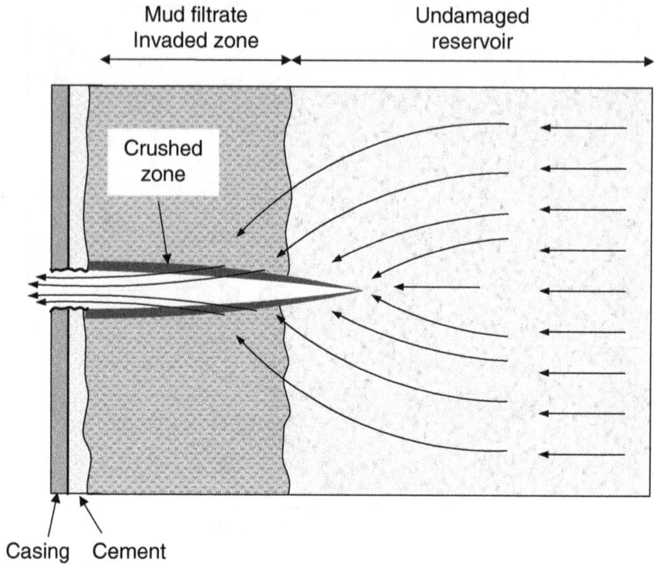

FIGURE 2.3 Flow into a perforated cased hole completion.

the reservoir and divides into separate flow streams that converge to a certain perforation (Fig. 2.3). Pressure losses associated with this flow convergence are a function of permeability impairment in the mud filtrate invaded zone surrounding the well and of perforation effects including both perforation geometry and damage effects (McLeod, 1983). Perforation geometry is comprised of perforation length, perforation diameter, perforation phasing, and shot density. Damage may occur as a result of the perforating process and well conditions at the time of perforating. This type of damage is typically referred to as crushed zone damage due to the crushing of formation grains immediately adjacent to the perforation. The control parameters are permeability impairment due to mud filtrate invasion, perforation length, and crushed zone permeability impairment.

Cased Hole Gravel Packs

Pressure losses associated with this type of completion are quite complex. Flow through the far-field reservoir can be idealized as radial flow with undisturbed parallel streamlines (Fig. 2.4). As flow approaches to the wellbore it deviates from the undisturbed radial flow pattern in the reservoir and converges to gravel-filled perforations. From the perforation, flow accelerates into the small area provided by the gravel-filled perforation tunnel and passes through the cement sheath and casing to the interior of the well. Once inside the casing, the flow spreads out into the larger area provided by the

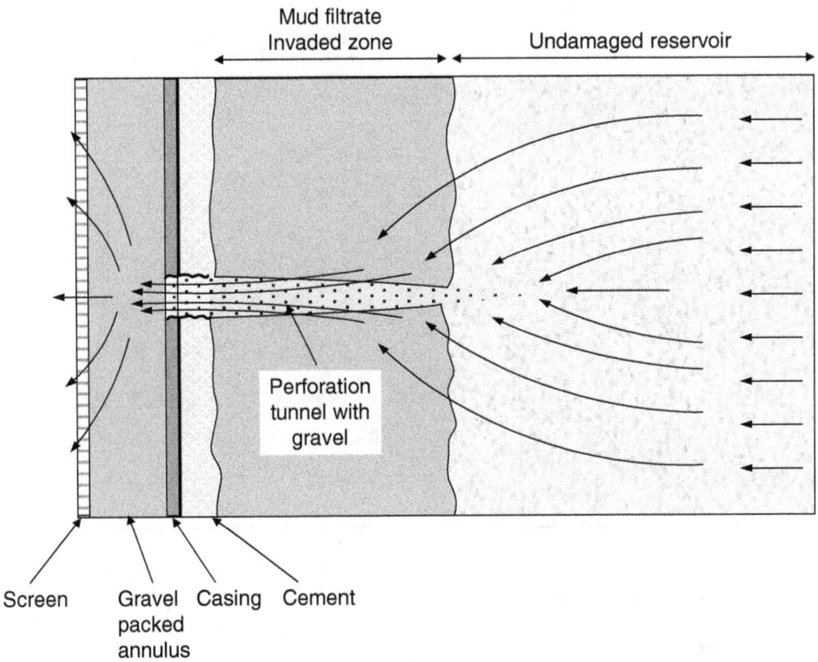

FIGURE 2.4 Flow into a perforated, gravel packed, cased hole completion.

casing/screen annulus, and moves across the gravel-filled annulus and through the screen to be produced to surface.

Typical skin effects for a cased hole gravel-packed completion are permeability impairment due to mud filtrate invasion and perforation geometry effects. Because of the low formation strengths typically encountered in gravel packed wells, the crushed perforation zone generally collapses onto the perforation, and flows into the wellbore. This can be removed during the gravel packing process.

Furthermore, mud filtrate invasion and perforation geometry skins are generally low and far outweighed by the much higher skin effects which occur as a result of high flow velocities occurring in the perforation tunnels (Burton, 1999). For this reason, the dominant control parameters for cased hole gravel-packed completions are perforation tunnel permeability and inflow area. It should be noted that the flow area provided by the perforation tunnels in a well-designed cased hole with gravel pack is low, typically less than 3% of the equivalent openhole area. This is true even for high shot density perforating using big-hole charges. The need for high inflow area becomes even more necessary when non-Darcy flow effects become significant in high-rate oil and gas wells.

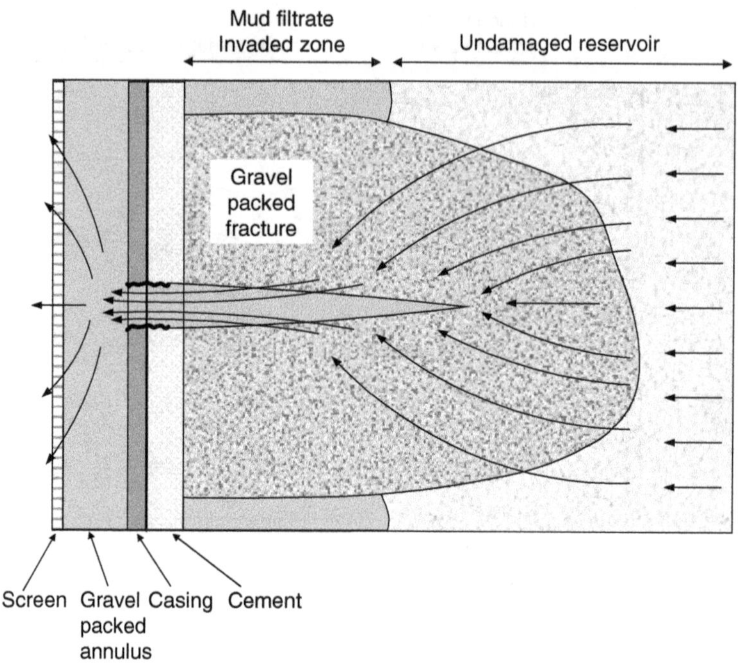

FIGURE 2.5 Flow into a frac-pack completion.

Cased Hole Frac-Packs

Frac-packing is a completion technique combining a short, wide fracture with a cased hole gravel pack (Fig. 2.5). The fracture is intended to improve cased hole gravel packed completion performance by providing greater inflow potential in the near-well region. For most high-permeability reservoirs the inflow improvement due to fracturing may not be the dominating advantage. Frac-packing has, however, been shown to provide a significant greater gravel placement than is routinely achieved in conventional cased hole gravel packs. In designing frac-pack completions, precautions must be taken to first achieve a solid gravel pack.

The primary control parameters are perforation tunnel permeability and inflow area (Burton, 1999). Procyk et al. (2009) and Nozaki et al. (2017) presented that perforation tunnel permeability estimated from flow tests is much smaller than gravel pack permeability, which causes high skin factors observed from high-rate frac-pack wells. Flow velocity is also a consideration as lower velocity improves sand control efficiency. In most high-permeability reservoirs ($k > 200$ mD), the productivity improvement by fracturing is moderate; for lower-permeability reservoirs fracturing offers stimulation benefit and improves overall well inflow performance. The impact of fracturing will be discussed in Chapter 4. In addition to fracturing, placing gravel, bypassing damage, and controlling sand production are major advantages of gravel-pack completion.

Openhole Gravel Packs and Openhole Completions with Stand-Alone Screens

Formation damage and hole/screen annular permeability skin effects tend to dominate overall skin factor in openhole gravel packs and in openhole completions with stand-alone screens (Fig. 2.6). The primary control parameters for these types of completions are mud cake removal and hole screen annular permeability (Burton and Hodge, 1998). The mud cake should be designed to be acid-soluble and should be removed by acidizing during completion operations. This will eliminate mud cake skin effects leaving only hole/screen annular flow skin as a potential flow restriction.

Matrix Acidized Wells

In sandstone reservoirs, matrix acidizing will improve performance by removing the damage in the near-wellbore region, but will not overcome the completion skin effects. On the other hand, in carbonate

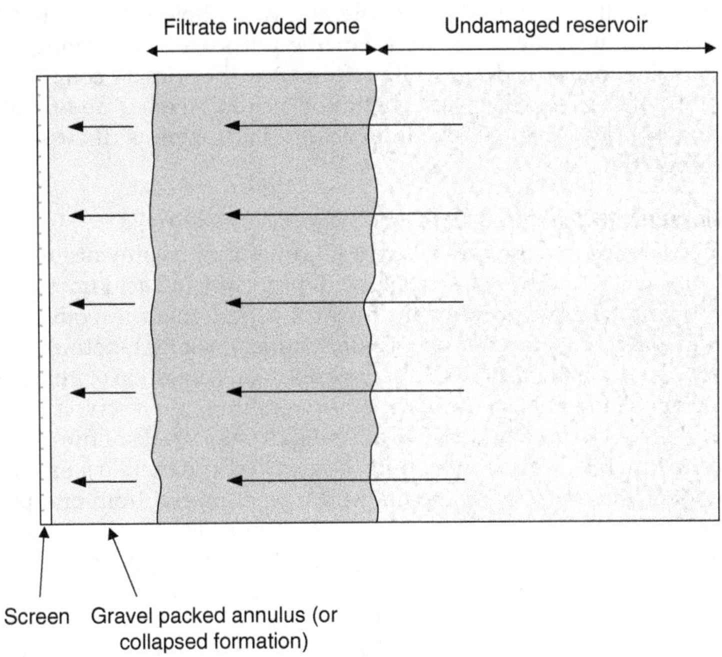

FIGURE 2.6 Flow into an openhole gravel packs and openhole completions with stand-alone screens.

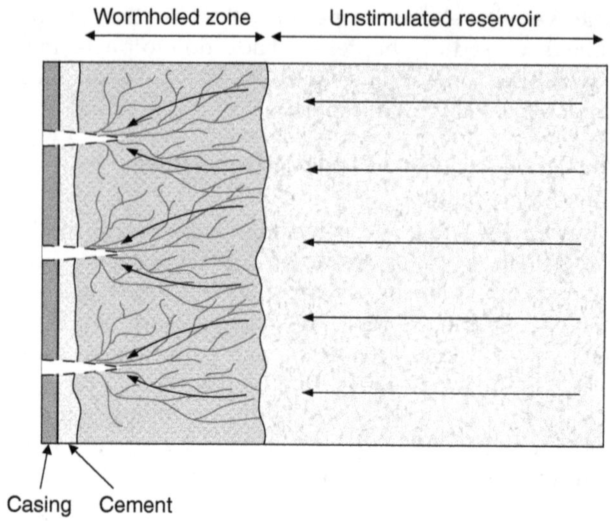

Figure 2.7 Flow into a matrix acidized well in carbonate formation.

reservoirs (Fig. 2.7), acid penetration and permeability improvement will reduce the completion skin effects as a result of wormhole formation (Ali et al., 2016). Therefore, wormhole skin effects tend to dominate overall skin in both openhole and cased hole matrix acidized wells for carbonate formations. For the purposes of designing and analyzing matrix acidized well performance, the primary completion control parameters are acid penetration, acid placement, and permeability improvement. The skin effect after acidizing is discussed in Chapter 5 in detail.

Hydraulically Fractured Wells

There are two major types of hydraulic fracturing treatments used in the industry. These are fracturing with proppant and acid fracturing.

Hydraulic fracturing with proppant is performed in both sandstone and carbonate reservoirs. These treatments use high-permeability granular proppant to hold the fracture open after cessation of pumping operations to establish conductive passages from the reservoir to the well (Fig. 2.8). The key variables affecting fractured well inflow performance are fracture geometry and conductivity (Prats, 1961). Fracture conductivity under ideal conditions can be estimated from proppant loading and proppant permeability data. These ideal fracture conductivity values must be modified for the effects of proppant crushing, proppant embedment, and gel damage inside the fracture pack. This is usually done by running laboratory tests or estimating conductivity loss with a commercially available program. Propped fracture half-lengths can be estimated from design programs. However, the best results are attained when design program results are correlated to the

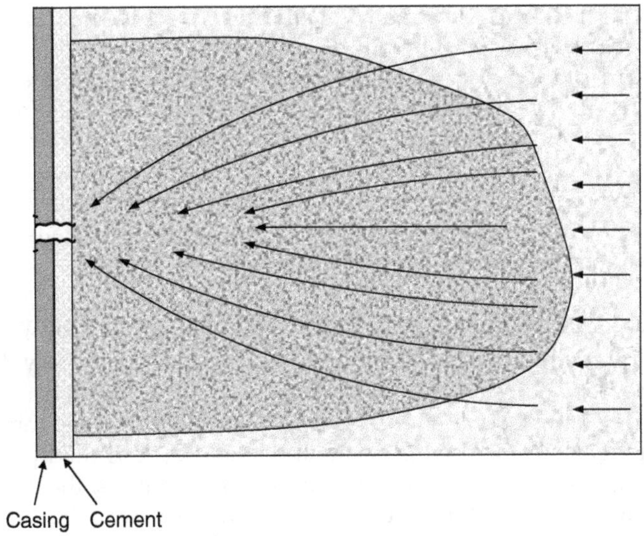

FIGURE 2.8 Flow into a hydraulically fractured well.

results of fracture half-length generated by pressure build-up tests from the field.

In carbonate formations, the reservoir rock can be dissolved with hydrochloric acid (HCl). As a result, these formations can also be acid-fractured. In acid fracturing, HCl acid is pumped along a hydraulic fracture to etch the formation face. After pumping operations are completed, the fracture will close up because of the closure stress applied to the fracture faces. Acid reaction/etching with formation rocks is a statistical process related to the permeability and mineralogy distribution of the formation rock. As a result, the etched fracture surfaces become uneven. Wormholes created during acid pumping, either into the formation by leakoff or along the fracture faces, promote the roughness of the fracture faces. These surface features are the source of conductivity after closure for acid fractures. Key control parameters are fracture half-length and fracture conductivity for this type of completion. Fracture conductivity under ideal conditions can be estimated from rock removal calculations or laboratory test data. These ideal fracture conductivity values must be adjusted for the effects of non-uniform etching and closure stress (Nierode and Kruk, 1973). Acid fracture half-lengths can be estimated from design programs. Similar to propped fracturing, the designed fracture half-length should be correlated to the test results using pressure build-up measured fracture half-lengths. Without proper calibration, fracture design programs overestimate fracture half-lengths.

Acid fracturing is extremely ineffective in sandstone formations because of the low reaction rate. It is not recommended.

Comparative skin factor analysis helps engineers select and design a well completion for particular applications. When forecasting completion performance for well design and/or economic studies using these skin equations, it is essential that the applicable control parameters be correctly identified and that realistic values be used to effectively bound the range of likely outcomes. The details of skin calculations will be discussed in Chapter 3.

2.1.3 Well (Borehole and Casing) Mechanical Stability

Underground formations are always in a stressed state, mostly due to overburden and tectonic stresses. When a well is drilled into a formation, rock material is removed. The borehole wall is then supported by its strength and the fluid pressure in the hole. As this fluid pressure generally does not match the in-situ formation stresses, the borehole deforms. If the rock strength of the formation is sufficient, the well may be completed as an openhole completion. On the other hand, if the rock strength is not sufficient to support near-wellbore stress levels, this may lead to rock failure creating borehole breakouts. The severity of these effects and subsequent hole failure depend on the magnitude of the formation stresses, the stress differences in different orientations, and the mechanical properties of the formation. In a formation where borehole breakouts are expected to be severe, openhole screen or liner completions can be used. In this type of completion, casing is set above the producing zone, and an uncemented screen and liner assembly are installed across the pay section. If the well requires selective stimulation or control of excessive gas or water production, cased, cemented, and perforated completions may have to be used.

Another significant well problem associated with a compacting reservoir is casing damage. Casing damage refers to as a casing being deformed or bent (by creation of dog-legs). Casing failure causes operational problems for workovers and recompletions. This necessitates costly drilling of a side-track well. The main mechanisms responsible for casing damage are compressional, shear, tensile, and bending deformations (Veeken et al., 1994). The mechanisms are illustrated in Fig. 2.9. The risk of casing buckling is the highest in unsupported intervals, for instance, near sand producing perforations where the pressure drawdown is also most severe, wherever the cement gives poor support, or at lower completion. Casing failure can be a result of well stimulation. In high porosity chalk formation, excess volume in acid stimulation can cause axial compression failure in both reservoir and casing/liner (Furui et al., 2010). Deformation of the casing cross-section due to non-uniform lateral loading is referred to as crushing or radial compression. This is mainly a problem in reservoir zones with large depletion, and where the formation is extremely soft. Under these

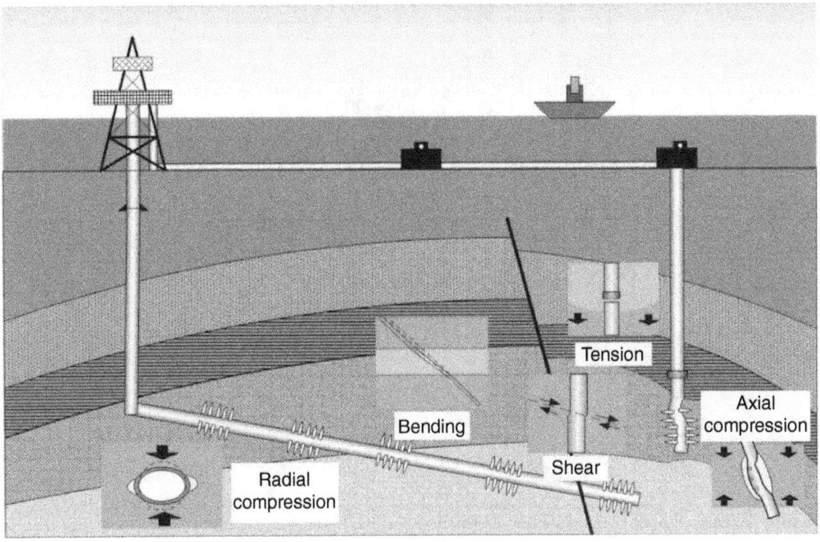

FIGURE 2.9 Casing failure scenarios.

conditions, proper casing grade and thickness (or proper diameter-to-thickness ratio) should be used to prevent premature casing collapse.

2.1.4 Sand (Solids) Control

Sand production is a leading operational problem in many oil and gas reservoirs worldwide. When hydrocarbons are produced from a reservoir, solid particles can follow the reservoir fluid into the well. This unintended byproduct of hydrocarbon production is called solids production. It is generally accepted that sand production cannot occur from intact rock, even if it is poorly consolidated, since a flowing fluid does not normally provide sufficient force to pull sand grains out of an intact rock. Sand production may thus occur only if the rock in the vicinity of the producing cavity is unconsolidated, or has been failed. The loose sand grains created at the cavity face are mobilized at certain pressure-drawdown levels, fluid velocities, and fluid viscosities. Once produced into the wellbore, these particles can create operational troubles downstream. Sand production may erode the tubing or surface valves and flowlines. In addition, sand could accumulate in downhole equipment, preventing the equipment working appropriately and creating problems in wireline servicing. Figure 2.10 illustrates a typical sand production event observed in a producing well.

There are many ways to avoid or minimize sand production. Sand control methods can be divided into three techniques: (a) Sand management techniques, (b) formation failure prevention techniques, and (c) sand exclusion techniques.

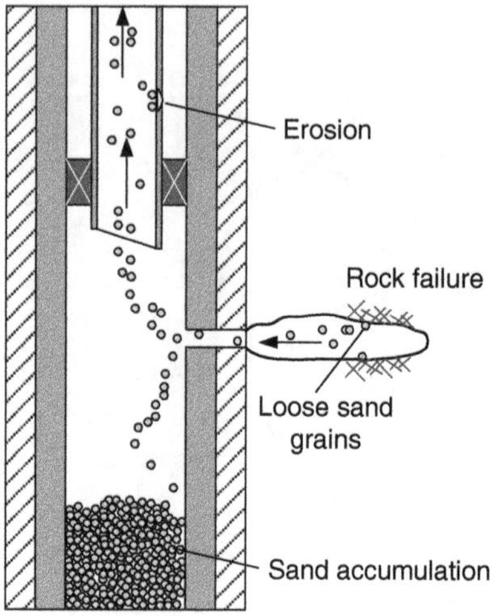

FIGURE 2.10 Sand production mechanism.

Sand management techniques allow wells to produce acceptable levels of sand and then handle solids with surface facilities. A safe and effective plan requires erosion control and satisfactory sand disposal. Wells that exceed acceptable levels of sand production are choked back or shut-in. This strategy is most effective in liquid-producing wells because the lower fluid velocity results in less erosion to the well systems when compared with gas wells.

Formation failure prevention techniques prevent formation failure through application of rock mechanics principles. Typical failure prevention techniques include perforating only high strength zones (i.e., selective perforating), fracturing from high strength zones, stress-oriented perforating (i.e., oriented perforating) (Tronvoll et al., 1993), increasing reservoir pressure, and increasing near-well rock strength with resin. This strategy is more effective in reservoirs with well-defined stress or strength contrasts.

Sand exclusion techniques prevent larger formation sand grains from flowing through the completion and damaging downhole and/or surface equipment. Typical sand exclusion techniques include open and cased hole gravel packs and frac-packs, stand-alone screens in openhole, and surface sand exclusion devices. It is important to remember that smaller grains should be allowed to flow to surface to preventing severe plugging of screens and gravels. The sand exclusion technique is capital-intensive but very reliable.

2.1.5 Selective Zonal Flow Control

A wide range of downhole equipment has been designed and manufactured to meet the needs of more complex well completions.

In situations where multiple reservoirs cannot be commingled, the zones are separated with isolation packers. Packers are devices that are run on, or in conjunction with, a string of tubing. The packer has a rubber element that can be extruded by compression to form a seal between the tubing and the casing. Packers are used for a variety of reasons in well completions.

Another component that has become an integral part of well completions is the sliding sleeve. In case that a reservoir consists of multiple zones being waterflooded, a sliding sleeve can be used to selectively shut-in the water producing zone while keeping the other zones open to produce hydrocarbon. The sleeve is opened or closed through the use of wireline servicing methods. Figure 2.11 shows an example of isolation packers and sliding sleeves for production control in a multi-zone water flooding reservoir. These devices are only a few examples of an extensive range of equipment that enables engineers to control fluid flow selectively and to stimulate producing reservoirs. Chapters 5 and 6 will discuss more advanced flow control and monitoring devices, referred to as intelligent well completions, a recently developed technology for real-time zonal flow control and monitoring purposes.

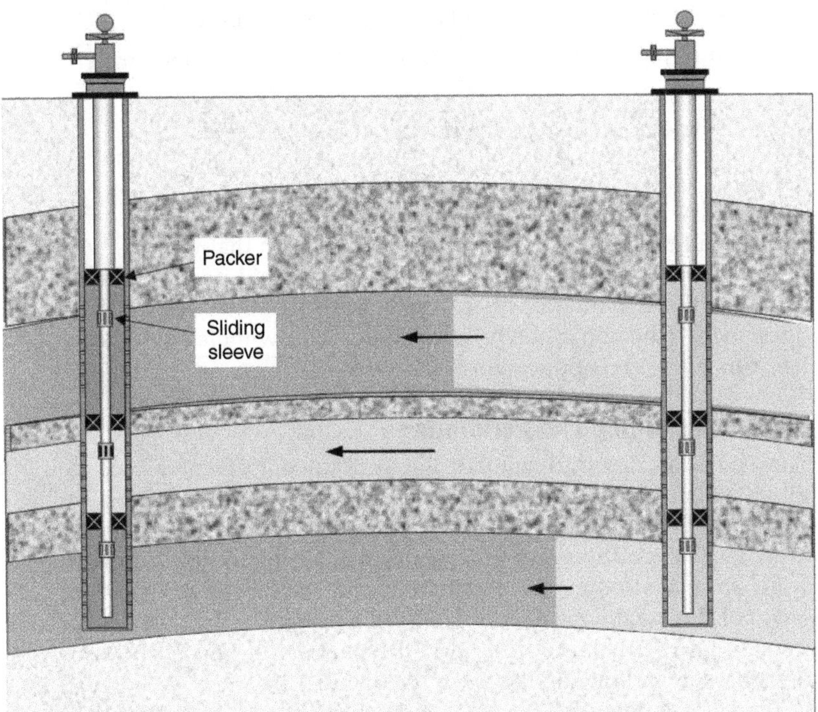

FIGURE 2.11 Selective flow control example in a water-flooded reservoir.

2.2 Completion Components

During the early time of petroleum industry development, oil and gas wells were commonly completed with only a single string of cemented casing. As wells are drilled deeper, multiple reservoirs with higher pressure and temperature were encountered. It was recognized that such completions imposed limitations on well servicing and control, and that downhole designs would need to be upgraded to meet increasing needs for safety, zonal isolation, selectivity, and flow and solids control.

Completions are often divided into lower and upper completions. The lower completions are the connection between the reservoir and the well, including packers, sliding sleeves, and flow control devices. The upper completions are the conduit from the lower completion to surface facilities including subsurface safety valves and tubing hangers. Here, we focus mainly on the lower completion equipment. Major design decisions associated with the lower completion are:

- Well trajectory and inclination
- Open hole versus cased hole
- Wellbore integrity
- Sand control requirement and type of sand control
- Stimulation (fracturing, proppant, or acid; and matrix acidizing)
- Single or multi-zone (commingled or selective)

Today, conventional and unconventional oil and gas wells are completed with a variety of downhole devices. The devices specified depend on the characteristics of the formation that the well will produce from/inject to the producing/injecting fluid properties, and other functional requirements of the completion. This section provides an introduction to the principal types of downhole equipment, their functions, and applications.

2.2.1 Cementing and Perforating

Many wells are cased, cemented, and perforated. Cement is used to hold the casing in place and to prevent fluid migration between subsurface formations. Cementing is the process of mixing a slurry of cement, cement additives, and water, and pumping it down through casing to critical points in the annulus around the casing or in the open hole below the casing string. The two principal functions of the cementing process are to restrict fluid movement between different payzones and to bond and support the casing.

Zonal isolation must be performed effectively to allow production or stimulation operations to be conducted. The success of a well depends on this primary operation. In addition to isolating oil-, gas-, and water-producing zones, cement also aids in (1) protecting the casing from corrosion, (2) preventing blowouts by quickly forming a seal, (3) protecting the casing from shock loads in deeper drilling, and (4) sealing off zones of lost circulation or thief zones.

Perforating is a process used to establish a flow path between the near reservoir and the wellbore after cementing. It normally involves initiating a hole from the wellbore through the casing and any cement sheath into the producing zone.

The perforating string contains a cable head, a correlation device, a positioning device, and the perforation guns. The cable head connects the string to a wireline, and meanwhile provides a weak point at which to disconnect the cable if problems arise. The correlation device is used to identify the exact position with a previously run correlation log, and frequently locates casing collars. The positioning device orients the shots on the perforation guns toward the casing for more optimum perforation geometry. The perforating guns are loaded with shape charges. Each shape charge consists of a case, explosive material, and a liner, as shown in Fig. 2.12. An electric current initiates an explosive wave. Typical perforations have diameters between 0.25 and 0.4 in and lengths between 6 and 12 in. Larger perforation holes can be created with higher power perforating guns.

Where cased hole gravel packs and frac packs are considered for as completion methods, perforation diameter becomes more critical. Altering the explosive charge design and materials creates a larger diameter entrance hole on the perforation while reducing the depth of penetration. For cased hole gravel packs and frac packs employed, particularly in high permeability formations, big hole (1.0-in diameter range) charges at high shot-density (12 to 18 SPF) are highly recommended to minimize gravel pack skin factor.

2.2.2 Slotted/Pre-Perforated Liners

Slotted liners are made from tubulars by saw-cutting slot configurations, as shown in Fig. 2.13. Slot widths are often referred to as in terms of gauge. Slot or screen gauge is simply the width of the opening in inches multiplied by 1000. For instance, a 12-gauge screen has openings of 0.012 in. The machining consists of cutting rectangular openings with small rotary saws. Typical slot widths are about 0.030 in but can be adjusted for specific uses.

Figure 2.13 displays different slot patterns for slotted liner design. The single-slot staggered, longitudinal pattern is generally preferred because the strength of the un-slotted pipe is preserved. The staggered pattern also gives a more uniform distribution of slots over the surface area of the pipe. The single-slot staggered pattern is slotted with an even number of rows around the pipe with a typical 6-in longitudinal spacing of slot rows. Pre-perforated liners are also

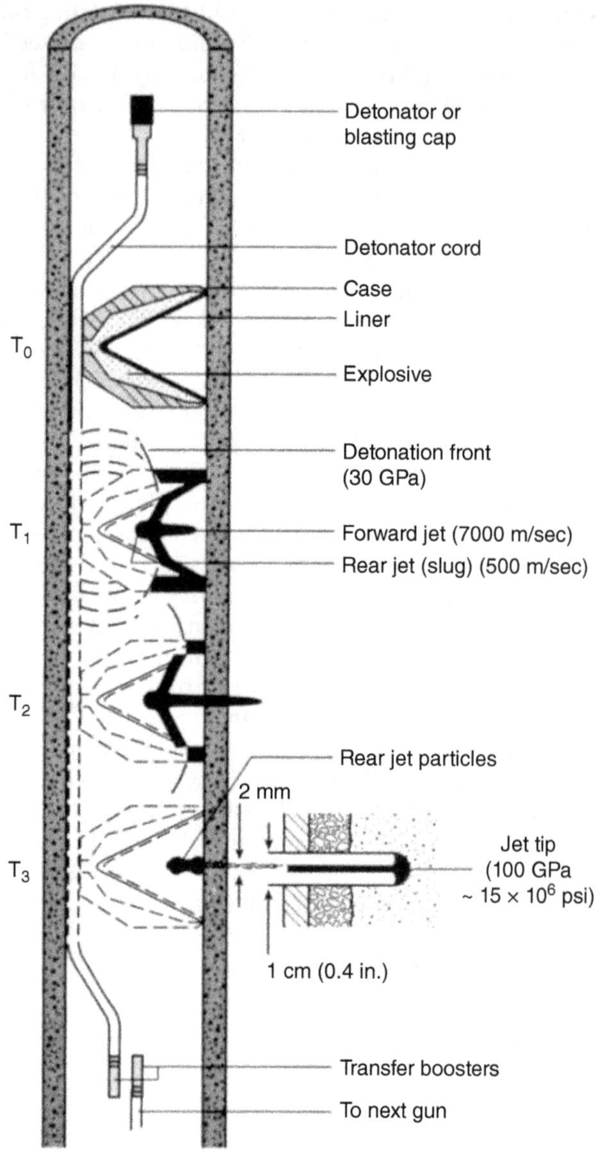

FIGURE 2.12 Detonation process of a shaped charge (Economides et al., 2012). *(Reprinted with permission from Pearson Education, Inc., NY.)*

commonly used for openhole completions. The hole size and phasing pattern can be easily adjustable and the risk of hole plugging is thought to be lower than slotted liners due to larger opening areas.

The slots can be straight or keystone-shaped, as illustrated in Fig. 2.14. The keystone slot is narrower on the outside surface of the

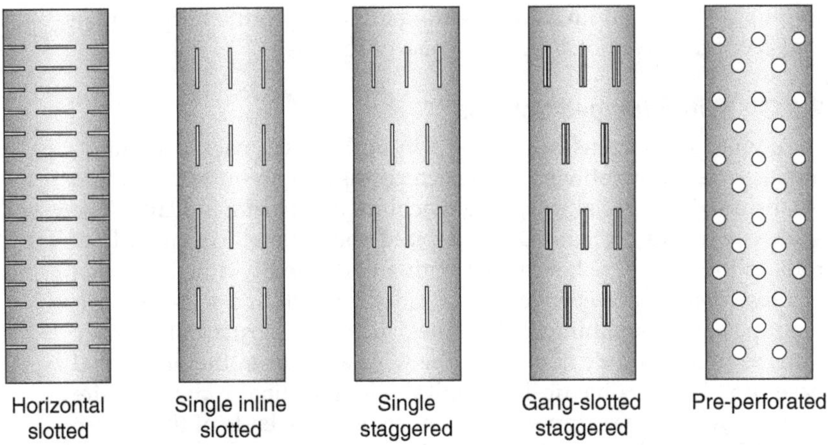

FIGURE 2.13 Various liner geometries.

pipe than on the inside. Slots formed in this way have an inverted V cross-sectional area and are less prone to plugging because any particle passing through the slot at the outside opening of the pipe will continue to flow through, rather than lodging within the slot. While the slotted liners usually cost less compared with screens, they have smaller inflow areas and experience higher pressure drops during production (i.e., higher skins).

The purposes of slotted liner completion are:

- Stop gross hole collapse.
- Allow zonal isolation packers to be deployed within the reservoir completion for upfront or later isolation.
- Allow the deployment of intervention tool strings such as production logs (PLTs). However, given that much of the flow is behind the pipe, interpreting such logs is normally difficult in high-angle wells unless proper isolation is placed behind liners.

Slotted liners are not normally a method for sand control as it is hard to make the slots small enough to stop sand. Where slots are manufactured with such a small aperture (cut with a laser), the flow area through the liner is too small such that they become susceptible

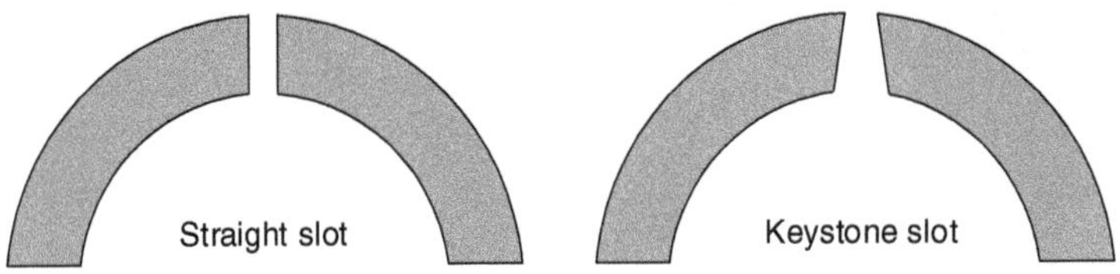

FIGURE 2.14 Straight and keystone-shaped slots.

to plugging. Slotted liners are used where well productivity is small and economics cannot support the use of screens.

2.2.3 Wire-Wrapped Screens

Wire-wrapped screens can be used in both gravel pack and standalone screen completions. The screen consists of an outer jacket that is fabricated on special wrapping machines that resemble a lathe. The shaped wire is simultaneously wrapped and welded to longitudinal rods to form a single helical slot with any desired width. The jacket is subsequently placed over and welded at each end to a supporting pipe base (containing drilled holes) to provide structural support. The longitudinal rods help offset the wire wrap from the pre-drilled base pipe holes. Some designs omit the longitudinal rods. The wire is either welded or gripped by a connector at the ends of the screen. Depending on the metallurgy, welding the screen to the base pipe can cause complication when compared with no-welding designs because of quality of welding and therefore mechanical integrity of completion. The problem can be avoided if operate correctly. This is a standard-commodity design manufactured by several companies. A schematic of the screen construction is shown in Fig. 2.15. The keystone (wedge) shape of the wire ensures that particles bridge off against the wire or pass right through and are produced.

Wire-wrapped screens have substantially more inflow area than a slotted liner but still have a relatively low inflow area when compared with premium screens (this is discussed in Section 2.2.5). The inflow area of screens varies from about 6% to 12%, depending on the slot opening. Screens with the smallest slot openings are typically

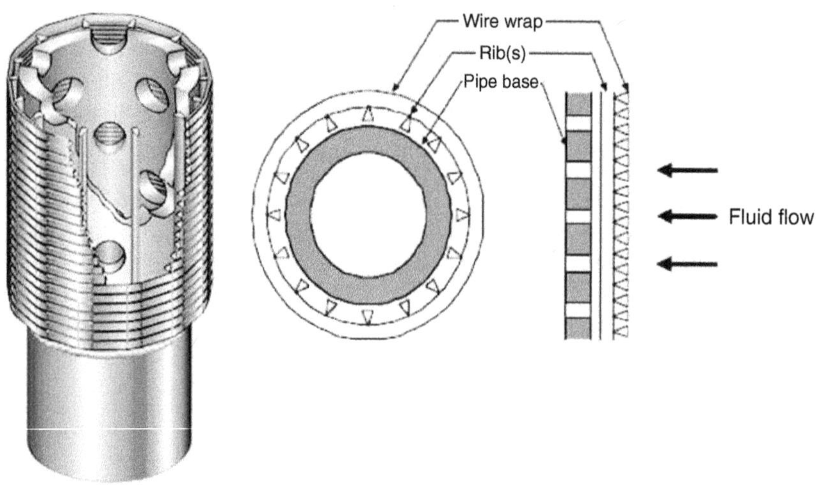

Figure 2.15 Wire-wrapped screens (*http://www.sand-screen.com*).

6 gauge (0.006 in). For large gravels, for example, 10 to 20 mesh, the screen slot openings are about 18 gauge (0.018 in).

The base pipe is normally the same metallurgy as the tubing (e.g., 13Cr). Base pipe failures are rare, but collapse failures have been reported when the screen is plugged.

2.2.4 Pre-Packed Screens

Pre-packed screens are constructed in a similar manner to wire-wrapped screens, but with two screens. The pre-packed screens consist of a standard screen assembly with a layer of resin-coated gravel (consolidated) placed around it in an annular ring supported by a second screen (dual-screen prepack) or outer shroud (single-screen prepack). The resin coating is a partially cured phenolic plastic. Being dry, the resin-coated gravel can be handled like ordinary gravel. After prepacking the screen, the complete unit is heated to cure and harden the resin. The thickness of the gravel layer can be varied to meet special needs. The screens with the lowest profiles are those that contain an annular pack between the jacket and the base pipe. This screen has a thin lattice screen wrapped around it to prevent gravel from flowing through the holes in the base pipe before consolidation. It is tempting to consider pre-packed screens as a pre-built gravel pack. However, the fundamental advantage of gravel packs is that they remove the annulus between the screen and formation, and thus prevent formation failure and sand transport, which a pre-packed screen cannot accomplish.

Examples of pre-packed screens are shown in Fig. 2.16. Pre-packed screens have been used with gravel packs instead of standard wire-wrapped screens, and also in stand-alone applications in

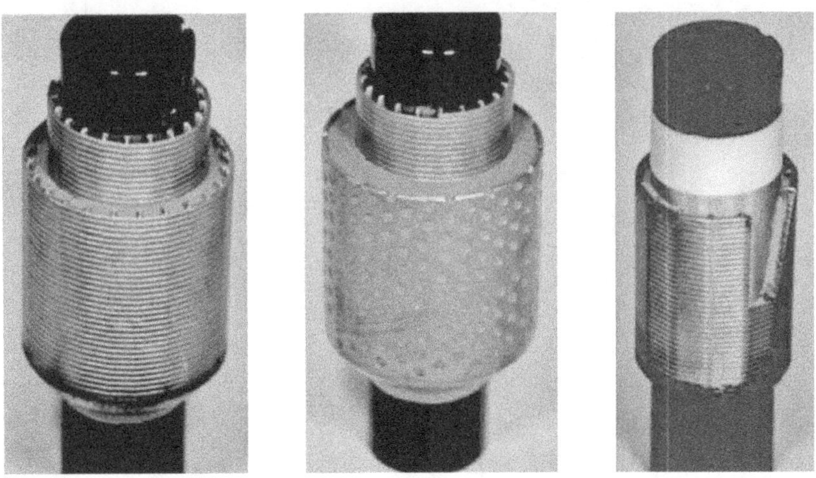

Dual-screen prepack | Single-screen prepack | Low-profile prepack

FIGURE 2.16 Types of pre-packed screens (*PetroWiki*).

horizontal wells. While the pre-packed screens have been used in stand-alone service, pre-packed screens can be prone to plugging, consequently restricting production. Also, pre-packed screens have no advantage at resisting jetting of sand when compared with wire-wrapped screens. The inflow area of these screens is typically about 4% to 6% of the surface area, which is significantly smaller than premium screens (see the following section) and wire-wrapped screens.

2.2.5 Premium Screens

In most critical sand production cases, either openhole or cased hole, premium sand-control screens are used as completion methods to minimize the chance of sand-production-related failures. These types of screens are generally constructed with a metal-mesh sand retention media that offers superior sand control characteristics over traditional wire-wrapped screens or pre-packed screens. These premium screens are typically run in long-interval openhole stand-alone screen and openhole gravel-pack completions to minimize the change of mud plugging and subsequent sand-control failure. These screens are also used in long-interval cased-hole gravel-pack and frac-pack completions to minimize the risk of completion sand-control failure in multisand reservoirs with a high potential for voids in the annular pack. In both openhole and cased-hole completions, the sand-control screen provides the final design barrier in the sand-control completion; if it fails, the completion fails.

Typical premium screen components are shown in Fig. 2.17 (Adam et al., 2009). The base pipe provides a continuous inner structure on which to attach the various drainage and sand-control layers.

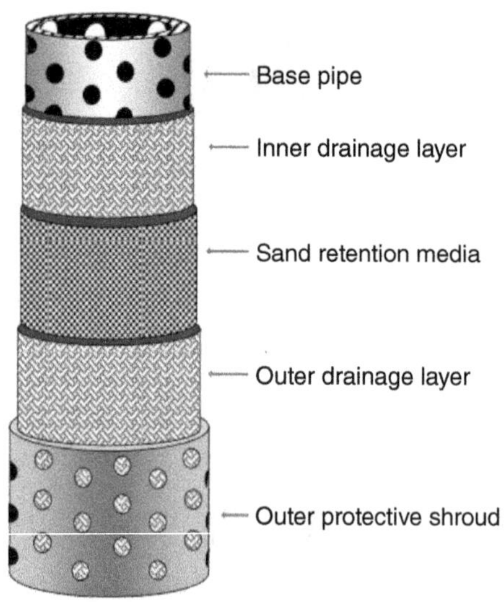

FIGURE 2.17 Typical premium sand-screen components (Adam et al., 2009).

The base pipe has holes drilled on it to provide desired flow paths from the various drainage and sand-control layers to the interior of the well. The inner-drainage layer consists of a high-flow-capacity, wire-wrapped screen or wire-mesh layer that distributes flow from the sand retention layer to the holes in the base pipe. This inner-drainage layer also provides support to the sand-retention layer from collapse by the loads imposed by fluid flow and/or the formation stress. The sand-retention layer typically consists of a thin woven-wire or distributed metal-fiber mesh sized to prevent formation sand grains from flowing through the screen. Some screen designs have an outer-drainage layer that is placed outside of the sand-retention layer to distribute flow from the shroud to the sand-retention layer. When used, this drainage layer typically consists of a woven-wire mesh. The woven design has non-uniform apertures. The outer protective shroud generally consists of a thin casing with a high-density of small holes or slots on it. This outer shroud provides protection to the thin sand-retention media during run-in. It also provides support to the sand-retention layer in the event of burst loads being applied by internal pressure during installation, or in the event of screen plugging. Premium screens typically have an inflow area of around 30%.

2.2.6 Gravel-Pack Completion Equipment

Figure 2.18 illustrates typical gravel packs for cased and openhole completions. These crossover gravel packing equipment are considered state-of-the-art in the industry today. Washdown and reverse circulation methods are other alternatives that are less expensive and are to be used when costs do not support crossover equipment.

The first step in installing a gravel-pack completion is to establish a base on which the screen will rest. In cased-hole completions, the most common type base is a sump packer. The sump packer is normally run into the well on an electric wireline before perforating and is set at a specified distance (5 to 10 ft) below the lowest planned perforation. The distance below the perforations must accommodate the length of the seal assembly and production screen overlap.

Although sump packers are the preferred gravel-pack base, other options such as a bridge plug or cement plug can be used. In openhole completions, provisions for a debris sump or logging access can be achieved, but these are not routine and may not be feasible in other situations. Therefore, the gravel-pack base is normally a bull plug on the bottom of the screen. The types of common gravel-pack bases are illustrated in Fig. 2.19.

The seal assembly is required to establish a seal in the bore of the sump packer to prevent gravel-pack sand from filling the bottom of the well during gravel packing. In the case of multiple gravel packs, the seal also provides zonal isolation.

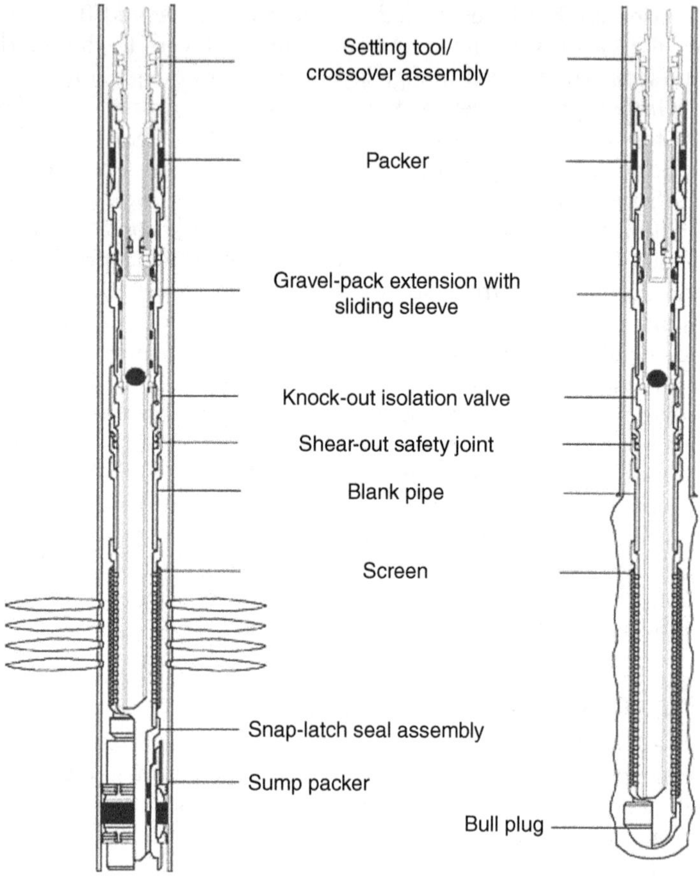

FIGURE 2.18 Typical gravel-pack completion equipment in cased and openhole completions (*PetroWiki*).

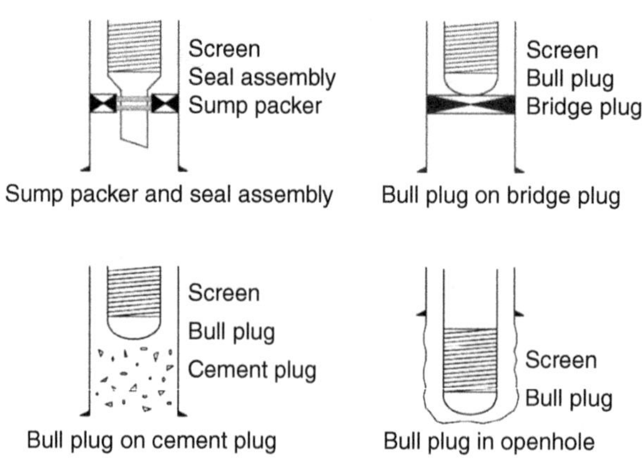

FIGURE 2.19 Types of gravel-pack bases (*PetroWiki*).

The purpose of blank pipe is to provide a reservoir of gravel-pack sand above the screen to ensure that the screen remains completely packed in the event of pack settling. During gravel-pack operations, it is possible for minor voids in the annulus pack to occur. In fact, gravel packing with viscous gel transport fluids commonly produces voids, particularly opposite to the short lengths of blank pipe between screen joints. Depending on deviation angle, pack settling shortly after gravel placement may fill the voids. It is important to have sufficient reserve of gravel-pack sand available for this process to occur to avoid uncovering the top of the screen.

A long-standing guideline for gravel reserve has been to maintain a minimum of 30 ft of packed gravel in the blank pipe above the top of the screen when packing with brine. When viscous fluids are used, blank lengths may be as much as twice the screen length for short completion intervals. This allows for additional settling with these fluids when the gel breaks.

A shear-out safety joint is located just above the blank pipe. It consists of a top and bottom subconnected by shear screws. This device is incorporated in most gravel-pack completion assemblies to allow retrieval of the gravel-pack packer and the gravel-pack extension independently of the blank pipe and screen for emergence situations.

A knock-out isolation valve is a mechanical device that prevents completion fluid loss and subsequent damage to the formation after install in a gravel-pack completion. The downward closing flapper in the valve is held open by the gravel-pack service tools (normally the washpipe) during the gravel pack. When the service tools are removed from the valve, the flapper closes, preventing fluid loss to the formation. The gravel-pack service tools can be removed from the well and the completion tubing run. When the well is producing, the flapper will open. Alternatively, a flapper made of breakable material can be used. The flapper is broken either hydraulically or mechanically before producing the well.

Gravel-pack extensions are used with the gravel-pack packer, which is at the top of the gravel-pack assembly to provide a flow path from the tubing above the packer and to the screen/casing annulus below the packer via perforated pup or sliding sleeve.

2.2.7 Gravel Pack Service Tools

Gravel-pack service tools are the equipment necessary to perform the gravel pack. They are removed from the well after gravel packing. In most cases, the type of gravel-pack equipment used dictates the service tools required for a gravel pack.

A hydraulic setting tool is a hydraulic piston that generates the force required to set gravel-pack packers. It is attached to the top of a crossover tool and has a sleeve shouldered against the setting sleeve of the packer. A setting ball is dropped to the ball seat in the crossover tool to plug off the ID of the work string. Applied pressure to the work string acts on the piston in the hydraulic setting tool to force the sleeve down to compress the slips and packing element of the packer. Special versions of setting tools are available, which allow for rotation and high-circulating rates while running the gravel-pack assembly.

Gravel pack crossover tools create the various circulating paths for fluid flow during gravel packing. The crossover tool consists of a series of molded seals surrounding a gravel-pack port midway down the tool and a return port near the top of the tool. A concentric tube (washpipe) design in the crossover tool along with the gravel-pack packer and gravel-pack extension allow the fluid pumped down the work string above the packer to "cross over" to the screen/casing annulus below the packer. Similarly, return fluids flowing up the washpipe and below the packer can "cross over" to the work string/casing annulus above the packer. Gravel-pack crossover tools typically have squeeze, circulating, and reverse circulating positions, as illustrated in Fig. 2.20.

The squeeze position is located by positioning to seal the return ports. The squeeze position allows all fluids pumped down the work string to be forced into the formation. It is used to perform squeeze gravel-pack treatments and/or inject acid treatments into the formation. The circulating position is located by picking the crossover tool up approximately 18 in above the squeeze position. The circulating position works with a properly sized washpipe to provide a flow path to circulate gravel-pack sand to completely fill the screen/casing annulus. The fluids flow down the work string into the crossover tool, out the gravel-pack extension, down the screen/casing annulus into the screen, up the washpipe into the crossover tool again, and up the work string/casing annulus.

Washpipe is run below the gravel-pack crossover tool inside the blank pipe and screen to ensure that the return circulation point for the gravel-pack carrier fluid is at the bottom of the screen. The washpipe assists placing gravel-pack sand at the bottom of the screen and packing from the bottom up. The end of the washpipe should be as close to the bottom of the screen as possible. When gravel packing is completed, the gravel pack crossover tool is set in the reverse circulating position to clean any remaining gravel out of treating string.

2.2.8 Expandable Techniques

Expandable technology has become an important component in different stages of a well life. It can be used for drilling, permanent completion (openhole or cased hole), workover, and stimulation. There are two groups of expandable completion hardware, expandable casings/liners and expandable screens.

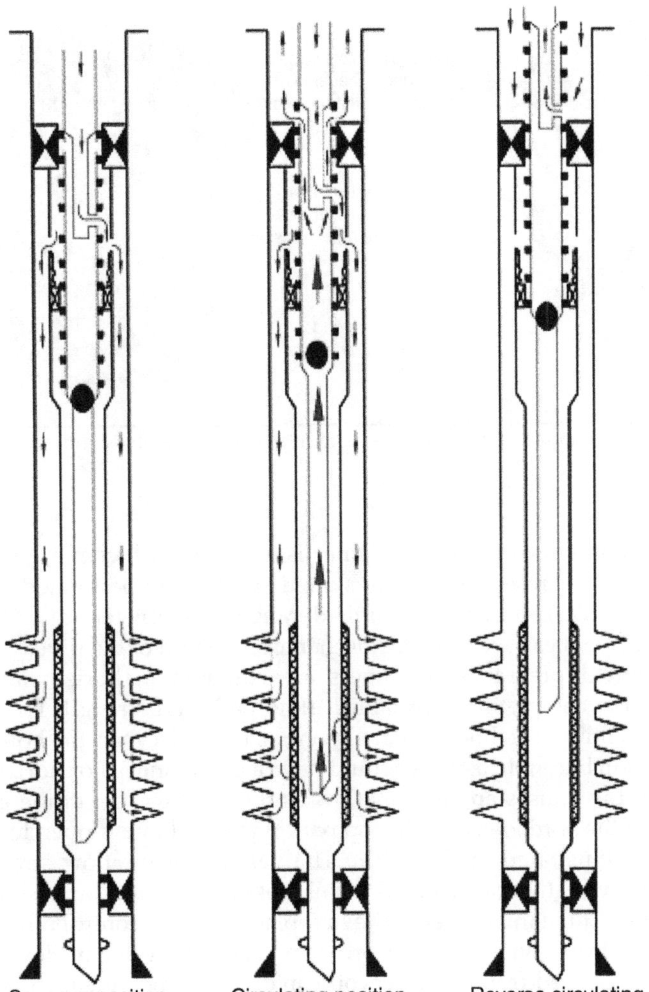

FIGURE 2.20 Gravel-pack crossover tool positions (*PetroWiki*).

An expandable tubular is a steel pipe, either liner or screen, that has the ability to have an enlarged diameter at downhole conditions. The enlargement can be accomplished by either hydraulic force with high pressure or mechanical force by a piston-like plunger. Screens are easier to expand, and enlarged size is more flexible. Figure 2.21 shows a picture of an expandable screen before and after the diameter is enlarged (Mohd Ismail and Geddes, 2013). Solid casings

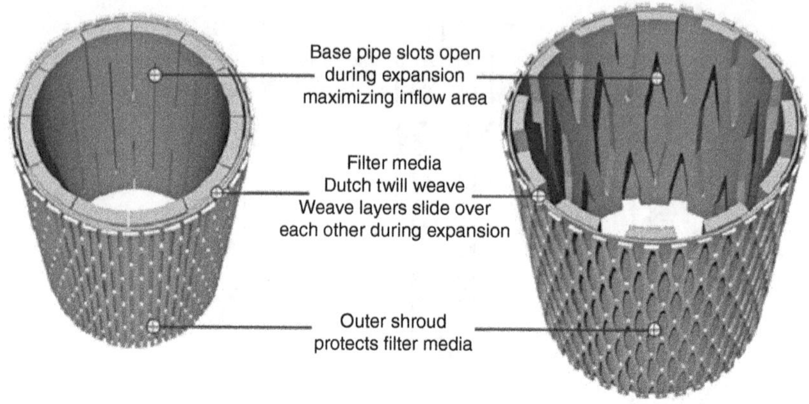

Figure 2.21 A composite liner before (*a*) and after (*b*) expansion (Mohd Ismail and Geddes, 2013).

and communicating liners (a liner is a casing with communication path with the reservoir, such as slotted liners or pre-perforated liners) can be enlarged within a limited range of diameter ratio. Over-enlargement can weaken the mechanical strength of the pipe.

Slotted/pre-perforated liners are common in horizontal well openhole completion to establish flow path and strength for borehole stability. The liner is usually one size smaller than the hole-size (e.g., 6-in borehole usually would have a 4½-in liner). For horizontal wells, this causes an annulus flow as the liner will set at the lower part of the borehole, and the top part of the wellbore is open to flow. This is illustrated in Fig. 2.22*a*. The velocity of the flow stream is generally high because of the small cross-section of flow area in the annulus. If the liner is set in the first place for sand control purposes, this could become a severe problem. The high-velocity flow with sand produced from the reservoir can quickly erode the liner. When using expandable liners in this case, the liner is enlarged to fill the borehole space, eliminating the annulus flow (Fig. 2.22*b*).

2.2.9 External Casing Packers

An external casing packer (ECP) is a casing string which has a part of it made with an inflatable or expandable material (Fig. 2.23). The expandable ECP expands when it contacts certain surrounding fluids (for example, oil or cementing fluid). A swell packer is an example of an ECP. An inflatable ECP is set by hydraulic forces. The expanding section can be several feet up to 40-ft long. The main function of an ECP is to isolate annular flow. It also centralizes the liner. This completion component is widely used for zonal control in horizontal wells. The ECP design depends on the purpose of the ECP and the type of formation. Most ECPs can be used in irregular hole shapes,

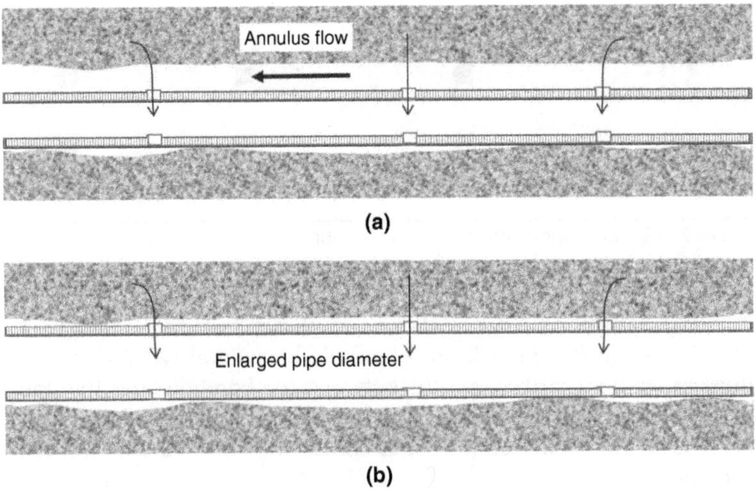

FIGURE 2.22 Expandable screens for sand control.

and some expandable ECPs have the ability to stand up to 10,000 psi pressure differential, providing reliable isolation. One thing that is important to remember is that the high pressure-differential packers are hard, if not impossible, to retrieve. If a packer sets at a wrong place, it cannot be released. The consequence is to drill/mill out the completion string to correct the mistake.

Expandable (swellable) packers and inflatable packers consist of a piece of synthetic material that expands as the surrounding environment changes. Swell packers are self-setting. Once the surrounding condition reaches the designed condition (for example, pH value), the packer expands into the void space between the liner and the borehole. It is simple from an operational point of view. Inflatable packers are set when the packing material is squeezed by hydraulic force. The setting time for inflatable ECPs is shorter and more controllable when compared with swelling packers.

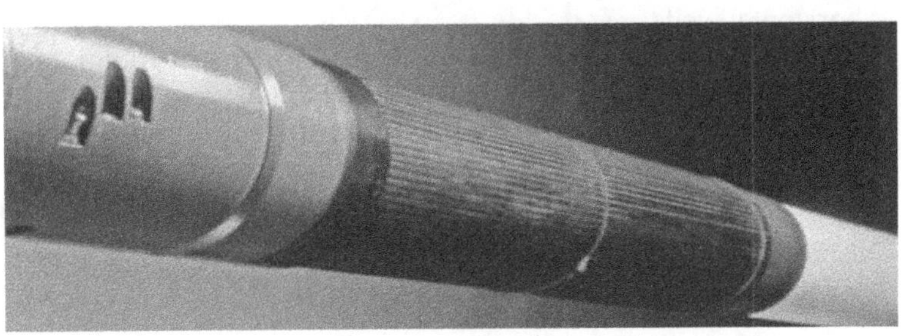

FIGURE 2.23 External Casing Packers. (*Courtesy of Baker Hughes Oilfield Operations, LLC.*)

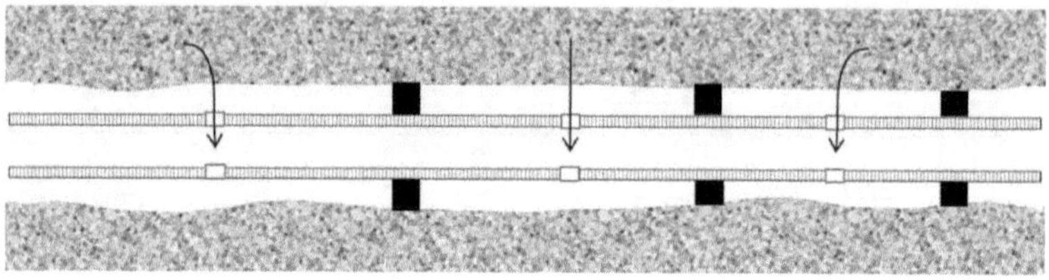

FIGURE 2.24 ECP application of zonal isolation.

One example of using ECPs for completion/production purposes is to install such a liner in an openhole horizontal well. This stops fluid movement in the annulus between the borehole and the liner, and forces the fluid entering the well and flowing inside the liner only. With this completion, zonal control of inflow along wellbore becomes possible.

Another popular application of ECP is for multiple staged well stimulation treatments in openhole horizontal wells. The packers can be set along a liner several hundred feet apart, depending on the design requirements, to isolate the wellbore into segments, and well stimulation, either by acid injection or by hydraulic fracturing, is delivered one stage at a time through the flow paths (sliding sleeves, slots, or pre-perforated holes) along the liner. The ECPs isolate the annular communication between stages. Figure 2.24 illustrates such a case. This will be further discussed in the later chapters.

2.2.10 Inflow Control Device

Inflow control devices (ICDs) are designed to regulate inflow distribution along a wellbore, either vertical or horizontal, and to control the effect of inflow profile on well performance. For vertical wells, ICDs can be used in multiple permeable zones; and for horizontal wells, ICDs can correct the negative impact of heterogeneous permeability along horizontal laterals, or eliminate the effect of toe-heel unbalanced production because of the wellbore frictional pressure drop inside the wellbore. The technology is applied more commonly in horizontal wells.

The principle of ICD design is to use flow-created additional pressure drop to balance the reservoir drawdown along a wellbore. The flow-created pressure drop can be either frictional pressure drop or kinetic pressure drop. For frictional pressure drop,

$$\Delta p_f = \frac{2 f_f u^2 L}{g_c D} \tag{2.2}$$

where f_f is frictional factor that is related to the Reynold's number and the roughness of the pipe, u is the velocity of fluid, L is the length of the flow path, and D is the diameter of the flow path. L and D are the design parameters for ICDs. And for kinetic pressure drop,

$$\Delta p_{KE} = \frac{\rho(u_2^2 - u_1^2)}{2g_c} \tag{2.3}$$

where ρ is the density of fluid. Kinetic pressure drop is created when the flow velocity is changed. Notice that both pressure drops are proportional to the squared flow velocity u^2, which is related to the flow rate q, and the flow cross-sectional area A, as

$$u = \frac{q}{A} \tag{2.4}$$

This indicates that the higher the flow rate or the smaller the flow cross-sectional area, the higher the pressure drop an ICD will create. Along a wellbore, when inflow profile is not evenly distributed because of reservoir heterogeneity or well structure design, production can be problematic. The examples of the production problems include early breakthrough of unwanted fluids at the high-permeability locations along the wellbore, sand production, or losing a part of production intervals toward the toe of a well when the frictional pressure drop is too high. ICDs can be installed to redistribute the flow along the wellbore. For example, at a high-permeability location of a horizontal well, since the reservoir has less resistance to flow, the inflow rate at the location would be higher than the locations that have lower permeability. If an ICD is installed, additional pressure will be created at the wellbore when the fluid enters the wellbore. The additional pressure (or additional resistance) will reduce the inflow rate at that location.

In summary, an ICD has a function to choke back inflow from the reservoir to the wellbore, and therefore it can generate a balanced flow rate distribution along the wellbore. It should be noted that ICDs do not increase the total volumetric flow rate. Based on this principle, there are different types of ICD designs.

ICD Design Using Frictional Pressure Drop

This type of ICD is one of the earliest types of ICDs that uses surface friction to generate pressure drop. Pre-cut channels with designed diameters and lengths around a base pipe can generate additional frictional pressure drop. Figure 2.25 shows a schematic of a channel-type ICD. The ICD design uses channels around a base-pipe combined with a screen wrapped outside the base pipe to regulate flow distribution in addition to sand control. The arrows indicate the flow path when entering the wellbore. Longer and smaller flow path results in lower flow velocities, which reduces the chances of erosion and plugging. However, a disadvantage of using friction to generate pressure drop is that frictional pressure drop increases as viscosity increases. For the

Figure 2.25 Channel-type ICD. (*Courtesy of Schlumberger.*)

wells that the oil phase has higher viscosity when compared with the water phase, ICDs choke more oil-phase flow than water-phase flow, which is not desirable. The viscosity-dependent ICDs should not be used in such wells.

ICD Design Using Kinetic Pressure Drop

This type of ICDs use flow restriction to generate a desired kinetic pressure drop. Fluid is forced through a pre-configured set of small-diameter nozzles or orifices before entering to the producing liners. The velocity change going through the restrictions helps balance the flow distribution. Figure 2.26 is a picture of a nozzle-type ICD and Fig. 2.27 shows an orifice ICD. An orifice ICD applies the same concept as a surface choke that is used to control flow rate at the wellhead. The pressure drop across this type of ICDs occurs instantaneously, and it is more dependent on the fluid density and velocity but less on the viscosity. This can be an advantage for the wells with produced hydrocarbon having higher viscosity than produced water. On the contrary, being dependent on fluid velocity makes the ICD prone to erosion from sand particles.

Autonomous ICD

Autonomous ICD is also referred to as AICD. The device is designed for controlling light-phase flow rate in a multiphase flow condition. The device contains a movable or rotatable disk inside, an inlet flow seat and an outlet flow seat. Figure 2.28 shows the picture of one type design of AICD, and illustrates how this AICD works. The rotation or motion of the disk is driven by the flow velocity. When the fluid velocity reaches a threshold value, the disk starts rotating, generating

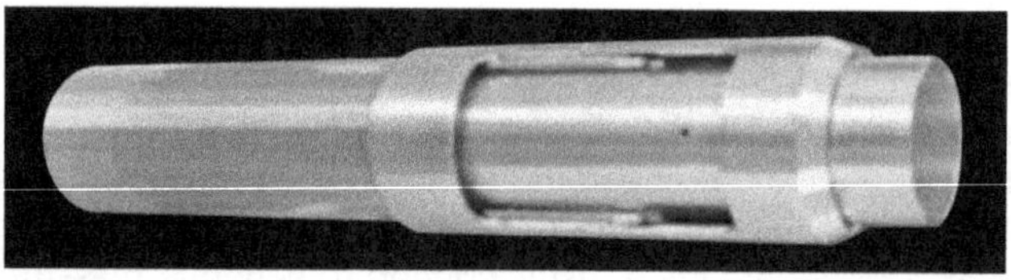

Figure 2.26 Nozzle-type ICD (Zhu, 2011–2012).

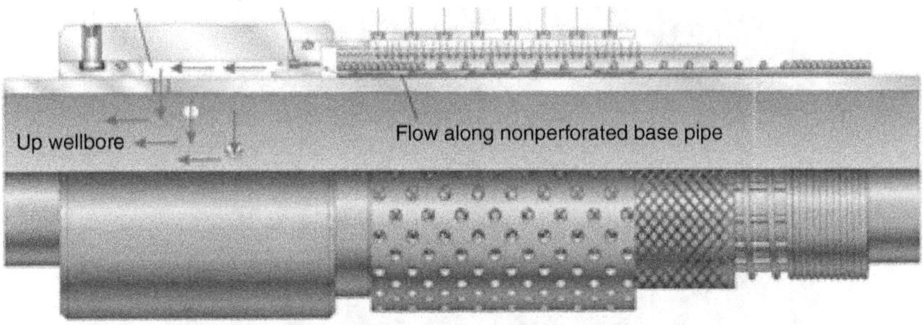

Figure 2.27 Orifice-type ICD. (Adopted and modified from Birchenko et al., 2010.)

additional pressure drop to the flow path. The additional pressure drop makes the pressure on the inlet side of the disk lower the back-side pressure. This pressure difference will start moving the disk toward the outlet seat to close up the flow path. AICD allows higher viscosity fluid flow through with less pressure drop (disk does not move at low velocity condition). It is suitable for heavy-oil wells that need to control water production, or oil wells that need to control gas production from the gas cap above. An advantage of this type of ICD is that the restriction only applies after the unwanted fluid breakthrough. Principally speaking, it does not generate additional pressure drop when only high-viscosity fluid is flowing. Sufficient viscosity difference is required for AICDs to function correctly.

Other ICD Types

There are many configurations of ICDs of channel-type, nozzle-type, orifice-type ICDs, or AICDs. More commonly used designs are a combination of channel, orifice, and nozzle. Efforts have been made to more evenly distribute the additional pressure drop to avoid point-restriction. An example of such an ICD is simply packing fine fiber material between two perforated or slotted base pipes (Fig. 2.29). The texture of the packing material determines the level of

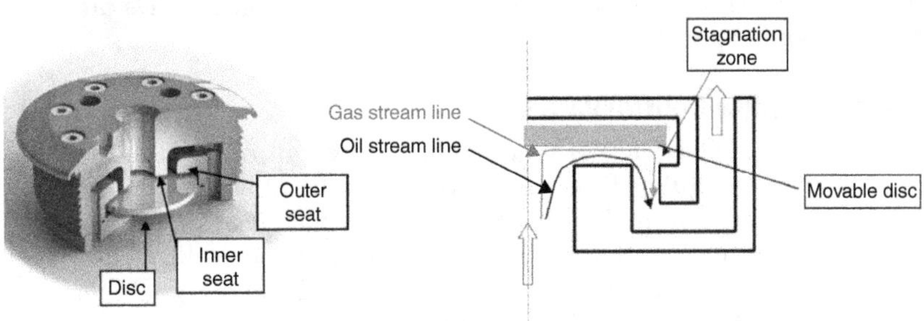

Figure 2.28 AICD mechanism. (*a*) Structure (Least et al., 2012) and (*b*) flow path (Halvorsen et al., 2012).

Figure 2.29 Fiber-filled ICD for more evenly distributed additional pressure drop. (*Courtesy of Schlumberger.*)

the additional pressure that can be created when flowing through the section. This can also serve as a sand-control device.

It is important to remember that an ICD is a permanent hardware installed upon completion of a well based on the initial reservoir condition evaluation and simulation prediction of reservoir performance. Most of these types of hardware are self-regulated, un-adjustable, and hard to retrieve. Once installed in the well, the device will function as designed and remains in the well throughout the life of the well. Because the reservoir condition is dynamic, choosing of the correct ICDs for flow control can be difficult. Changes in fluid viscosity, density, and velocity occur with time, and ICDs should be designed considering these conditions to avoid creating negative effects on production. Even though the restriction to flow by ICDs is linearly proportional to the flow rate to the second power ($Dp \propto q^2$), and as production rates decline, the restriction becomes smaller, it is clear that ICDs are the downhole chokes in the well system, and over-designed ICDs can be detrimental to production.

2.2.11 Inflow Control Valves

Inflow control valves (ICVs) are downhole flow control devices that are designed to regulate flow along a vertical or horizontal wellbore. It is a component of intelligent completions, which uses downhole monitoring and downhole flow control to optimize well performance. ICVs are fundamentally different from ICDs in design. ICVs usually consist of sliding sleeves and balanced piston that can be operated by hydraulic control lines from the surface. Most of ICVs commercially available also allow mechanical shifting by a slickline

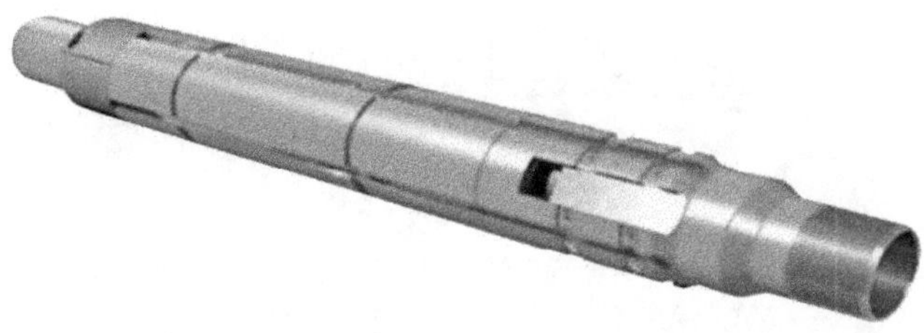

FIGURE 2.30 Inflow control valve. (*Courtesy of Halliburton Completion Tools.*)

tool in case the valve is unable to move hydraulically. Figure 2.30 shows an ICV.

ICVs can be used in openhole completions along with ECPs (external casing packers) to isolate the sections along a wellbore. ICVs are installed between two ECPs to regulate the flow from each section individually. Figure 2.31 illustrates a configuration of ECP and ICV completion for multilateral well production control. Certain section can be shutoff when unwanted fluid starts entering the wellbore. Compared with ICDs, ICVs are a more positive control method for downhole flow regulation.

2.2.12 Mechanical Sliding Sleeves

Mechanical sliding sleeves are used to establish communication between the tubing string and the casing annulus. The inner pipe with slots can be moved inside the outer liner by a slickline tool and/or ball. When the slots on the inner pipe are aligned with the slots on the outer liner, the communication is established between the annular

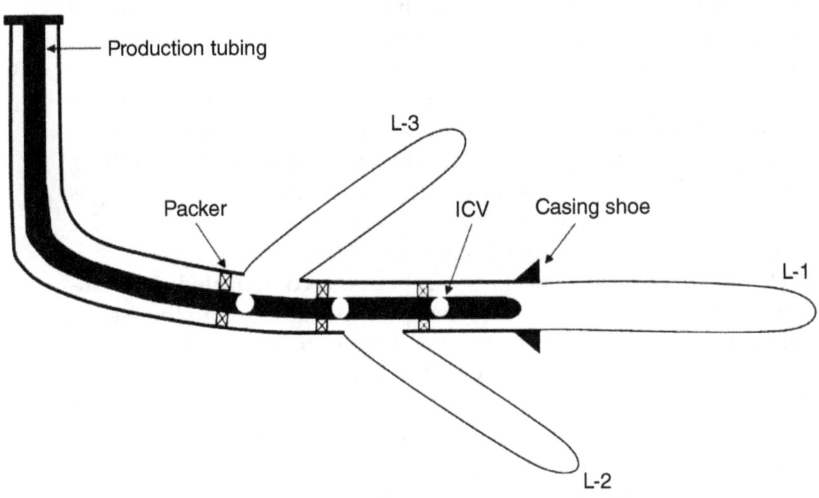

FIGURE 2.31 A multilateral ICV well completion with ECPs.

FIGURE 2.32 Ball-activated sliding sleeve. *(Courtesy of oilpro.com. Reprinted with permission of Completions.)*

and the producing tubing. This open position can be reversed to close the communication when local shut-in is desired. An example of this kind of valves is fracture sleeves used in multiple-stage hydraulic fracture treatments. Figure 2.32 illustrates how a ball-activated valve works. The valve is opened by a ball pumped down from the surface. When the ball reaches the seat, it seals off the upstream flow. Further pumping increases the pressure inside the liner, and triggers the movement of the sleeve to an open position. Communication between the liner and formation is established this way.

2.3 Completion Design Philosophy

Based on the discussions of the previous sections of this chapter, we understand the functions and hardware structures of well completions, and how each completion is designed for well production. The completion should be selected based on the characteristics of the reservoirs in which wells are going to be drilled and produced, and cover broad considerations of the purpose of the well (producing, injecting, or possibly alternating), the formation rock properties (mineralogy, competency, sand production, depth, etc.), the reservoir flow properties (permeability, porosity, heterogeneity, temperature, pressure, and thickness), and fluid properties (viscosity, density, gas cap, water aquifer). From the functions of each completion, Table 2.2 suggests a preliminary guideline for completion selection.

In the real world, problems can be more complicated and usually are not isolated. When designing completions before the well is put on production/injection, or even before the well is drilled, it is difficult to identify all challenges that may be associated with well flow in the future. For example, reservoir heterogeneity plays an important role in well completion design, but heterogeneity can only be assessed in a statistical manner, and high uncertainty is always in the assessments. When multiple problems are anticipated in the target reservoir, a combination of different types of completions is

Function	Problem	Completion
Borehole stability	Unconsolidated formation, incompetant formation, washout/cavity	Cased/cemented/perforation Openhole slotted/pre-perforated liners
Well flow efficiency	Low-permeability formation Large skin	Acid fracturing/Proppant fracturing Matrix acidizing Hydraulic fracturing
Flow control	Vertical barriers in reservoir, Horizontal well zonal control, Multilateral well layer control, Water/gas conning, Crossflow	Cased/cemented and perforated Openhole liners with ECP, Openhole liners with ICD/ICV/AICD, Cased/cemented completion with ICD/ICV/AICD
Sand control	Sand production	Openhole gravel packs Openhole stand-alone screens Cased hole gravel packs Cased hole frac packs

TABLE 2.2 Completion Design Philosophy

commonly deployed. Each completion, when installed in a well, adds restriction to the reservoir flow. Unnecessary over-powered completion design will result in negative impact on well production, and should be avoided.

2.4 New Challenges in Unconventional Reservoirs

As the oil and gas industry moved toward the 21st century, the resources for hydrocarbon production became much more complex compared with the early time development. Unconventional reservoirs with low flow abilities that were defined as non-economic reserves for development became the new targets for oil and gas production around the world. This revolution was made possible because of the advances in drilling, completion, and stimulation.

The word "unconventional" has been used by the industry for tight sand and shale formations. In this book, *unconventional* is used relative to the common understanding of *conventional*. From Eq. (2.1), the group of reservoir properties, kh/μ, often referred to as transmissibility of the reservoir, and has strong impact to the flow rate, q_o. It is obvious that the flow rate, q_o, is linearly proportional to transmissibility, kh/μ. Unconventional resources are defined as the formations with low permeability (a permeability range from hundreds of nano-Darcy to 0.1 mD). Unconventional reservoirs can also refer to as the reservoirs with thin payzones (reservoir thickness ranges 10 to 40 ft), or the reservoirs that have fluid viscosity of hundreds or even up to thousands of centipoise (heavy oil). For unconventional resources, the well configuration and completion design plays an important role in successful development. In this section, we will discuss some new challenges of each unconventional case.

2.4.1 Tight Sand and Shale Formation

Tight sand production has a relatively long history comparing with shale gas and oil productions. With the breakthrough in the Barnett shale gas reservoir in the United States in 1990s, drilling, completion, and stimulation activities in shale formations quickly became an important portion of the oil and gas industry. This is not limited to the United States.

For low-permeability reservoirs, conventional vertical well structure drains reservoir fluids extremely slow. The key to successful development is to create extended reservoir contact with the wellbore. This can be accomplished by either drilling vertical wells and hydraulically fracture the wells; drilling horizontal or multilateral wells; or drilling horizontal wells and creating multiple fractures along the wellbores. For moderately low-permeability (~0.01 to 0.1 mD) formations, drilling vertical wells with hydraulic fracturing or drilling horizontal wells may be sufficient to solve the production problem.

When matrix permeability of the target reservoir is of the order of 10^{-3} to 10^{-4} mD (most shale formations), fractured vertical wells or horizontal wells alone are not sufficient to make economic wells. The combination of horizontal drilling and multiple-stage fracturing along the horizontal wellbore is the most commonly used method in tight sand and shale formations. The surface area of the created fractures is directly related to the well production. The lower the permeability of the reservoir, the higher the fracture area is desired. Economically producing from shale formations is mainly limited by how to efficiently place fractures in a reservoir according to the rock mechanical and fluid properties. The completion is the key to successfully developing the shale formations. The details of multi-stage fracturing of horizontal wells are discussed in Chapter 4.

2.4.2 Heavy Oil Reservoirs

Heavy oil reservoirs were produced by thermal injection with patterned arrangements of vertical injectors and producers at early times. As we are targeting the reservoirs that have fluid viscosity in the range of hundreds or thousands of centipoise, standard vertical injector/producer approach becomes extremely inefficient. Newer technologies in heavy oil recovery include steam-assisted gravity draining (SAGD) by drilling two horizontal laterals, with the upper horizontal well or lateral as the injector, and the lower one as the producer. Drilling multilateral wells with many short laterals to produce oil without thermal assistance is another option. In either case, steam and fluid distribution along the horizontal laterals is an issue to successful production. Inflow control completions (ICVs or ICDs) are used to evenly distribute the flow. Most heavy oil reservoirs do not suffer from low-permeability problems, but have sand production problems. Sand control is an important issue in completion design. It should be kept in mind that when design completions for flow distribution and sand control are added to the well, it restricts fluid flow also, and this results in reduced production. For heavy oil reservoirs, production is more sensitive to completion than for conventional reservoirs, and any restriction by pressure drops related to fluid viscosity should be avoided in completion design.

2.4.3 Thin Oil Rims

Thin oil rim reservoirs are reservoirs that have small payzone thickness, and there are strong gas caps above and/or aquifers below the payzone. Because gas and water could have high mobility compared with oil, early breakthrough of gas or water can jeopardize well production. Many thin-oil-rim wells have to sacrifice oil production rate because of the limited capacity of handling produced gas. To produce from such a well, reservoir drawdown has to be controlled at a small value, and therefore the oil production rate is limited. Horizontal wells are good candidates to produce from thin oil-rim reservoirs. It increases the contact between formations and wellbores, and thus improves production performance. Even with horizontal wells, early breakthrough of gas or water is still unavoidable, and most likely happens at the heel of a wellbore. The completion, in this case, can be designed to help to delay the breakthrough at the heel. ICDs or ICVs are installed along wellbores with a stronger choking effect toward the heel. AICD works well in thin oil-rim reservoirs because the device chokes the light phase (gas phase), and it only acts when the light phase breakthrough happens. It has less impact to the higher-viscosity phase.

2.4.4 Offshore Development

For offshore reservoirs, especially in deepwater fields, drilling and completion are challenged by the harsh environment, high cost and risk, and limited operating space. Deepwater wells are expensive to drill and complete, often costing tens to hundreds of millions dollars per well. Most of these costs are associated with the rig time; so, preventing operational problems is critical. Because of the limitation to access offshore locations (platforms or subsea templates), horizontal or highly-deviated wells are commonly planned to save the drilling and completion cost, and access more reserve with limited surface space. Two challenges are commonly faced in offshore completions, sand production and zonal isolation for inflow control and water management.

Industry experiences have shown that frac-pack completions can provide effective production stimulation and reliable sand control for unconsolidated sand reservoirs. One of the major considerations in planning sand control completions offshore is operational efficiency. In reservoirs that consist of multiple target zones, completing each zone independently with a stack-pack completion can be cost-prohibitive under offshore environments. A single trip multizone frac-pack system enables several intervals of a well to be isolated and treated with high-rate frac pack during a single trip of the workstring leading to significant cost saving due to reduced rig time.

Because of the high cost environment, offshore wells require large reservoir contact, high rates, and good reliability in order to be economic. This results in many offshore completions incorporating ICVs and downhole gauges. ICVs can shut off unwanted fluids (water or gas) immediately without having to mobilize intervention equipment, and in worst cases, a rig. Having ICVs allows for more proactive and regular reservoir management and hence can increase hydrocarbon reserves.

Nomenclature

A = flow cross-section area, ft^2, m^2
A = drainage area, ft^2, acres, m^2
B = formation volume factor, res bbl/STB, m^3/Sm^3
B_o = oil formation volume factor, res bbl/STB, m^3/Sm^3
c_t = total system compressibility, psi^{-1}, Pa^{-1}
C_H = geometric factor, dimensionless
D = diameter, ft, m
D = non-Darcy coefficient, (STB/d)$^{-1}$, (m^3/s)$^{-1}$
f_f = friction factor, dimensionless
g_c = unit conversion factor
h = reservoir thickness, ft, m
I_{ani} = index of permeability anisotropy, dimensionless
J = productivity index, STB/d/psi, m^3/s/Pa
k = permeability, md, m^2
k_o = oil permeability, md, m^2
k_y = permeability in the y direction, md, m^2
k_z = permeability in the z direction, md, m^2
L = pipe length, ft, m
L = horizontal well length, ft, m
$m(p)$ = real gas pseudo-pressure, psi^2/cp, Pa/s

$\bar{p}$ = average reservoir pressure, psi, Pa
p_e = reservoir outer boundary pressure, psi, Pa
p_i = initial reservoir pressure, psi, Pa
p_{wf} = flowing bottomhole pressure, psi, Pa
q = production rate, STB/d, m^3/s
q_o = oil flow rate, STB/d, m^3/s
$q_{o,max}$ = absolute open flow potential, STB/d, m^3/s
Q = production rate, STB/d, m^3/s
r_e = drainage radius, ft, m
r_w = well radius, ft, m
s = skin effect, dimensionless
s_R = skin resulting from partial penetration, dimensionless
t = time, h, s
T = temperature, °F, °C
u = velocity, ft/s, m/s
y_b = reservoir drainage width, ft, m
Z = gas compressibility (gas deviation factor), dimensionless
ϕ = porosity, fraction
Δp_f = frictional pressure drop, psi, Pa
ρ = density, lb/ft^3, kg/m^3
μ = viscosity, cp, Pa·s
μ_o = oil viscosity, cp, Pa·s

References

Adams, P. R., Davis, E. R., Hodge, R. M., Burton, R. E., Ledlow, L. B., Procyk, A. D., and Crissman, S. C. (2009, September 1). "Current State of the Premium Screen Industry; Buyer Beware. Methodical Testing and Qualification Shows You Don't Always Get What You Paid For," *Society of Petroleum Engineers*. doi:10.2118/110082-PA.

Ali, S. A., Falfayan, L., and Montgomery, C. (2016). "Acid Stimulation. SPE Monograph Series," Vol. 26, ISBN: 978-1-61399-426-9. *Society of Petroleum Engineers*.

Birchenko, V. M., Murradov, K. M., and Davies, D. R. (2010) "Reduction of the Horizontal Well's Heel-Tow Effect with Inflow Control Devices," *Journal of Petroleum Science and Engineering*. Elsevier.

Burton, R. C., Rester, S., and Davis, E. R. (1996, January 1). "Comparison of Numerical and Analytical Inflow Performance Modelling of Gravelpacked and Frac-Packed Wells," *Society of Petroleum Engineers*. doi:10.2118/31102-MS.

Burton, R. C., and Hodge, R. M. (1998, January 1). "The Impact of Formation Damage and Completion Impairment on Horizontal Well Productivity," *Society of Petroleum Engineers*. doi:10.2118/49097-MS.

Burton, R. C. (1999, December 1). "Use of Perforation-Tunnel Permeability to Assess Cased Hole Gravelpack Performance," *Society of Petroleum Engineers*. doi:10.2118/59558-PA.

Carter, N., Hembling, D., Salamy, S. P., Qahtami, A., and Jacob, S. (2006, January 1). "Swell Packers: Enabling Openhole Intelligent and Multilateral Well Completions for Enhanced Oil Recovery," *Society of Petroleum Engineers*. doi:10.2118/100824-MS.

Economides, M. J., Hill, A. D., Ehlig-Economides, C., and Zhu D. (2012). *Petroleum Production Systems*, 2nd ed., ISBN: 0-13-703158-0, Prentice Hall, NJ.

Furui, K., Zhu, D., and Hill, A. D. (2005, August 1). "A Comprehensive Model of Horizontal Well Completion Performance," *Society of Petroleum Engineers*. doi:10.2118/84401-PA.

Furui, K., Fuh, G.-F., Abdelmalek, N. A., and Morita, N. (2010, December 1). "A Comprehensive Modeling Analysis of Borehole Stability and Production-Liner Deformation for Inclined/Horizontal Wells Completed in a Highly Compacting Chalk Formation," *Society of Petroleum Engineers*. doi:10.2118/123651-PA.

Halvorsen, M., Elseth, G., and Naevdal, O. M. (2012, January 1). "Increased Oil Production at Troll by Autonomous Inflow Control with RCP Valves," *Society of Petroleum Engineers*. doi:10.2118/159634-MS.

Least, B., Greci, S., Burkey, R. C., Ufford, A., and Wilemon, A. (2012, January 1). "Autonomous ICD Single Phase Testing," *Society of Petroleum Engineers*. doi:10.2118/160165-MS.

McLeod, H. O. (1983, January 1). "The Effect of Perforating Conditions on Well Performance," *Society of Petroleum Engineers*. doi:10.2118/10649-PA.

Mohd Ismail, I., and Geddes, M. W. (2013, September 30). "Fifteen Years of Expandable Sand Screen Performance and Reliability," *Society of Petroleum Engineers*. doi:10.2118/166425-MS.

Nozaki, M., Burton, R. C., Pandey, V. J., and Furui, K. (2017, October 9). "Lessons Learned from Evaluation of Frac-Pack Flow Performance Before and After Full Clean-up for Subsea Wells in High-Permeability Gas Reservoirs," *Society of Petroleum Engineers*. doi:10.2118/187227-MS.

Nierode, D. E., and Kruk, K. F. (1973, January 1). "An Evaluation of Acid Fluid Loss Additives Retarded Acids, and Acidized Fracture Conductivity," *Society of Petroleum Engineers*. doi:10.2118/4549-MS.

NORSOK Standard D-010 rev.3, Well Integrity in Drilling and Well Operations, August 2004.

Prats, M. (1961, June 1). "Effect of Vertical Fractures on Reservoir Behavior-Incompressible Fluid Case," *Society of Petroleum Engineers*. doi:10.2118/1575-G.

Procyk, A. D., Jamieson, D. P., Miller, J. A., Burton, R. C., Hodge, R. M., & Morita, N. (2009, December 1). "Completion Design for a Highly Compacting Deepwater Field," *Society of Petroleum Engineers*. doi:10.2118/109824-PA.

Tronvoll, J., Kessler, N., Morita, N., Fjær, A., and Santarelli, F. J. "The Effect of Anisotropic Stress State on the Stability of Perforation Cavities," *International Journal of Rock Mechanics and Mining Sciences & Geomechanics Abstracts*, Volume 30, Issue 7, 1993, Pages 1085–1089, ISSN 0148-9062, http://dx.doi.org/10.1016/0148-9062(93)90075-O.

Van Everdingen, A. F. "The Skin Effect and Its Influence on the Productivity Capacity of a Well," *Trans. AIME* (1953), 198.

Veeken, C. A. M., Wahleitner, L. P., and Keedy, C. R. (1994, January 1). "Experimental Modelling of Casing Deformation in a Compacting Reservoir," *Society of Petroleum Engineers*. doi:10.2118/28090-MS.

Zhu, D. (2011–2012). Understanding the Roles of Inflow-Control Devices in Optimizing Horizontal Well Performance. SPE Distinguished Lecturer Program.

CHAPTER 3

Completion Impact in Near-Wellbore Region to Inflow Performance

In Chapter 2, we introduced well completion hardware and their functions in assisting well production. When a well is drilled and completed, the flow from the reservoir to the wellbore follows the path defined by the completion. This path deviates from the assumed ideal flow streamlines that are used to predict flow behavior. For example, for a vertical well, the flow velocity is calculated by analytical models or simulated by numerical methods based on the assumption that the fluids follow radial flow streamlines from the reservoir drainage volume to the line source (the well) in the flow domain. But when the well is drilled, and completed partially along the payzone with perforations, the fluid will converge to the perforated section that is open to flow when getting closer to the wellbore, instead of following the ideal flow path. Figure 3.1 illustrates this phenomenon. An additional pressure drop results in when flow deviates from the assumed ideal flow path or when any permeability change occurs along the flow path. This additional pressure drop is captured by the concept of skin factor in petroleum engineering. This skin factor has a positive value.

On the other hand, when the permeability of a reservoir is increased from the original permeability in the near-wellbore region by stimulation (acidizing or fracturing), the improved flow ability of a well can be represented with a negative skin factor. Deviated wells have larger contact area with the reservoir compared with true vertical wells, and this increased contact increases the flow rate for a defined pressure difference. This effect can also be captured by a negative skin factor.

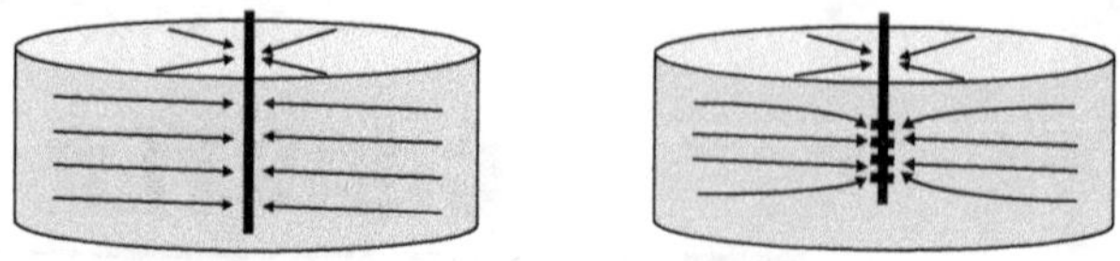

FIGURE 3.1 An example of flow path change by completion that results in additional pressure drop.

Skin factor is an important concept for completion and production engineers to understand well flow behavior after a well is drilled and completed, and is the starting point to bring a well to its optimal operating condition.

There are two types of skin effects: formation damage skin effects and mechanical skin effects. Formation damage skin captures the permeability changes in the near-wellbore region during drilling, completion, production, or even stimulation. Mechanical skin factors are mainly caused by flow path changes. This can be either by the structure of a well (partially completed payzone, for example), or by the individual completion (when perforating a well, the flow path is limited to the perforations only). Figure 3.2 summarizes all possible skin factors associated with a well during drilling, completion, and operation.

Skin factors are commonly estimated from well testing. Well testing analysis uses transient pressure behavior instigated by a flow rate change (increase, decrease, or shut-in) to estimate reservoir parameters such as permeability, initial reservoir pressure, and skin factor. When estimating skin factor from well testing, the skin factor is a combination of all skin effects. Understanding the causes of skin factors is the first step of well performance diagnosis. In this chapter, we will first show how to calculate skin factor for each component, and then discuss the interactive impact of skin factors on well performance.

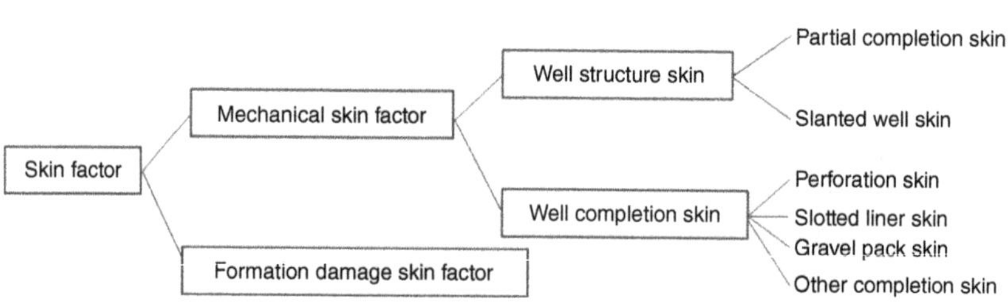

FIGURE 3.2 Different skin factors caused by permeability changed in near-wellbore region or well constructions.

3.1 Formation Damage Skin

Formation damage skin describes the phenomenon of permeability reduction in the near-wellbore region. Examples of causes of this permeability reduction include drilling or completion fluid invasion, filter cake build-up from drilling or completion fluid, fines migration from well production, or solids precipitation. Formation damage can happen in any wells, with or without completion effects. In an openhole completion, the flow path does not change from the ideal, and the only possible skin factor is the formation damage skin factor.

3.1.1 Damage Skin Factor for Vertical Wells

Hawkins formula (Hawkins, 1956) is commonly used to estimate damage skin factor. If a well with a wellbore radius of r_w is drilled in a reservoir that has a drainage radius of r_e, and the formation permeability is altered from the original permeability, k, to a reduced permeability, k_s, in a region that has a radius of r_s (Fig. 3.3), then damage skin factor is calculated by

$$s_d = \left(\frac{k}{k_d} - 1\right) \ln\left(\frac{r_s}{r_w}\right) \tag{3.1}$$

The two parameters in the above equation that describe the level of damage, damaged permeability, k_s, and damaged zone radius, r_s are difficult to measure or estimate, especially before a well is drilled. Logging tools that have multiple sensors with different sensing ranges can measure resistivity at different depth into the formation. If a damaged zone is in the sensing range of the shallow sensor, the resistivity change can be detected, and thereafter the damaged depth may be estimated. Calibration of damage skin factor from production tests also helps to estimate the level of damage. Other methods such as using transient temperature behavior interpret damaged zone radius and damaged permeability (Sui et al., 2008).

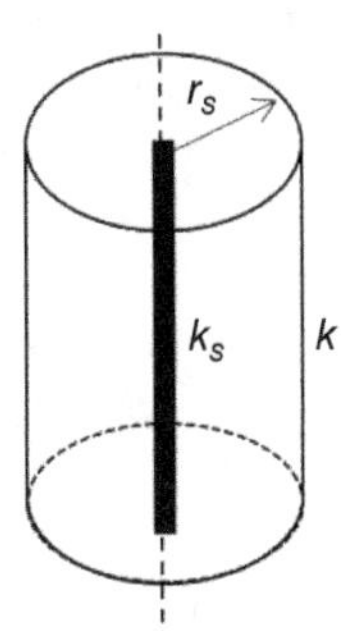

FIGURE 3.3 Illustration of formation damage parameters.

Example 3.1 A damaged zone is created after drilling a well that has a wellbore radius of 0.4 ft. If we assume that the permeability in the damaged zone k_s is reduced to 10% of the original permeability k and the damaged zone extends from the wellbore wall into the formation a distance of 1 ft, what is the damage skin factor?

Solution In the Hawkins' equation, r_s, is the radius of damaged zone, and it should be counted from the center of the well system,

$$r_s = \text{depth of damage} + r_w = 1 + 0.4 = 1.4 \text{ (ft)} \tag{3.2}$$

$$s_d = \left(\frac{k}{k_s} - 1\right) \ln\left(\frac{r_s}{r_w}\right) = \left(\frac{k}{0.1k} - 1\right) \ln\left(\frac{1.4}{0.4}\right) = 11.3 \tag{3.3}$$

3.1.2 Damage Skin Factor for Horizontal Wells

For horizontal wells, damage skin factor is a function of the location along the wellbore. This is because drilling mud is exposed to the formation longer at the heel of a wellbore compared with toward the toe. The damaged radius is more likely larger at the heel of a horizontal well compared with toward the toe. As mud cake is built up on the borehole surface, it slows down the filtration into the formation. Because of a much longer contact with reservoirs for horizontal wells compared with vertical wells, heterogeneity also plays a role for non-constant damage along a wellbore. Damage is deeper where permeability is higher. The distribution of the damaged zone radius is not easy to describe, especially when permeability along the wellbore is not a constant. What complicates the situation even more is that damage profiles of horizontal wells are also affected by the permeability anisotropy in the horizontal and vertical directions. The invasion of drilling and completion fluids tends to be deeper in the direction that the permeability is higher, which generally is the horizontal direction.

To develop a method for damage skin factor calculation for horizontal wells, we make some assumptions to simplify the damage distribution along a horizontal wellbore. Assuming a homogeneous but anisotropic formation, in the cross-sectional area perpendicular to the wellbore, the damaged zone is assumed to have an elliptical shape, with the long axis aligned in the higher permeability direction. Along the wellbore, the damage zone has a cone shape. Figure 3.4 illustrates the concept of formation damage in a horizontal well.

Defining the anisotropic ratio of permeability I_{ani} as

$$I_{\text{ani}} = \sqrt{\frac{k_H}{k_V}} \tag{3.4}$$

For homogeneous reservoirs, assuming that permeability along the wellbore is a constant, then the damage skin factor for horizontal wells is (Furui et al., 2003)

$$s = \left[\frac{k_H}{k_{sH}} - 1\right] \ln\left\{\frac{1}{I_{\text{ani}} + 1}\left[\frac{r_{sH}}{r_w} + \sqrt{\frac{r_{sH}^2}{r_w^2} + I_{\text{ani}}^2 - 1}\right]\right\} \tag{3.5}$$

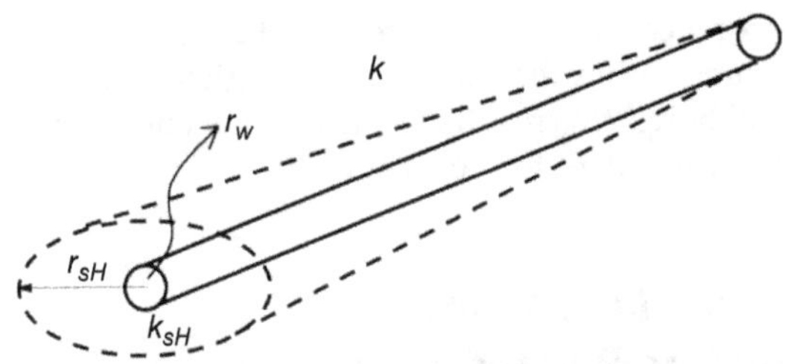

FIGURE 3.4 Damage zone in a horizontal well has an elliptical shape and the long axis is aligned with higher permeability direction.

Notice that for an isotropic reservoir, where k_H is equal to k_V (I_{ani} is equal to one), Eq. (3.5) reduces to Eq. (3.1). Similar to the vertical well situation, k_{sH} and r_{sH} are difficult to measure in the field.

Equation (3.5) presents the simplest estimate of damage skin factor, assuming that all parameters that describe the damage are constant. For heterogeneous reservoirs, the permeability, both original and damaged, and the damaged radius are functions of locations along a wellbore. A more general equation for the local damage skin factor is

$$s(x) = \left[\frac{k}{k_{sH}(x)} - 1\right] \ln\left[\frac{1}{I_{ani}+1}\left(\frac{r_{sH}(x)}{r_w} + \sqrt{\left(\frac{r_{sH}(x)}{r_w}\right)^2 + I_{ani}^2 - 1}\right)\right] \tag{3.6}$$

and the equivalent skin over a wellbore that has a length of L is

$$s_{eq} = \frac{L}{\int_0^L \{\ln[h/(2r_w)] + s(x)\}^{-1}\, dx} - \ln\left(\frac{h}{2r_w}\right) \tag{3.7}$$

Example 3.2 For a horizontal well, if the permeability anisotropy ratio is 3.16 ($k_H/k_V = 10$), r_{sH} is 1.4 ft, the undamaged permeability to damaged permeability ratio k_H/k_{sH} is 10 and r_w is 0.4 ft, what is the skin factor in this case?

Solution With a consistent unit of ft for both r_{sH} and r_w, using Eq. (3.5),

$$s = [10-1]\ln\left\{\frac{1}{3.16+1}\left[\frac{1.4}{0.4} + \sqrt{\frac{1.4^2}{0.4^2} + 3.16^2 - 1}\right]\right\} = 6 \tag{3.8}$$

From this example, it shows that at the same level of damage, the horizontal well skin factor is smaller than the vertical well skin factor (Example 3.1, $s = 11.3$). This is contributed by the anisotropy of permeability in the horizontal and vertical directions. In a general case, the damage radius in the vertical direction r_{sV} is smaller than r_{sH} (see Fig. 3.4), indicating shallower damage invasion in the vertical direction when compared with the invasion in the horizontal direction.

This is only for local skin. We also realize that contact lengths of horizontal wellbores with reservoir are much higher than contact lengths of vertical wellbores, this results in a less impact of skin factor on flow for horizontal wells than for vertical wells.

3.2 Mechanical Skin Factor: Well Structure-Related Skin

Well structure-related skin factors have two components: partial completion skin and deviated well skin (slanted well skin). Partial completion skin occurs when the completion does not cover the entire payzone. A positive skin factor is resulted in this case. Slanted skin accounts for a well that is deviated from vertical, which is a negative skin factor because more reservoir is contacted when a well is deviated.

3.2.1 Partial Completion Skin Factor

When a well is partially completed, the fluid flows through a converged path to the completed portion of the payzone when approaching the wellbore. This convergence causes two alterations to the original ideal path. First of all, the fluid flowing toward the portion where there is no completion has to turn to the portion that is open to flow. This results in a longer flow path compared with if the entire payzone is open. Secondly, when the flow converges to the open portion of the wellbore, vertical flow happens that involves both horizontal permeability and vertical permeability instead of just horizontal permeability in the case of a fully opened completion. Since vertical permeability tends to be smaller than horizontal permeability in general, this makes flow more difficult. For a partially completed well, both flow path change and permeability alteration require additional pressure drop to flow the same rate compared with a fully completed well. Partial completion skin factor is developed in this case to describe the effect of additional pressure drops.

Papatzacos (1987) presented a model for a partially completed well with anisotropic reservoir permeability. Figure 3.5 shows the geometry used to develop the model. The partial completion skin factor is calculated by

$$s_c = \left(\frac{1}{h_{wD}} - 1\right) \ln \frac{\pi}{2r_D} + \frac{1}{h_{wD}} \ln \left(\frac{h_{wD}}{2+h_{wD}} \sqrt{\frac{A-1}{B-1}}\right) \quad (3.9)$$

where the parameters A and B in the above equation are defined as

$$A = \frac{1}{h_{1D} + \frac{h_{wD}}{4}} \quad (3.10)$$

and

$$B = \frac{1}{h_{1D} + \frac{3h_{wD}}{4}} \tag{3.11}$$

and the dimensionless variables are

$$h_{wD} = \frac{h_w}{h} \tag{3.12}$$

$$r_D = \frac{r_w}{h}\sqrt{\frac{k_V}{k_H}} \tag{3.13}$$

and

$$h_{1D} = \frac{h_1}{h} \tag{3.14}$$

In the above equations, r_w is wellbore radius, h is payzone thickness, h_w is the thickness of completed zone, k_H is horizontal permeability, and k_V is vertical permeability. Notice that h_1 is the distance from the top of completed zone to the top of payzone (see Fig. 3.5). The above equations should be used in consistent units.

The model is straightforward. The completed zone can be located anywhere in the payzone. The partial completion skin model is independent of the type of completion. Regardless a well is completed with cementing and perforating, or slotted liners, as long as the completion does not cover the entire payzone, partial completion skin applies. Partial completion skin can be reduced/removed by increasing the completion interval or fully completing the payzone. Sometimes partial completion is necessary based on the reservoir characteristics. For example, when a reservoir has a strong bottom aquifer, the completion tends to be located toward the top of the payzone to avoid early water breakthrough. Even though partial completion reduces the production rate, sometimes partial completion is designed (and very commonly applied in the field) for better production management.

FIGURE 3.5 Partially completed well configuration.

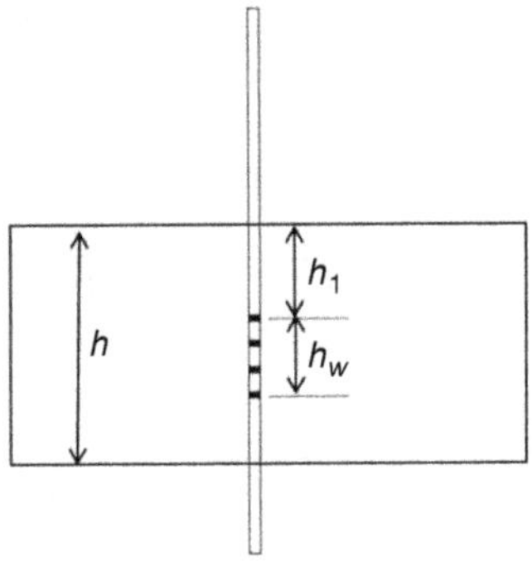

When planning to partially complete a well, a skin calculation should be conducted to make sure that the restriction by the completion will not jeopardize the well production.

Example 3.3 A wellbore with a radius of 0.4 ft is placed in a 150-ft thick formation. The completion covers 80 ft of the payzone and its location starts from the top of the formation. The permeability ratio in horizontal and vertical directions, k_H/k_V, is 10. Calculate the partial completion skin factor.

Solution

$$h_{wD} = \frac{80 \text{ ft}}{150 \text{ ft}} = 0.53 \tag{3.15}$$

$$r_D = \frac{0.4 \text{ ft}}{150 \text{ ft}} \sqrt{\frac{1}{10}} = 0.00084 \tag{3.16}$$

Because the completion is from the top of the payzone, $h_1 = 0$ ft, then

$$h_{1D} = \frac{0}{150 \text{ ft}} = 0 \tag{3.17}$$

With the dimensionless parameters,

$$A = \frac{1}{0 + \frac{0.53}{4}} = 7.6 \tag{3.18}$$

$$B = \frac{1}{0 + \frac{(3)(0.53)}{4}} = 2.5 \tag{3.19}$$

And the partial completion skin factor for this completion is

$$s_c = \left(\frac{1}{0.53} - 1\right) \ln \frac{\pi}{2 \times 0.00084} + \frac{1}{0.53} \ln\left(\frac{0.53}{2 + 0.53} \sqrt{\frac{7.6 - 1}{2.5 - 1}}\right) = 5.1 \tag{3.20}$$

3.2.2 Slanted Well Skin Factor

When a wellbore is drilled with a deviation angle α through a permeable zone, the contact of the well with the payzone increases compared with if the well is drilled vertically. The impact of this increased contact to the flow from reservoir to wellbore can be included in the inflow relationship through a slanted well skin factor. There are several models that can calculate the slanted well skin factor, we use the model developed by Besson (1990) to illustrate the concept of slanted well skin. For an isotropic reservoir,

$$s_\theta = \ln\left(\frac{4r_w \cos\alpha}{h}\right) + \cos\alpha \ln\left(\frac{h}{4r_w\sqrt{\cos\alpha}}\right) \tag{3.21}$$

For the anisotropic reservoir with horizontal permeability of k_H and vertical permeability of k_V,

$$s_\theta = \ln\left\{\frac{1}{I_{\text{ani}}\gamma}\left(\frac{4r_w \cos\alpha}{h}\right)\right\} + \frac{\cos\alpha}{\gamma} \ln\left\{\frac{2I_{\text{ani}}\sqrt{\gamma}}{1 + \frac{1}{\gamma}}\left(\frac{h}{4r_w\sqrt{\cos\alpha}}\right)\right\} \tag{3.22}$$

where

$$\gamma = \sqrt{\frac{1}{I_{ani}^2} + \cos^2\alpha\left(1 - \frac{1}{I_{ani}^2}\right)} \tag{3.23}$$

and I_{ani} is defined by Eq. (3.4).

Example 3.4 For the well in Example 3.2, if the well is 30° deviated from vertical, what is the slanted well skin?

Solution For an anisotropic reservoir, slanted skin is calculated by Eq. (3.22). We first calculate the parameter γ by Eq. (3.23)

$$\gamma = \sqrt{\frac{1}{10} + \cos^2(30)\left(1 - \frac{1}{10}\right)} = 0.88 \tag{3.24}$$

From the permeability ratio,

$$I_{ani} = \sqrt{10} = 3.16 \tag{3.25}$$

The slanted skin factor is

$$s_\theta = \ln\left\{\frac{1}{3.16(0.88)}\left(\frac{4(0.4)\cos(30)}{150}\right)\right\} + \frac{\cos(30)}{0.88}\ln\left\{\frac{2(3.16)\sqrt{0.88}}{1+\dfrac{1}{0.88}}\left(\frac{150}{4(0.4)\sqrt{\cos(30)}}\right)\right\}$$
$$= -0.2 \tag{3.26}$$

As mentioned before, slanted skin factor is always negative, but the magnitude is usually small unless the deviation angle becomes really large. The two structural skin factors can be simply added when both exist in a completion. For Examples 3.3 and 3.4, the combined skin factor is 4.9 (5.1–0.2). Table 3.1 shows the values of skin factor as the completion zone and well deviation angle change for Examples 3.3 and 3.4. The completion zone varies from 50 to 150 ft

	Angle			
Location	**0°**	**15°**	**30°**	**45°**
50/top	11.2	11.1	11	10.7
50/middle	10.23	10.19	10.07	9.79
100/top	2.89	2.85	2.73	2.45
100/middle	2.73	2.69	2.56	2.28
150	0	–0.04	–0.17	–0.44

Table 3.1 Partial Completion and Slanted Well Skin Factor

(fully completed), the location changes from top to middle, and the deviation angle changes from vertical to 45°. From the table we can see that for the same formation, the combined skin factor can change from 11.2 to −0.44, depending on how the completion is selected. 50-ft completion at the top or bottom of a vertical well ($\alpha = 0°$) has the highest skin factor. The partial completion skin model does not consider gravitation effects, and the partial completion skin factor is symmetrical in the vertical direction. In another words, 50 ft at the top yields the same skin factor as 50 ft at the bottom of the payzone.

3.3 Mechanical Skin Factor: Completion-Related Skin

Any completion-related skin can have two parts, the flow-related skin factor and flow-independent skin factor. For most oil wells, the flow-related skin factor can be small and flow-independent skin factor is the dominant component. When the fluid velocity for a well is high (high flow rate oil wells and gas wells), flow-dependent skin factor can be significant. The flow-dependent skin factor is most likely a nonlinear function of flow velocity (flow rate), thus making the flow problem more complicated to model. In this section, we will focus on flow-independent skin factors for different completions. For flow-dependent skin factor, the readers can refer to Furui's work (Furui, 2004).

3.3.1 Cased, Cemented, and Perforated

When a wellbore is cased, cemented, and perforated, the flow path changes; so does, most likely, the permeability in the completion region because of perforating. The parameters that describe a perforation completion include perforation length, l_{perf}, perforation radius, r_{perf}, perforation density, SPF, and perforation phasing, θ. Perforation density, SPF, is also the inverse of vertical distance between perforations, h_{perf}, if the perforations are aligned. The geometry and its parameters of a perforation completion are shown in Fig. 3.6.

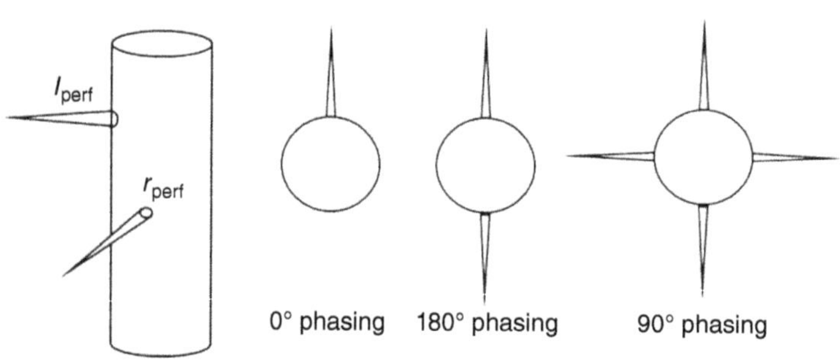

Figure 3.6 Completion geometry parameters.

Vertical Well Model

The skin factor for a perforated vertical well consists of three components: the flow convergence in the horizontal plane, s_H, convergence in the vertical plane, s_V, and the convergence caused by the wellbore blockage, s_{wb}. Perforation skin factor is expressed as (Karakas and Tariq, 1991)

$$s_p = s_H + s_V + s_{wb} \tag{3.27}$$

Each component in the above equation is calculated as follows. The horizontal skin component is

$$s_H = \ln \frac{r_w}{r'_w(\theta)} \tag{3.28}$$

where $r'_w(\theta)$ is the effective wellbore radius and is a function of the phasing angle θ

$$r'_w(\theta) = \begin{cases} \dfrac{l_{\text{perf}}}{4} & \text{for } \theta = 0 \\ a_\theta (r_w + l_{\text{perf}}) & \text{for } \theta \neq 0 \end{cases} \tag{3.29}$$

To calculate s_V, two dimensionless variables are defined based on the perforation parameters:

$$h_D = \frac{h_{\text{perf}}}{l_{\text{perf}}} I_{\text{ani}} \tag{3.30}$$

and

$$r_D = \frac{r_{\text{perf}}}{2h_{\text{perf}}} \left(1 + \frac{1}{I_{\text{ani}}}\right) \tag{3.31}$$

I_{ani} in the above equations is defined by Eq. (3.4). The vertical skin component is

$$s_V = 10^a h_D^{b-1} r_D^b \tag{3.32}$$

where

$$a = a_1 \log r_D + a_2 \tag{3.33}$$

and

$$b = b_1 r_D + b_2 \tag{3.34}$$

For the calculation of s_{wb}, a dimensionless parameter, r_{wD}, is defined first:

$$r_{wD} = \frac{r_w}{l_{\text{perf}} + r_w} \tag{3.35}$$

Perforation Phasing	a_θ	a_1	a_2	b_1	b_2	c_1	c_2
0° (360°)	0.250	−2.091	0.0453	5.1313	1.8672	1.6E−1	2.675
180°	0.500	−2.025	0.0943	3.0373	1.8115	2.6E−2	4.532
120°	0.648	−2.018	0.0634	1.6136	1.7770	6.6E−3	5.320
90°	0.726	−1.905	0.1038	1.5674	1.6935	1.9E−3	6.155
60°	0.813	−1.898	0.1023	1.3654	1.6490	3.0E−4	7.509
45°	0.860	−1.788	0.2398	1.1915	1.6392	4.6E−5	8.791

Source: From Karakas and Tariq (1991). Used with permission.

Table 3.2 The Constants in Perforation Skin Model

and, with this parameter,

$$s_{wb} = c_1 e^{c_2 r_{wD}} \tag{3.36}$$

In the above equations, the constants a_θ [Eq. (3.29)], a_1 and a_2 [Eq. (3.33)], b_1 and b_1 [Eq. (3.34)], c_1 and c_2 [Eq. (3.36)] are correlation constants as functions of perforation phasing, θ. The values of these constants are listed in Table 3.2.

Example 3.5 A well is completed with cementing and perforating. The perforation design has a phasing of 180°, perforation radius r_{perf} of 0.25 in, length l_{perf} of 8 in (0.667 ft), and the perforation density is 2 SPF ($h_{perf} = 0.5$). If the wellbore radius r_w is 0.4 ft and the permeability ratio in the horizontal and vertical directions, k_H/k_V, is 10, what is the skin factor caused by perforations?

Solution From Table 3.2, for perforation phasing equal to 180°, we have $a_\theta = 0.5$, $a_1 = -2.025$, $a_2 = 0.0943$, $b_1 = 3.0373$, $b_2 = 1.8115$, $c_1 = 0.026$, $c_2 = 0.532$. The horizontal skin factor is

$$r_w'(\theta) = 0.5(0.4 + 0.667) = 0.53 \tag{3.37}$$

$$s_H = \ln\frac{0.4}{0.53} = -0.29 \tag{3.38}$$

For s_v calculation, we calculate the two dimensionless parameters first,

$$h_D = \frac{0.5}{0.667}(3.16) = 2.4 \tag{3.39}$$

$$r_D = \frac{0.25}{2(0.5)(12)}\left(1 + \frac{1}{3.16}\right) = 0.027 \tag{3.40}$$

$$a = -2.025\log(0.027) + 0.0943 = 3.27 \tag{3.41}$$

$$b = 3.0373(0.027) + 1.8115 = 1.89 \tag{3.42}$$

thus

$$s_V = 10^{3.27} 2.4^{0.89} 0.027^{1.89} = 4.3 \tag{3.43}$$

To calculate s_{wb}, we calculate the dimensionless parameter first:

$$r_{wD} = \frac{0.4}{0.667 + 0.4} = 0.375 \tag{3.44}$$

The wellbore blockage skin factor is

$$s_{wb} = 0.026e^{4.532(0.375)} = 0.14 \tag{3.45}$$

The perforation skin factor in this case is

$$s_p = -0.29 + 4.3 + 0.14 = 4.2 \tag{3.46}$$

Notice that the vertical convergence effect, s_v, is the largest skin component in perforation skin factor. Perforation skin factor can be positive or negative. Longer perforations, larger perforation radius, and higher perforation density lead to lower perforation skin factor, or even a negative skin factor. Sometimes perforation design is limited by production operation strategies. For example, in well stimulation, injection velocity is the key for successful treatments, both acidizing and fracturing. The concept of limited entry is used which suggests small perforation densities (as small as 0.1 SPF). When design perforation, comprehensive considerations of operations (completion, production/injection, and stimulation) of a well should be included, and perforation skin factor should be checked to make sure the optimized performance can be achieved.

Horizontal Well Model

For horizontal wells, the orientation of perforations has an effect on perforation skin factor if the permeability field is anisotropic. If we assume that horizontal permeability is usually higher than vertical permeability, Fig. 3.7 illustrates that the vertically oriented perforations allow the fluid flow through the better permeability direction (in the horizontal permeability direction), and should be preferred when design perforation completion for horizontal wells. This is only significant when the perforation phasing is 0° or 180° (one or two perforations in the cross-sectional area of the wellbore). The perforation orientation effect diminishes as more perforations are evenly distributed around the circumference of the wellbore.

Furui et al. (2008) modified Karakas and Tariq's model for horizontal wells with a new parameter α to count for the perforation orientation effect. When the perforation is oriented horizontally, α equals to zero (0). The perforation skin model for horizontal wells is

$$s_p = s_{2D} + s_{wb} + s_{3D} \tag{3.47}$$

In the above equation, s_{2D} accounts for the flow convergence in the y-z plane. This skin factor can be negative or positive depending on

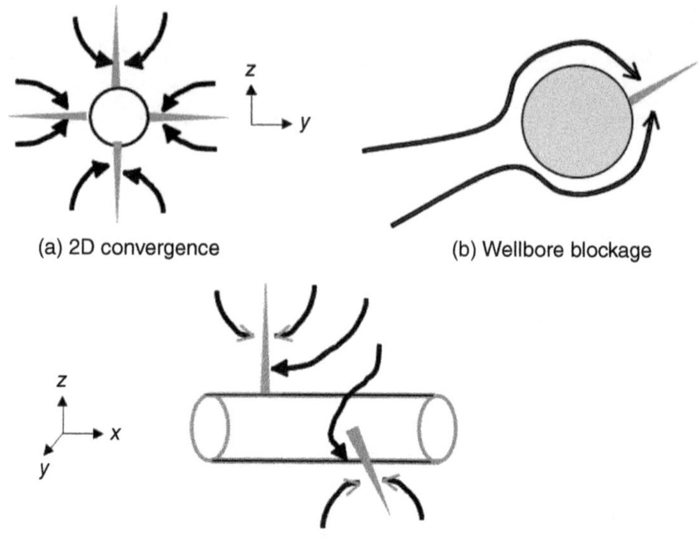

FIGURE 3.7 2D flow convergence, wellbore blockage, and 3D flow convergence skin components for perforation skin factor for horizontal wells.

the perforation conditions and the reservoir anisotropy. For one or two perforations (perforation phasing is 0° or 180°),

$$s_{2D} = a_m \ln\left(\frac{4}{l_{pD}}\right) + (1-a_m)\ln\left(\frac{1}{1+l_{pD}}\right) + \ln\left[\frac{\sqrt{k_y/k_z}+1}{2(\cos^2\alpha + (k_y/k_z)\sin^2\alpha)^{0.5}}\right] \tag{3.48}$$

For perforation phasing of 120° or 90°,

$$s_{2D} = a_m \ln\left(\frac{4}{l_{pD}}\right) + (1-a_m)\ln\left(\frac{1}{1+l_{pD}}\right) \tag{3.49}$$

where

$$l_{pD} = l_p / r_w \tag{3.50}$$

The wellbore blockage skin s_{wb} accounts for the effect of the wellbore existence (the fluid has to flow around the wellbore to find the perforations) and it is positive for any perforation condition. The wellbore blockage skin equation is

$$s_{wb} = b_m \ln\{c_m / l_{pD,\text{eff}} + \exp[-c_m / 1_{pD,\text{eff}}]\} \tag{3.51}$$

where effective dimensionless perforation length, $l_{pD,\text{eff}}$, is a function of perforation phasing. For perforation phasing equal to 0°,

$$l_{pD,\text{eff}} = l_{pD}\left[\frac{(k_y/k_z)\sin^2\alpha + \cos^2\alpha}{(k_y/k_z)\cos^2\alpha + \sin^2\alpha}\right]^{0.675} \tag{3.52}$$

if perforation phasing is 180°, then

$$l_{pD,\text{eff}} = l_{pD}\left[\frac{1}{(k_y/k_z)\cos^2\alpha + \sin^2\alpha}\right]^{0.625} \tag{3.53}$$

and for 120° and 90° phasing,

$$l_{pD,\text{eff}} = l_{pD} \tag{3.54}$$

The 3D convergence skin correlation equation accounts for the reservoir anisotropy and the perforation orientation, α,

$$s_{3D} = 10^{\beta_1} h_D^{\beta_2-1} r_{pD}^{\beta_2} \tag{3.55}$$

where

$$\beta_1 = d_m \log r_{pD} + e_m \tag{3.56}$$

$$\beta_2 = f_m r_{pD} + g_m \tag{3.57}$$

For 0° and 180° phasing,

$$h_D = \frac{h}{l_p\sqrt{(k_x/k_z)\sin^2\alpha + (k_x/k_y)\cos^2\alpha}} \tag{3.58}$$

$$r_{pD} = \frac{r_p}{2h}\left[\cos(\alpha'' - \alpha')\sqrt{\left(\frac{k_x}{k_y}\sin^2\alpha + \frac{k_x}{k_y}\cos^2\alpha\right)} + 1\right] \tag{3.59}$$

where

$$\alpha' = \arctan\left[\sqrt{\frac{k_y}{k_x}}\tan\alpha\right] \tag{3.60}$$

$$\alpha'' = \arctan\left[\sqrt{\frac{k_z}{k_y}}\tan\alpha\right] \tag{3.61}$$

And for 120° and 90° phasing,

$$h_D = \frac{h}{l_p}\left(\frac{\sqrt{k_y k_z}}{k_x}\right)^{0.5} \tag{3.62}$$

$$r_{pD} = \frac{r_p}{2h}\left[\left(\frac{k_x}{\sqrt{k_y k_z}}\right)^{0.5} + 1\right] \tag{3.63}$$

m	a_m	b_m	c_m	d_m	e_m	f_m	g_m
1	1	0.9	2	−2.091	0.0453	5.1313	1.8672
2	0.45	0.45	0.6	−2.025	0.0943	3.0373	1.8115
3	0.29	0.2	0.5	−2.018	0.0634	1.6136	1.777
4	0.19	0.19	0.3	−1.905	0.1038	1.5674	1.6935
>4	0	0	0				

TABLE 3.3 The Constants in the Perforation Skin Factor Model for Horizontal Wells

The numerical values of a_m, b_m, c_m, d_m, e_m, f_m, and g_m in Eqs. (3.47) to (3.57) are listed in Table 3.3. The perforation phasing is presented in the term of number of perforations per cross-sectional area of the wellbore. For example, 0° phasing corresponds to the number of perforations, m, of 1, 180° phasing sets m equals to 2, and so on.

3.3.2 Slotted Liner Completions

Slotted liner skin factor is calculated based on the geometry of the slots. Slot style is characterized by the arrangement of the slots around the circumference of a liner (see Fig. 3.7). Slots can be arranged with several in a group (multiple) or individually (single) as a unit. Slot units can be aligned (inline) or offset to each other (staggered). The most commonly used slot pattern is staggered slots with multiple slots in a unit (multiple staggered). From a flow convergence point of view, a multiple-staggered design yields the smallest skin factor.

For inline slotted liners, the slotted liner skin factor, s_{SL}, is a function of slot width, w_s; slot length, l_s; number of slots (or slot units) around the circumference of the liner, m_s; slot penetration ratio, λ (defined as the length of slots per unit length of pipe), the number of slots in the slot unit, n_s; the width of a slot unit, w_{sj}; and wellbore radius, r_w. All of these parameters are marked in Fig. 3.8.

The skin factor for slotted liner completions is a result of the radial convergence of flow from reservoir to wellbore, and the liner convergence flowing through the slots. The skin factor can be expressed as

$$s_{SL} = s_{SL,r} + s_{SL,l} \tag{3.64}$$

The subscripts, l and r, denote the linear flow through the slots and the radial flow outside the liner, respectively.

To calculate the radial flow components of slotted liner skin factor $s_{SL,r}$, we first define a group of dimensionless variables. The slot penetration ratio, λ, is the ratio of slot length to slot unit length,

$$\lambda = \frac{l_s}{l_u} \tag{3.65}$$

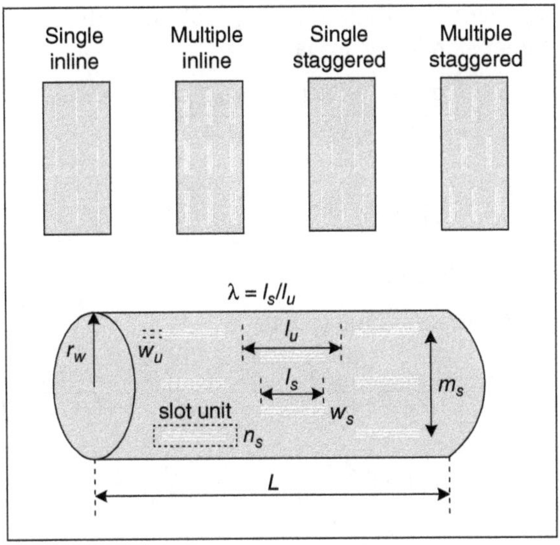

FIGURE 3.8 The geometry and parameters for slotted liner completion.

The dimensionless slot unit length γ is

$$\gamma = \frac{l_u}{2r_w} \tag{3.66}$$

The dimensionless slot length l_{sD} is

$$l_{sD} = \frac{l_s}{r_w} \tag{3.67}$$

The dimensionless slot width w_{sD} is

$$w_{sD} = \frac{w_s}{r_w} \tag{3.68}$$

The dimensionless unit width w_{uD} is

$$w_{uD} = \frac{w_u}{r_w} \tag{3.69}$$

and the geometry parameter υ is defined as

$$\upsilon = \begin{cases} \sin\left(\dfrac{\pi}{m_s}\right) & \text{when } m_s \neq 1 \\ 1.5 & \text{when } m_s = 1 \end{cases} \tag{3.70}$$

Then for high slot penetration ratio ($\gamma < \upsilon$), the radial component of the slotted liner skin factor $s_{SL,r}$ is

$$s_{SL,r} = \left(\frac{2}{n_s m_s \lambda}\right)\ln\left(\frac{1-\lambda+2l_{sD}/w_{sD}}{1-\lambda+n_s l_{sD}/w_{uD}}\right) + \left(\frac{2}{m_s}\right)\left[\frac{1}{\lambda}\ln(1-\lambda+l_{sD}/w_{uD})+\ln\left(\frac{2\lambda\upsilon}{l_{sD}}\right)\right] - \ln(1+\upsilon) \tag{3.71}$$

For low slot penetration ratio ($\gamma > \upsilon$),

$$s_{SL,r} = \left(\frac{2}{n_s m_s \lambda}\right)\ln\left(\frac{1-\lambda+2l_{sD}/w_{sD}}{1-\lambda+n_s l_{sD}/w_{uD}}\right) + \left(\frac{2}{m_s \lambda}\right)\ln\left[\frac{1-\lambda+l_{sD}/w_{uD}}{1-\lambda+l_{sD}/(2\upsilon)}\right] + \left(\frac{l_{sD}/\lambda}{l_{sD}-2(1-\lambda)}\right)\ln\left\{\left(\frac{\lambda+l_{sD}/2}{1+\upsilon}\right)\left[1+\frac{2\upsilon(1-\lambda)}{l_{sD}}\right]\right\} - \ln\left(1+\frac{l_{sD}}{2\lambda}\right) \tag{3.72}$$

Example 3.6 A slotted liner design has 4 sets of inline multiple slots around the pipe circumference ($m_s = 4$), and each set has 3 slots ($n_s = 3$). The slot length l_s is 2 in, the slot opening w_s is 0.02 in, and the slot unit length l_u is 4 in, which gives a penetration ratio λ of 0.5. w_u is 0.16 in. The pipe radius is 0.3 ft. Calculate the skin factor for this slotted liner design.

Solution We first calculate the dimensionless variables,

$$\gamma = \frac{4 \text{ in}}{2(0.3 \text{ ft})(12 \text{ in/ft})} = 0.56 \tag{3.73}$$

$$l_{sD} = \frac{2}{(0.3)(12)\left[\frac{\text{in}}{\text{ft}}\right]} = 0.56 \tag{3.74}$$

$$w_{sD} = \frac{0.02}{0.3(12 \text{ in/ft})} = 0.0056 \tag{3.75}$$

$$w_{uD} = \frac{0.16}{0.3(12 \text{ in/ft})} = 0.044 \tag{3.76}$$

and the geometry parameter,

$$\upsilon = \sin\left(\frac{\pi}{4}\right) = 0.707 \tag{3.77}$$

Because in this case $\gamma < \upsilon$, the radial component skin factor for this slotted liner design is

$$s_{SL,r} = \left(\frac{2}{3\times4\times0.5}\right)\ln\left(\frac{1-0.5+\frac{2\times0.56}{0.0056}}{1-0.5+\frac{3\times0.56}{0.044}}\right) + \frac{2}{4}\left[\frac{1}{0.5}\ln\left(1-0.5+\frac{0.56}{0.044}\right)+\ln\left(\frac{2\times0.5\times0.707}{0.56}\right)\right] - \ln(1+0.707) = 2.7 \tag{3.78}$$

n_s	$m_s = 2$	$m_s = 4$
1	10.5	4.9
2	7.1	3.2
3	6.1	2.7

TABLE 3.4 Radial Component of Skin Factor for Slotted Liners

Table 3.4 lists the slotted liner skin factors for radial flow for different liner designs. It shows that skin factor is more than tripled when the slot design changed from four units ($m_s = 4$) with multiple-staggered 3 slots ($n_s = 3$) to 2 units with single slot, indicating that completion design for slotted liners is critical to well performance.

For the linear flow convergence skin component, letting k_l be permeability of slots, the linear flow components are given by

$$s_{SL,l} = \left(\frac{2\pi}{n_s m_s w_{sD} \lambda}\right)\left(\frac{k}{k_l}\right) t_{sD} \tag{3.79}$$

where t_{sD} is the dimensionless thickness of a liner or the depth of partial plugging in slots, and is defined as

$$t_{sD} = \frac{t_s}{r_w} \tag{3.80}$$

For unplugged slots ($k_l >> k$), $s_{SL,l}$ can be neglected. If slots are plugged, the liner flow convergence skin can increase dramatically as a function of k_l and terminate the flow through the slots. Slot plugging can be a fatal problem in slotted liner completions. Once it happens, it is very difficult to clean the liner because of the huge pressure drop required to flow through the plugged slots. This situation becomes worse when the slot width is small. If a slotted liner is designed for sand control, the design should be compromised between sand production and liner plugging.

For staggered slots, we expect slightly smaller skin factors than inline slots. Staggering of slots is characterized by the distances between slots. The number of slots around the circumference of the liner may be effectively doubled as l_{uD} ($= l_{sD}/\lambda$) approaches 0. With these observations, we introduce the following correlation equation to obtain the effective slot angular distribution m'_s as

$$m'_s = m_s(1 + e^{-m_s l_{sD}/\lambda}) \tag{3.81}$$

The skin factors for staggered slots can be obtained by simply replacing m_s by m'_s in Eqs. (3.70) to (3.72).

Example 3.7 If in Example 3.6, instead of inline slots, the slots are staggered. Recalculate skin factor.

Solution For staggered slots,

$$m_s' = 4\left(1+\exp\left(-\frac{4\times 0.56}{0.5}\right)\right) = 4.05 \tag{3.82}$$

This changes γ in Eq. (3.73) to 0.57, and the impact on skin factor is insignificant.

Unlike cased and perforated wells, the effect of the formation anisotropy in permeability is not significant for slotted liner completion with four or more slot units around the circumference of a liner. The liner orientation relative to the permeability field has no significant effect on skin factor. Using the coordinate transformation into the equivalent isotropic system, the dimensionless slot length can be obtained by a function of the formation permeabilities as

$$l_{sD} = \frac{2l_s}{r_w}\frac{1}{\left(\sqrt{\frac{k_x}{k_z}}+\sqrt{\frac{k_x}{k_y}}\right)} \tag{3.83}$$

In Example 3.6, for a permeability anisotropic ratio of 3.16, l_{sD} become 0.13, leads to a slotted liner skin factor of 2.1, instead of 2.7 for the isotropic case.

3.3.3 Perforated Liner Completions

The flow geometry of perforated liners is similar to that of slotted liners. The convergent flow to perforations becomes more likely to be hemispherical flow rather than radial flow. Replacing l_s with a perforation diameter, $2r_p$, in Eq. (3.66), γ for perforated liners can be given by

$$\gamma = \frac{r_{pD}}{\lambda} \tag{3.84}$$

The skin factor for perforated liner completions is obtained as

For $\gamma < \upsilon$

$$s_{PL} = \left(\frac{2}{m_p\lambda}\right)\left(\frac{3}{2}-\lambda\right)+\left(\frac{2}{m_p}\right)\ln\left(\frac{\upsilon\lambda}{r_{pD}}\right) - \ln(1+\upsilon) \tag{3.85}$$

For $\gamma > \upsilon$

$$s_{PL} = \left(\frac{2}{m_p\lambda}\right)\left(\frac{3}{2}-\frac{r_{pD}}{\upsilon}\right)+\left(\frac{r_{pD}/\lambda}{r_{pD}+\lambda-1}\right) \times \ln\left\{\left(\frac{\lambda+r_{pD}}{1+\upsilon}\right)\left(1+\frac{\upsilon(1-\lambda)}{r_{pD}}\right)\right\} - \ln(1+r_{pD}/\lambda) \tag{3.86}$$

Example 3.8 A perforated liner has a phasing of 90°, and perforation radius of 0.4 in. The penetration ratio λ is 0.5. Calculate the skin factor for this completion. The pipe radius is 0.3 ft (3.6 in).

Solution Calculate the dimensionless variable first,

$$r_{pD} = \frac{r_p}{r_{\text{pipe}}} = \frac{0.4}{3.6} = 0.11 \tag{3.87}$$

$$\gamma = \frac{0.11}{0.5} = 0.22 \tag{3.88}$$

For $\gamma < \upsilon$

$$s_{PL} = \left(\frac{2}{4 \times 0.5}\right)\left(\frac{3}{2} - 0.5\right) + \left(\frac{2}{4}\right)\ln\left(\frac{0.707 \times 0.5}{0.11}\right) - \ln(1 + 0.5) = 1.2 \tag{3.89}$$

The perforated liner has smaller skin factor than what the slotted liner has when the ratio of open area to the total surface area is the same. Notice that when the perforation radius is 0.4 in, it does not have much function for sand control. This is a typical base pipe structure to strength the completion mechanical strings.

3.3.4 Gravel Pack Completions

As mentioned in Chapter 2, gravel pack completions are more commonly used in high permeability formations with sand production problems. Gravel packs can be installed in both openhole completions and cased/perforated completions. The mechanical skin factor caused by flow through these types of completions is presented by Furui (2004). Flow through an openhole gravel pack is modeled as flow through a series of three permeability regions—the gravel, a possible damaged zone in the formation, and the undamaged zone away from the well. The skin factor for this system can be easily obtained as

$$s_{GO} = \frac{k}{k_g}\ln\frac{r_w}{r_g} + \left(\frac{k}{k_s} - 1\right)\ln\frac{r_s}{r_w} \tag{3.90}$$

This equation shows that when the permeability of gravel, k_g, is significantly higher than the reservoir permeability, the gravel pack has insignificant effect on flow (the first term on the right-hand side is very small). Even when the gravel pack permeability, k_g, is comparable with the formation permeability k an openhole gravel pack has small restriction to production.

For a cased-hole gravel pack, the flow path is changed from a radial flow in openhole completion to a combination of reservoir flow converging to the perforations, through the gravel-filled perforations, and through the gravel packed annulus between the casing and the screens or liners, as shown in Fig. 3.9.

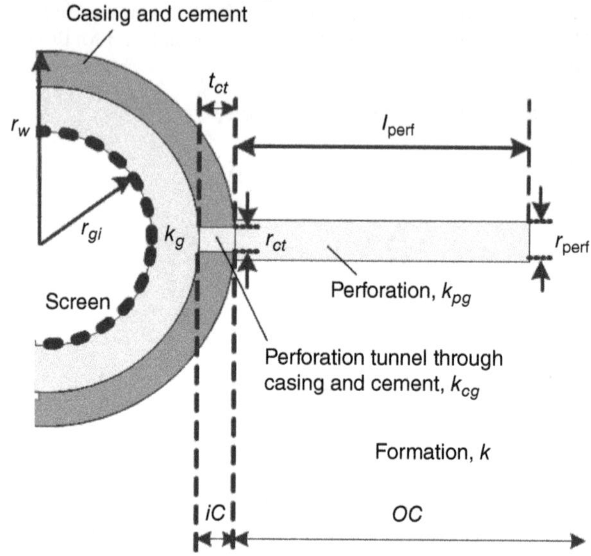

FIGURE 3.9 Cased-hole gravel pack geometry.

If the permeability of the gravel in the annulus between the screen and the casing, k_g, is large, the cased-hole gravel pack skin factor can be separated into two components: the skin factor component related to the flow through the perforation tunnel of cement and casing and the skin component related to the flow within the perforation cavity extending into the formation

$$s_{CG} = \left(\frac{2h_{pD}}{r_{tD}^2}\right)\frac{l_{tD}}{k_{gD}} + \left[\left(1-k_{gD}^{-0.5}\right)s_p + k_{gD}^{-0.5}s_{pl}\right] \tag{3.91}$$

The first term on the right-hand side of the above equation represents the flow convergence in the perforation tunnels through casing and cementing, and the second term is for the flow through the gravel-filled perforations. The dimensionless variables in Eq. (3.90) are defined as,

$$r_{tD} = \frac{r_{ct}}{r_w} \tag{3.92}$$

$$h_{pD} = \frac{h_{perf}}{r_w} \tag{3.93}$$

$$l_{tD} = \frac{l_{ct}}{r_w} \tag{3.94}$$

and

$$k_{gD} = \frac{k_g}{k} \tag{3.95}$$

In the above equations, r_{ct} is radius of the perforation tunnel, l_{ct} is length of the tunnel through the casing and cement, h_{perf} is perforation spacing (inverse of perforation density), and k_g is permeability of the gravel-packed tunnel. The flow to the gravel-packed cavity extending into the formation is analogous to the flow to a perforated completion or to the flow to a perforated liner completion, depending on the lengths of perforations and the permeability of the gravel packed in perforation tunnels. If the perforations are very short or the permeability of the gravel in the perforation tunnels is low, the completion will behave essentially like a perforated liner; longer perforations containing high permeability gravel result in the completion behavior of a cased, perforated well. s_p in Eq. (3.91) is the standard perforation skin factor as discussed in Sec. 3.3.1 and s_{pl} is perforated liner skin factor,

$$s_{pl} = \frac{3h_{pD}}{2r_{pD}} + \ln\left[\frac{\upsilon^2}{h_{pD}^2(1+\upsilon)}\right] - 0.61 \tag{3.96}$$

and

$$\upsilon = \begin{cases} 1.5 & \theta = 360°,\ 0° \\ \sin\left(\dfrac{\pi}{360/\theta}\right) & \theta \neq 360°,\ 0° \end{cases} \tag{3.97}$$

Example 3.9 A gravel pack design has r_{ct} of 0.4 in, t_{ct} of 1.2 in, r_w of 6 in, and the gravel permeability, k_g is 500,000 mD filled both in the casing/cement tunnels and in the perforations. The perforation has a density of 10 SPF (h_{perf} = 0.1), 90° phasing, the perforation radius is 0.4 in, and the perforation length is 6 in. The reservoir has a permeability of 500 mD. Calculate gravel pack skin factor for k_g = 500,000 mD (the original design) and k_g = 500 mD.

Solution We calculate the dimensionless parameters first,

$$r_{tD} = \frac{0.4}{6} = 0.067 \tag{3.98}$$

$$h_{pD} = \frac{1}{(SPF)r_w} = \frac{1}{10 \times 0.5} = 0.2 \tag{3.99}$$

$$l_{tD} = \frac{1.2}{6} = 0.2 \tag{3.100}$$

and

$$k_{gD} = \frac{500{,}000}{500} = 1000 \tag{3.101}$$

From the perforation skin model, we can calculate the perforation skin factor s_p. In this case,

$$s_p = -0.298 + 0.128 + 0.08 = -0.16 \tag{3.102}$$

For perforated liner,

$$\upsilon = \sin\left(\frac{\pi 90}{360}\right) = 0.707 \tag{3.103}$$

$$s_{pl} = \frac{3(0.12)}{2(0.067)} + \ln\left[\frac{0.707^2}{0.2^2(1+0.707)}\right] - 0.61 \tag{3.104}$$
$$= 5.86$$

For the gravel permeability of 500,000 mD,

$$s_{CG} = \left(\frac{2\times 0.2}{(0.067)^2}\right)\frac{0.2}{1000} + [(1-1000^{-0.5})\times(-0.16) + 1000^{-0.5}\times 5.86] \tag{3.105}$$
$$= 0.05$$

If k_g is 500 mD,

$$s_{CG} = \left(\frac{2\times 0.2}{(0.067)^2}\right)\frac{0.2}{1} + (5.86) \tag{3.106}$$
$$= 23.7$$

The example shows that if the gravel pack permeability is reduced during production (e.g., fine migration that plugs up the gravel pack), the gravel pack itself can generate a large restriction to flow.

3.3.5 Inflow Control Device

Inflow control devices generate a skin factor to balance the flow distribution along a horizontal wellbore. Unlike other completion, the sole purpose of using inflow control devices (ICDs) in well completion is to generate this skin factor. The skin factor for ICDs depends on the type of ICDs. Since all ICDs generate flow-related pressure drop, most likely, the pressure drop is linearly proportional to q^2,

$$\Delta p_{ICD} = Cq^2\rho_m \tag{3.107}$$

where ρ_m is density of the fluid. The constant C in the above equation is determined by the ICD structures. For example, for a channel-type ICD,

$$\Delta p_{\text{ICD, channel}} = \frac{2f_f\rho_m u^2 L}{g_c D} \tag{3.108}$$

or in terms of flow rate q in the field units (Economides et al., 2012),

$$\Delta p_{\text{ICD, channel}} = \frac{7.35\times 10^{-7} f_f\rho_m q^2 L}{D^5} \tag{3.109}$$

where f_f is friction factor, ρ_m is fluid density, L is the length of flow path, and D is the channel diameter. L and D are design parameters. For a channel that wraps around a base pipe,

$$L = n_c \pi D_{bc} \tag{3.110}$$

In the above equation, n_c is the number of times the channel is wrapped around the base pipe, and D_{bc} is the base pipe diameter. We can see that the pressure drop across an ICD is structure dependent. Once L, D, D_{bs}, and n_c is defined, C in Eq. (3.107) can be determined,

$$C = \frac{7.35 \times 10^{-7} f_f L}{D^5} \tag{3.111}$$

Example 3.10 For a channel ICD with a channel that is 0.2 in diameter, and wraps around a 4½-in base pipe two times, generate a plot of Δp_{ICD} versus q^2. Assume that the friction factor is 0.001 and the averaged density of the fluid is 58 lb/ft³.

Solution The length of channel path is calculated as

$$L = 2 \times 2 \times 3.14 \times 4.5 = 56.5 \text{ [in]} \tag{3.112}$$

Friction factor is a function of Reynold's number, which is also related to flow rate. Thus the pressure drop can be calculated by

$$\Delta p_{\mathrm{ICD,\,channel}} = \frac{7.35 \times 10^{-7} \times 0.001 \times 58 \left(\dfrac{56.5}{12 \text{ [in/ft]}} \right)}{0.2^5} q^2 \tag{3.113}$$

Figure 3.10 shows the plot of flow rate versus Δp_{ICD}. The additional pressure of an ICD is self-regulated by the flow rate. At low flow rate, pressure drop is small. Pressure drop increases greatly as flow rate increases.

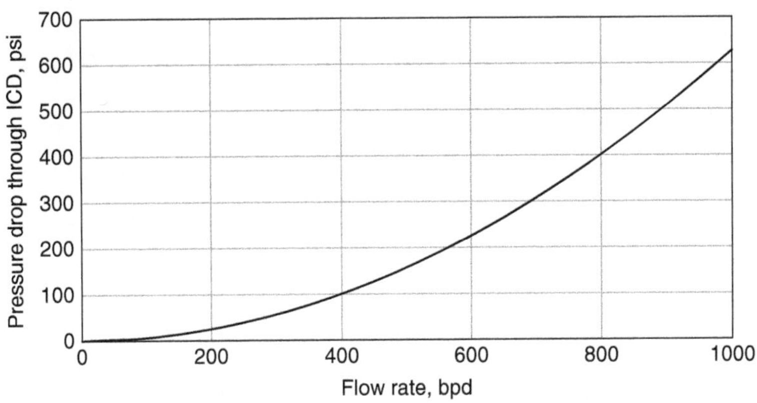

FIGURE 3.10 Pressure drop for ICD.

3.4 Combined Skin Effect

When formation damage is present in near-wellbore regions, completion skin effects can be compounded to a much higher level for some completions. This is because the fluid not only flows through a deviated path from the ideal path, it also flows through a lower permeability region compared with the original permeability. For some completions, this compound skin factor is much larger than the sum of damage skin and completion skin.

3.4.1 Combined Effect of Damage Skin with Perforation Skin

When a damage zone exists in a perforated completion, the compound skin factor is determined by if the perforations are extended outside the damaged zone or not. Figure 3.11 shows the two different cases for perforation in damaged zone. The skin factor is calculated by

$$(s_p)_d = \begin{cases} (s_d)_0 + \dfrac{k}{k_s} s_p & \text{when } (l_{\text{perf}} + r_w) < r_s \\ s'_p & \text{when } (l_{\text{perf}} + r_w) > r_s \end{cases} \tag{3.114}$$

where s'_p is the equivalent perforation skin factor and it is calculated by the same perforation skin equations with the new parameter l'_{perf} and r'_w

$$l'_{\text{perf}} = l_{\text{perf}} - \left(1 - \frac{k_s}{k}\right)(r_s - r_w) \tag{3.115}$$

and

$$r'_w = r_s - \frac{k_s}{k}(r_s - r_w) \tag{3.116}$$

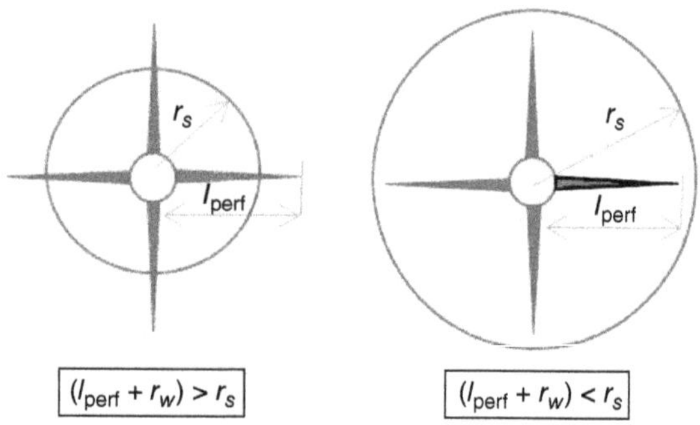

FIGURE 3.11 Two cases for perforation skin in damaged zones.

Example 3.11 If the perforation completion in Example 3.3 is in a damaged formation, and the damage is described as in Example 3.1, what is the compound skin factor because of damage and perforation? If the perforation length is extended to 18 in, what is the skin factor?

Solution From Example 3.1, the damaged zone radius is 1.3 ft and the damage skin factor is 11.3. From Example 3.5, the perforation length is 0.667 ft, and the perforation skin factor is 4.2. Using Eq. (3.114), the compound skin factor is

$$(s_p)_d = (11.3) + (10 \times 4.2) = 53.3 \tag{3.117}$$

If l_{perf} is 18 in,

$$l'_{\text{perf}} = 12 - (1 - 0.1) \times (1.4 - 0.4) = 11.1 \tag{3.118}$$

$$r'_w = 1.4 - 0.1 \times (1.4 - 0.4) = 1.3 \tag{3.119}$$

$$s_p = -2.7 + 0.35 + 0.03 = -2.3 \tag{3.120}$$

This example shows that when placing a perforated completion in a damaged zone, the resultant skin factor can be much larger than simply adding the two skin factors, s_d and s_p, together if the perforations are shorter than the damaged zone depth. On the other hand, if the perforations extend beyond the damaged zone, the compound skin factor is smaller than the addition of the two skin factors. The skin factor in the example changed from 53 to −2.3 just by reperforating the well with longer perforations. Understanding this phenomenon can help to design perforation completion more efficiently for better well flow performance.

3.4.2 Effect of Formation Damage on Slotted/Perforated Liner Completions

The skin caused by slots (perforations) and formation damage only influence the pressure drop in the region close to the wellbore. If formation damage exists around a slotted or perforated liner, the total skin effect is the combination of damage skin and slot/perforation skin.

For a slotted liner completion in a damaged zone, we follow a similar approach for perforation completion in a damaged zone (Karakas and Tariq, 1991) to derive a skin equation for slotted liner completions inside a damaged zone,

$$(s_{SL})_d = s_d + s_{SL,l} + \left(\frac{k}{k_s}\right) s_{SL,r} \tag{3.121}$$

Once again, similar to perforated wells in a damaged zone, an important fact is that the effect of formation damage on slotted liners is much worse than openhole completions with damage. The reduced permeability region magnifies the skin due to the convergent flow inside the damaged zone. The linear flow term in Eq. (3.121) does not have the multiplier, k/k_s, because the linear flow convergence happens inside the perforation tunnel in the casing and cement.

Example 3.12 For the slotted liner shown in Example 3.6, if the well is completed in a damaged zone with a damaged radius of 1 ft from the wellbore wall and a permeability reduction to 10% of the original in the damaged zone, what is the compound skin factor for this case?

Solution From Example 3.1, the damage skin, s_d, is 11.3, and from Example 3.6, the linear slotted liner skin, $s_{SL,l}$, is 10.9. If we neglect the linear skin for slotted liner, from Eq. (3.121),

$$(s_{SL})_d = 11.3 + 10 \times 10.9 = 120.3 \tag{3.122}$$

Skin factor at this level will tremendously reduce the flow rate from a well. The important fact is that even through the damage described here is moderate, the compound skin effect between damage and completion can be overwhelmingly detrimental.

3.5 Effect of Skin Factor on Production

Understanding how skin factors impact well flow behavior is important for well performance optimization. In the previous sections of this chapter, we have discussed how to estimate skin factors that are caused by drilling, completion and operation. Using the equations of inflow performance relationship (IPR), we can evaluate how skin effects change well flow rates for vertical wells and horizontal wells.

3.5.1 Skin Effect for Vertical Wells

For a single phase oil well under steady-state condition, the IPR equation is

$$q = \frac{kh(p_e - p_{wf})}{141.2\ B\mu\left(\ln\left(\frac{r_e}{r_w}\right) + s\right)} \tag{3.123}$$

If we assume that during drilling, completion and well operations, the changes of near-wellbore permeability and flow path can be captured by skin factors, and the rest of the overall reservoir properties in the above equation does not change, then Eq. (3.123) indicates that the impact of skin factor on flow rate is determined by the value of skin factor related to the term $\ln(r_e/r_w)$. Using the ratio of productivity index for original performance without skin J_{orig} and altered performance with skin J_{skin},

$$\frac{J_{\text{skin}}}{J_{\text{orig}}} = \frac{\dfrac{q_{\text{orig}}}{p_e - p_{wf}}}{\dfrac{q_{\text{skin}}}{p_e - p_{wf}}} = \frac{\ln\left(\dfrac{r_e}{r_w}\right)}{\ln\left(\dfrac{r_e}{r_w}\right) + s} \tag{3.124}$$

We can estimate the flow rate changes caused by skin factor.

Example 3.13 For the openhole completion with formation damage in Example 3.1, evaluate the skin impact on well performance.

Solution Example 3.1 shows a damage skin of 11.3. Using r_e of 790 ft (about 40 acre drainage radius) and r_w of 0.4 ft,

$$\frac{J_{\text{damage}}}{J_{\text{orig}}} = \frac{\ln\left(\frac{790}{0.25}\right)}{\ln\left(\frac{790}{0.25}\right)+11.2} = \frac{7.6}{7.6+11.2} = 0.40 \tag{3.125}$$

This simple calculation shows that a damage skin of 11.2 can reduce the well productivity to 40% of the undamaged production, a significant impact on well performance. In fact, since the two variables, r_e and r_w, that we use to evaluate skin impact are in a logarithm term, the change of this term is limited. Typical vertical wells have the value of this logarithm term ranging from 6 to 9. This tells us that once the skin value is in this range, the flow rate will reduce to about 50% of the original. Table 3.5 shows the productivity index ratios for other examples in this chapter.

It is clear from Table 3.4 that skin factor is an important concept to understand individual well performance. Well structure, well completion and formation damage can alter flow rate dramatically, especially when a completion is placed in a damaged zone. In Example 3.10, perforation design changes the well performance from extremely detrimental (87% reduction) to significant improvement (36% increasing). In Example 3.11, a plugged slotted liner in a damaged zone almost completely plugs the well flow (94% reduction). As we emphasize the important functions of well completions in Chapter 2, any overdesigned completion with unnecessary restriction to production will cause negative impact to the well performance, and should be avoided.

	Skin Factor	PI Ratio	Production Change, %
Formation damage, Example 3.1	11.3	0.40	−60
Partial completion, Example 3.3	5.1	0.59	−41
Perforation, Example 3.5	4.2	0.64	−36
Slotted liner, Example 3.6	10.9	0.41	−59
Short perforation in damage zone, Example 3.10	53.3	0.13	−87
Long perforation in damage zone, Example 3.10	−2.3	1.36	+36
Slotted Liner in damage, Example 3.11	120.3	0.06	−94

TABLE 3.5 Productivity Index Ratio for Examples 3.1, 3.3, 3.5, 3.6, 3.10, and 3.11

3.5.2 Skin Effects for Horizontal Wells

Using a steady-state equation for a horizontal well in a box-shaped reservoir, the IPR is (Furui et al., 2003)

$$q = \frac{kb(p_e - p_{wf})}{141.2\mu B_o\left(\ln\left[\frac{hI_{\text{ani}}}{r_w(I_{\text{ani}}+1)}\right] + \frac{\pi y_b}{hI_{\text{ani}}} - 1.224 + s + s_R\right)} \tag{3.126}$$

where b is the dimension of the reservoir along the well direction, y_b is the dimension of the reservoir perpendicular to the well direction, h is the third dimension of the reservoir, and it is also the pay-zone thickness. s_R, the denominator of the equation, is a skin factor accounting for the convergence when a well is not fully drilled through the reservoir ($L < b$). From the above equation, we can see that the impact of skin factor on well flow now is compared with several terms instead of just $\ln(r_e/r_w)$, as in for a vertical well case. The terms py_b/hI_{ani} and s_R sometimes can be large, which reduces the impact of skin factor on flow rate.

Example 3.14 Evaluate the damage skin effect calculated from Example 3.2 on flow rate. Use reservoir dimension of 5000 ft in length, 2000 ft in width, and 150 ft in thickness. Assume the partial penetration skin factor s_R is 3.

Solution If the reservoir width is 2000 ft, y_b is 1000 ft. Using the productivity index ratio, we can evaluate skin impact on well performance

$$\begin{aligned}\frac{J_{\text{damage}}}{J_{\text{orig}}} &= \frac{\left(\ln\left[\frac{hI_{\text{ani}}}{r_w(I_{\text{ani}}+1)}\right] + \frac{\pi y_b}{hI_{\text{ani}}} - 1.224 + s_R\right)}{\left(\ln\left[\frac{hI_{\text{ani}}}{r_w(I_{\text{ani}}+1)}\right] + \frac{\pi y_b}{hI_{\text{ani}}} - 1.224 + s + s_R\right)} \\ &= \frac{\left(\ln\left[\frac{150\times 3.16}{0.4\times(3.16+1)}\right] + \frac{3.14\times 1000}{150\times 3.16} - 1.224 + 3\right)}{\left(\ln\left[\frac{150\times 3.16}{0.4\times(3.16+1)}\right] + \frac{3.14\times 1000}{150\times 3.16} - 1.224 + 6.1 + 3\right)} \\ &= 0.7\end{aligned} \tag{3.127}$$

In this specific case, formation damage effect reduces production to 38%, and it is less compared with the same damage condition for the vertical well (0.4 PI ratio or 60% rate reduction). This is not always true. For fully penetrated wells (well length is equal to reservoir length) with small width to thickness ratio (y_b/h), skin factor can have significant effect on well performance.

The equations for skin factor calculations for different completion structures are summarized in Table 3.6.

	Parameters
<u>Barefoot completions</u> $s_{fo} = \left(\frac{k}{k_s} - 1\right)\ln\frac{r_s}{r_w}$ for vertical wells $s_{fo} = \left(\frac{k}{k_s} - 1\right)\ln\left\{\frac{1}{I_{ani}+1}\left[\frac{r_{sH}}{r_w} + \sqrt{\left(\frac{r_{sH}}{r_w}\right)^2 + I_{ani}^2 - 1}\right]\right\}$ for horizontal wells	
<u>Openhole with slotted liners</u> $s_{sl} = s_{fo} + s_{sl,l}^0 + \frac{k}{k_s}s_{sl,r}^0$ where $s_{sl,l}^0 = \left(\frac{2\pi}{n_s m_s w_{sD}\lambda}\right)\left(\frac{k}{k_l}\right)t_{sD}$ $s_{sl,r}^0 = \left(\frac{2}{n_s m_s\lambda}\right)\ln\left(\frac{1-\lambda+2l_{sD}/w_{sD}}{1-\lambda+n_s l_{sD}/w_{uD}}\right)$ $+\left(\frac{2}{m_s}\right)\left[\frac{1}{\lambda}\ln(1-\lambda+l_{sD}/w_{uD})+\ln\left(\frac{2\lambda\upsilon}{l_{sD}}\right)\right]$ $-\ln(1+\upsilon)$ for $\gamma < \upsilon$ $s_{sl,r}^0 = \left(\frac{2}{n_s m_s\lambda}\right)\ln\left(\frac{1-\lambda+2l_{sD}/w_{sD}}{1-\lambda+n_s l_{sD}/w_{uD}}\right)$ $+\left(\frac{2}{m_s\lambda}\right)\left[\ln\left(\frac{1-\lambda+l_{sD}/w_{uD}}{1-\lambda+l_{sD}/(2\lambda)}\right)\right]$ $+\left[\frac{l_{sD}/\lambda}{l_{sD}-2(1-\lambda)}\right]\ln\left\{\left(\frac{\lambda+l_{sD}/2}{1+\upsilon}\right)\left[1+\frac{2\upsilon(1-\lambda)}{l_{sD}}\right]\right\}$ $-\ln\left(1+\frac{l_{sD}}{2\lambda}\right)$ for $\gamma > \upsilon$	$t_{sD} = t_s/r_w$ $w_{sD} = w_s/r_w$ $w_{uD} = w_u/r_w$ $l_{sD} = l_s/r_w$ $\lambda = l_s/l_u$ $\upsilon = (\sin\pi/m_s)$ $\gamma = l_{sD}/(2\lambda)$
<u>Openhole perforated liners</u> $s_p l = s_f o + k/k_s s_p l^0$ where $s_{pl}^0 = \left(\frac{2}{m_p\lambda}\right)\left(\frac{3}{2}-\lambda\right)+\left(\frac{2}{m_p}\right)\ln\left(\frac{\upsilon\lambda}{r_{pD}}\right)$ $-\ln(1+\upsilon)$ for $\gamma < \upsilon$	$r_{pD} = r_p/r_w$ $\gamma = r_{pD}/\lambda$

TABLE 3.6 Equations for Skin Factor Calculations

	Parameters
$s_{pl}^{0}=\left(\frac{2}{m_p\lambda}\right)\left(\frac{3}{2}-\frac{r_{pD}}{\upsilon}\right)+\left(\frac{r_{pD}/\lambda}{r_{pD}+\lambda-1}\right)$ $\times\ln\left\{\left(\frac{\lambda+r_{pD}}{1+\upsilon}\right)\left[1+\frac{\upsilon(1-\lambda)}{r_{pD}}\right]\right\}$ $-\ln\left(1+\frac{r_{pD}}{\lambda}\right)\text{for } \gamma>\upsilon$	
Openhole gravel packs $s_{og}=s_{fo}+\frac{k}{k_g}\ln\frac{r_w}{r_g}$	
Cased, perforated wells $s_p=s_H+s_V+s_{wb}$ where $s_H=\ln\frac{r_w}{r_w'(\theta)}$ $s_V=10^a h_D^{b-1} r_D^b$ s_{wb} $=c_1\exp\left[c_2\left(\frac{r_w}{l_{perf}+r_w}\right)\right]$	$r_w'(\theta)=\begin{cases}\frac{l_{perf}}{4} & \text{for } \theta=0\\ a_\theta(r_w+l_{perf}) & \text{for } \theta\neq 0\end{cases}$ $h_D=\frac{h_{perf}}{l_{perf}}\sqrt{\frac{k_H}{k_V}}$ $r_D=\frac{r_{perf}}{2h_{perf}}\left(1+\sqrt{\frac{k_V}{k_H}}\right)$
Cased hole gravel packs $s_{cg}=\left(\frac{2h_{Dp}}{r_{Dct}^2}\right)\frac{t_{Dct}}{k_{Dcg}}+\left(1-k_{Dpg}^{-0.5}\right)s_p+k_{Dpg}^{-0.5}s_{pl}$	$r_{Dct}=r_{ct}/r_w$ $h_{Dp}=h_p/r_w$ $t_{Dct}=t_{ct}/r_w$
Hydraulically fractured wells $s_{fr}=\ln\left(\frac{x_f/r_w}{\frac{\pi}{C_{fD}}+2}\right)$	$C_{fD}=\frac{k_f w}{k x_f}$
Matrix acidized wells $s_{wh}=\left(\frac{k}{k_{wh}}-1\right)\ln\frac{r_{wh}}{r_w}$	

TABLE 3.6 Equations for Skin Factor Calculations (*Continued*)

Nomenclature

a = parameter defined by Eq. (3.33)

a_m = constant used in the perforation skin model (Table 3.3), dimensionless

a_θ = constant used in the perforation skin model (Table 3.2), dimensionless

a_1 = constant used in the perforation skin model (Table 3.2), dimensionless

a_2 = constant used in the perforation skin model (Table 3.2), dimensionless

A = parameter defined by Eq. (3.10)

b = parameter defined by Eq. (3.34)

b_m = constant used in the perforation skin model (Table 3.3), dimensionless

b_1 = constant used in the perforation skin model (Table 3.2), dimensionless

b_2 = constant used in the perforation skin model (Table 3.2), dimensionless

B = parameter defined by Eq. (3.11)

B = formation volume factor, res bbl/STB, m^3/Sm^3

B_o = oil formation volume factor, res bbl/STB, m^3/Sm^3

c_m = constant used in the perforation skin model (Table 3.3), dimensionless

c_1 = constant used in the perforation skin model (Table 3.2), dimensionless

c_2 = constant used in the perforation skin model (Table 3.2), dimensionless

C = constant defined by Eq. (3.111)

d_m = constant used in the perforation skin model (Table 3.3), dimensionless

D = channel diameter, ft, m

D_{bc} = base pipe diameter, ft, m

e_m = constant used in the perforation skin model (Table 3.3), dimensionless

f_f = friction factor, dimensionless

f_m = constant used in the perforation skin model (Table 3.3), dimensionless

g_m = constant used in the perforation skin model (Table 3.3), dimensionless

h = reservoir thickness, ft, m

h_D = parameter defined by Eq. (3.30), dimensionless

h_D = dimensionless thickness, dimensionless [Eq. (3.58)]
h_1 = distance from the top of completed zone to the top of payzone, ft, m
h_{1D} = dimensionless distance defined in Eq. (3.14), dimensionless
h_{perf} = perforation spacing, ft, m
h_{pD} = dimensionless perforation spacing, dimensionless
h_{wD} = dimensionless thickness of completed zone, dimensionless [Eq. (3.12)]
I_{ani} = index of permeability anisotropy, dimensionless
J = productivity index, STB/d/psi, $m^3/s/Pa$
k = permeability, md, m^2
k_d = permeability in the damaged zone, md, m^2
k_g = permeability of gravel, md, m^2
k_{gD} = normalized gravel permeability, dimensionless
k_H = permeability in the horizontal direction, md, m^2
k_s = reduced permeability in the damage zone, md, m^2
k_{sH} = reduced permeability in the horizontal direction, md, m^2
k_V = permeability in the vertical direction, md, m^2
k_x = permeability in the x-direction, md, m^2
k_y = permeability in the y-direction, md, m^2
k_z = permeability in the z-direction, md, m^2
l_{ct} = length of the tunnel through casing and cement, ft, m
l_{pD} = dimensionless perforation length, dimensionless
$l_{pD,eff}$ = effective dimensionless perforation length, dimensionless
l_{perf} = perforation length, ft, m
l'_{perf} = effective perforation length calculated by Eq. (3.115), ft, m
l_s = slot length, ft, m
l_{sD} = dimensionless slot length, dimensionless
l_{tD} = dimensionless tunnel length, dimensionless
l_u = slot unit length, ft, m
L = length of the flow path, ft, m
m_s = number of slot units around the circumference of the liner
m'_s = number of effective slot angular distribution
n_c = number of times the channel is wrapped around the base pipe
n_s = number of slots in the slot unit
p_e = reservoir outer boundary pressure, psi, Pa
q = flow rate, STB/d, m^3/d
r_{ct} = radius of the perforation tunnel, ft, m
r_e = drainage radius, ft, m
r_D = dimensionless variable defined by Eq. (3.13), dimensionless

r_D = dimensionless variable defined by Eq. (3.30), dimensionless
r_g = radius of the gravel-packed zone, ft, m
r_p = perforation radius, ft, m
r_{pD} = dimensionless perforation radius, ft, m
r_s = radius of the damaged zone, ft, m
r_{sH} = damage zone radius in the horizontal direction, ft, m
r_{tD} = dimensionless perforation tunnel, dimensionless
r_w = wellbore radius, ft, m
r_{wD} = dimensionless parameter defined by Eq. (3.35)
r'_w = effective wellbore radius, ft, m
s = skin factor, dimensionless
s_c = partial completion skin, dimensionless
s_{CG} = skin factor for cased-hole gravel packed wells, dimensionless
s_d = damage skin factor, dimensionless
s_{eq} = equivalent skin, dimensionless
s_{GO} = skin factor for openhole gravel packed wells, dimensionless
s_H = flow convergence skin in the horizontal plane, dimensionless
s_p = perforation skin, dimensionless
s'_p = equivalent perforation skin, dimensionless
s_{PL} = perforated liner skin, dimensionless
s_R = skin resulting from partial penetration, dimensionless
s_{SL} = slotted liner skin, dimensionless
$s_{SL,r}$ = skin due to radial flow outside the liner, dimensionless
$s_{SL,l}$ = skin due to liner flow through the slots, dimensionless
s_V = flow convergence skin in the vertical plane, dimensionless
s_{wb} = wellbore blockage skin, dimensionless
s_θ = slanted well skin factor, dimensionless
s_{2D} = 2D flow convergence skin, dimensionless
s_{3D} = 3D flow convergence skin, dimensionless
w_s = slot width, ft, m
w_{sD} = dimensionless slot width, dimensionless
w_u = slot unit width, ft, m
w_{uD} = dimensionless slot unit width, dimensionless
x = x coordinate, ft, m
y_b = reservoir drainage width, ft, m
α = well deviation angle, degree
α' = parameter defined by Eq. (3.60)
α'' = parameter defined by Eq. (3.61)
β_1 = parameter defined by Eq. (3.56)
β_2 = parameter defined by Eq. (3.57)

γ = parameter defined by Eq. (3.23)
γ = dimensionless slot unit length, dimensionless
λ = slot penetration ratio, fraction
υ = geometric parameter defined by Eq. (3.70), dimensionless
θ = perforation phase angle, degree
Δp_{ICD} = pressure drop across ICD
μ = viscosity, cp, Pa·s
ρ_m = fluid density, lb/ft^3, kg/m^3

Subscripts

damage = damaged
orig = original performance
skin = altered performance with skin

References

Besson, J. (1990, January 1). Performance of Slanted and Horizontal Wells on an Anisotropic Medium. *Society of Petroleum Engineers*. doi:10.2118/20965-MS

Furui, K. (2004, May). "A Comprehensive Skin Factor Model for Well Completions Based on Finite Element Simulations," Ph.D. Dissertation, The University of Texas at Austin.

Furui, K., Zhu, D., and Hill, A. D. (2008, September 1). "A New Skin-Factor Model for Perforated Horizontal Wells," *SPE Drilling & Completion*, Vol. 23 (3), pp. 205–215. doi:10.2118/77363-PA.

Furui, K., Zhu, D., and Hill, A. D. (2003, August 1). "A Rigorous Formation Damage Skin Factor and Reservoir Inflow Model for a Horizontal Well," *SPE Production & Facilities*, Vol. 18 (3), pp. 151–157. doi:10.2118/84964-PA.

Hawkins, M. F. Jr. (1956). "A Note on the Skin Effect," *Transactions of AIME*, Vol. 207, pp. 356–357.

Karakas, M., and Tariq, S. M. (1991, February 1). "Semianalytical Productivity Models for Perforated Completions," *SPE Production Engineering*, Vol. 6 (1), pp. 73–82. doi:10.2118/18247-PA.

Papatzacos, P. (1987, May 1). "Approximate Partial-Penetration Pseudoskin for Infinite- Conductivity Wells," *SPE Reservoir Engineering*, Vol. 2 (2), 227–234. doi:10.2118/13956-PA.

Sui, W., Zhu, D., Hill, A. D., and Ehlig-Economides, C. (2008, January 1). "Model for Transient Temperature and Pressure Behavior in Commingled Vertical Wells," presented at SPE Russian Oil and Gas Technical Conference and Exhibition, Moscow, Russia, 28–30 Oct. doi:10.2118/115200-MS.

CHAPTER 4

Completions for Fracturing

Hydraulic fracturing, first applied in 1947 in the Hugoton Gas Field, Kansas (Gidley et al., 1989), has played a tremendous role in the history of oil and gas production, and is one of the most crucial technologies today for efficient and economic development of oil and gas reserves, especially in unconventional plays. Fracturing techniques have progressed from simply injecting high-pressure fluid into vertical wells to increase injectivity to today's multiple-stage fracturing of long horizontal wells with millions of pounds of proppant and millions of gallons of fracture fluid. The progression of hydraulic fracturing strongly depends on the advances in completion technologies. Many special completion methods have been developed to meet the challenges of current fracturing practices. In this chapter, we review some basics of hydraulic fracture design and completion requirements and discuss the completion technologies that are currently used in the field for fracturing.

4.1 Basics of Hydraulic Fracturing Stimulation

There are a plethora of literature regarding hydraulic fracturing technology in petroleum engineering, from rock mechanics, to fracture treatment design, to fracture fluid systems and proppant selection, to fracture diagnosis by pressure analysis, and fractured well performance modeling. This section only reviews the fundamentals of fracturing design that relates to completion selection and design. For the details of fracturing theories, design, and field executions, the following books are recommended: *Recent Advances in Hydraulic Fracturing* (Gidley et al., 1989); *Modern Fracturing: Enhancing Natural Gas Production* (Economides and Martin, 2007); *Hydraulic Fracturing* (Smith and Montgomery, 2015); *Essentials of Hydraulic Fracturing: Vertical & Horizontal Wellbores* (Veatch et al., 2017); *Petroleum Production Systems* (Economides et al., 2012); and for rock mechanics theory applied to fracturing, *Mechanics of Hydraulic Fracturing* (Yew and Wong, 2015).

4.1.1 In-Situ Stress of Formation and Its Impact on Hydraulic Fracturing

An element of formation rock is exposed to a three-dimensional (3D) in-situ stress field at the reservoir condition. Understanding in-situ stress is important in designing a fracture treatment because in-situ stresses determine what breakdown pressure is required to initiate a fracture, orientation and dimensions of fractures created, as well as the final conductivity of fractures after closure.

Figure 4.1 illustrates a rock element under 3D in-situ stress condition. The three principal stresses are vertical stress, σ_v, maximum horizontal stress, σ_H, and minimum horizontal stress, σ_h. In general, if the formation is not shallow, σ_v usually has the highest magnitude of the three principal stresses. As formation depth increases, the difference between vertical stress and horizontal stresses increases. Because of this, most fractures created by hydraulic fracturing are vertical fractures unless the formation is very shallow, and the minimum stress is the vertical stress (fracture is always perpendicular to the minimum stress).

There have been extended discussions on how to estimate the stress field (orientation and magnitude of principal stress components). Openhole logs, borehole breakout information, wellbore caliper logs, and rock mechanics lab measurements all provide information for stress field estimation. Since fracture propagation is dominated by the in-situ stresses, it is imperative to obtain a description of the stress field as accurate as possible before designing a fracture treatment.

To initiate a fracture, the injection pressure at bottomhole has to be higher than the formation breakdown pressure. The breakdown pressure is related to the rock mechanical properties of the formation. For a linear elastic solid, the breakdown pressure is calculated as (assuming that compression is positive),

$$p_{bd} = 3\sigma_h - \sigma_H + T_0 - p_p \tag{4.1}$$

where T_0 is tensile strength of the formation and p_p is pore pressure. Both are properties of the formation.

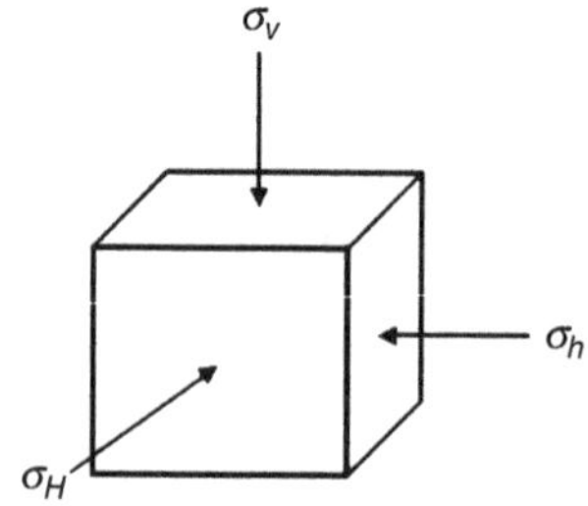

FIGURE 4.1 A rock element under 3D stresses at reservoir condition.

Example 4.1 Breakdown pressure calculation.
Estimate the breakdown pressure of the targeted fracture zone at a formation depth of 8000 ft. At that depth, the vertical stress is 9360 psi. The minimum horizontal stress estimated from vertical stress is 4829 psi, and the maximum stress is 6329 psi. Use the normal hydrostatic pressure gradient to estimate the pore pressure, and use 1200 psi for the tensile strength.

Solution We first calculate the pore pressure using the water pressure gradient,

$$p_p = 0.433\left[\frac{\text{psi}}{\text{ft}}\right] \times 8000[\text{ft}] = 3464 \text{ psi} \tag{4.2}$$

then the breakdown pressure is

$$p_{bd} = 3 \times 4829 - 6329 + 1200 - 3463 = 5894 \text{ psi} \tag{4.3}$$

4.1.2 Breakdown Pressure for Highly Deviated Wells

When a wellbore is deviated from the vertical direction with an inclination angle, the wellbore may no longer be aligned with the principal stress field, and the breakdown pressure is a dependent of deviation angle because the shear force becomes a part of the stress field that determines the breakdown pressure.

Yew et al. (1989) analyzed borehole breakdown pressure and orientation of fractures in deviated wells based on 3D elasticity theory. Figure 4.2 shows the coordinate systems. The coordinate system (1, 2, 3) is aligned with the principal stresses (σ_1, σ_2, σ_3), and in this case, they are (σ_h, σ_H, σ_V). The first step is to transfer the coordinate system (1, 2, 3) to a coordinate system that aligned with the deviated wellbore. Defining α as the wellbore azimuth angle from the axis-1 (aligned with σ_h) and β is the wellbore deviation angle from the horizontal plane, the coordinate transformation is achieved by rotating the axis-1 with the α angle, and the axis-2 with the β angle, as shown in Fig. 4.2. In the transferred new system, the x axis always intersects the highest point of the wellbore cross-sectional area, the y axis is perpendicular to the x axis, and the z axis is along the wellbore. The stress system from the general coordinate system (1, 2, 3) to the new coordinate system (x, y, z) can be established by

$$\begin{bmatrix} \sigma_{xx} & \sigma_{xy} & \sigma_{xz} \\ \sigma_{yx} & \sigma_{yy} & \sigma_{yz} \\ \sigma_{zx} & \sigma_{zy} & \sigma_{zz} \end{bmatrix} = A \begin{bmatrix} \sigma_1 & 0 & 0 \\ 0 & \sigma_2 & 0 \\ 0 & 0 & \sigma_3 \end{bmatrix} A^T \tag{4.4}$$

where the transformation matrix, A, is defined as

$$A = \begin{bmatrix} \cos(\alpha)\cos(\beta) & \sin(\alpha)\cos(\beta) & \sin(\beta) \\ -\sin(\alpha) & \cos(\alpha) & 0 \\ -\cos(\alpha)\sin(\beta) & -\sin(\alpha)\sin(\beta) & \cos(\beta) \end{bmatrix} \tag{4.5}$$

In Eq. (4.5), A^T is the transpose of matrix A. The stress distribution around the hole surface (at $r = r_w$) is a function of the location along

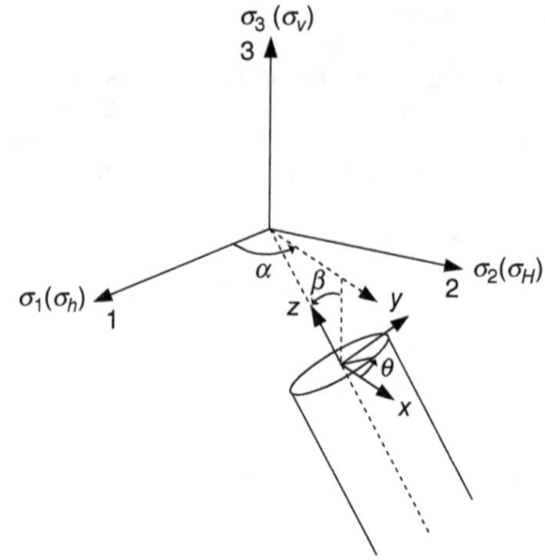

FIGURE 4.2 3D coordinate system for stress calculation of deviated wells.

the circumference of the borehole. If we define a new angle θ with θ equals to zero when we are at the highest point of the wellbore circumference (aligned with the x axis), the resultant stress can be calculated as follows:

$$\begin{aligned}
&\sigma_{rr} = -p_{bd} \\
&\sigma_{\theta\theta} = p_{bd} + \sigma_{xx}(1 - 2\cos 2\theta) + \sigma_{yy}(1 + 2\cos 2\theta) - 4\sigma_{xy}\sin 2\theta \\
&\sigma_{zz} = \sigma_{zz}^{\infty} + v[2(\sigma_{xx} - \sigma_{yy})\cos 2\theta + 4\sigma_{xy}\sin 2\theta] \qquad (4.6) \\
&\sigma_{r\theta} = \sigma_{rz} = 0 \\
&\sigma_{\theta z} = -2\sigma_{xz}\sin\theta + 2\sigma_{yz}\cos\theta
\end{aligned}$$

where v is the Poisson's ratio of the rock and σ_{zz}^{∞} denotes σ_{zz} in the (x, y, z) coordinate [from Eq. (4.4)]. Notice that Eq. (4.6) assumes that compression is negative. Angle θ is measured around the circumference of the hole (zero defined as the highest point of the hole). A fracture will be initiated on the borehole surface when the maximum (principal) stress on the surface reaches the tensile failure stress, σ_T (= $-T_0$). The maximum stress can be calculated by

$$\sigma_{max}(\theta) = \frac{\sigma_{zz} + \sigma_{\theta\theta}}{2} + \sqrt{\left(\frac{\sigma_{zz} - \sigma_{\theta\theta}}{2}\right)^2 + \sigma_{\theta z}^2} \qquad (4.7)$$

The hole breakdown pressure can be calculated by equating the maximum stress σ_{max} to σ_T. Notice that for homogeneous rock, this is likely happening at θ equals to 0° and 180°. The calculation can be done by first assuming a borehole pressure p_{bd} and carrying out computation iteratively until the calculated maximum stress from Eq. (4.7) equals to σ_T. The orientation of the fracture γ aligns with the borehole surface (θ-z plane) can be calculated from the following equation:

$$\gamma = \frac{1}{2}\tan^{-1}\left(\frac{2\sigma_{\theta z}}{\sigma_{\theta\theta} - \sigma_{zz}}\right) \tag{4.8}$$

Example 4.2 Calculate the breakdown pressure for Example 4.1 if the well is deviated 65° from vertical ($\beta = 65°$) but aligned with the principal stress in horizontal direction ($\alpha = 0°$).

Solution We first compute the *A* matrix and its transpose based on Eq. (4.5). Because the deviation angle is from the vertical axis, when we say the well is 65° deviated from vertical, β is 65° in this case. The *A* matrix is

$$A = \begin{bmatrix} \cos 0 \cos 65 & \sin 0 \cos 65 & \sin 65 \\ -\sin 0 & \cos 0 & 0 \\ -\cos 0 \sin 65 & -\sin 0 \sin 65 & \cos 65 \end{bmatrix} \tag{4.9}$$

$$A = \begin{bmatrix} 0.42 & 0 & 0.91 \\ 0 & 1 & 0 \\ -0.91 & 0 & 0.42 \end{bmatrix} \tag{4.10}$$

The effective stress is the difference between original absolute stress and pore pressure. Using 3464 psi as the pore pressure, the original effective stress is

$$\begin{bmatrix} \sigma_1' \\ \sigma_2' \\ \sigma_3' \end{bmatrix} = \begin{bmatrix} \sigma_1 + p_P \\ \sigma_2 + p_P \\ \sigma_3 + p_P \end{bmatrix}$$

$$= \begin{bmatrix} -4829 + 3464 & 0 & 0 \\ 0 & -6329 + 3464 & 0 \\ 0 & 0 & -9360 + 3464 \end{bmatrix} \tag{4.11}$$

$$= \begin{bmatrix} -1365 & 0 & 0 \\ 0 & -2865 & 0 \\ 0 & 0 & -5896 \end{bmatrix}$$

The resultant stress field after the coordinate transform by Eq. (4.6) is

$$\begin{bmatrix} \sigma_x \\ \sigma_y \\ \sigma_z \end{bmatrix} = \begin{bmatrix} 0.42 & 0 & 0.91 \\ 0 & 1 & 0 \\ -0.91 & 0 & 0.42 \end{bmatrix} \begin{bmatrix} -1365 & 0 & 0 \\ 0 & -2865 & 0 \\ 0 & 0 & -5896 \end{bmatrix} \begin{bmatrix} 0.42 & 0 & -0.91 \\ 0 & 1 & 0 \\ 0.91 & 0 & 0.42 \end{bmatrix}$$

$$= \begin{bmatrix} -577 & 0 & -5344 \\ 0 & -2865 & 0 \\ 1237 & 0 & -2499 \end{bmatrix} \begin{bmatrix} 0.42 & 0 & -0.91 \\ 0 & 1 & 0 \\ 0.91 & 0 & 0.42 \end{bmatrix} \tag{4.12}$$

$$= \begin{bmatrix} -5087 & 0 & 1735 \\ 0 & -2865 & 0 \\ 1735 & 0 & -2174 \end{bmatrix}$$

To obtain the breakdown pressure, we need to use Eqs. (4.7) and (4.8) to calculate σ_{max} with an assumed p_{bd}. With iteration, when σ_{max} is equal to the tensile failure stress (1200 psi), the fracture will initiate at the borehole location of $\theta = 0$. For the given condition, the calculation is as following. Assuming that the breakdown pressure is 4708 psi,

$$\begin{aligned}
\sigma_{rr} &= -4708 \text{ psi} \\
\sigma_{\theta\theta} &= 4708 - 5087(1 - 2\cos 0) - 2865(1 + 2\cos 0) - 0 = 1200 \text{ psi} \\
\sigma_{zz} &= -2174 - 0.2[2(-5087 + 2865)\cos 0 + 0] = -1286 \text{ psi} \\
\sigma_{r\theta} &= \sigma_{rz} = 0 \\
\sigma_{\theta z} &= -0
\end{aligned} \tag{4.13}$$

and

$$\sigma_{max}(\theta) = \frac{-1286 - 1200}{2} + \sqrt{\left(\frac{-1286 + 1200}{2}\right)^2 + 0} = 1200 \text{ psi} \tag{4.14}$$

As the failure criteria states that when σ_{max} equals to $-T_0$, which is -1200 psi for this example, fracture will initiate. This example shows that the effective breakdown pressure needs to be 2086 psi to initiate fractures. The breakdown pressure calculated is the effective breakdown pressure, and the absolute breakdown pressure in this case is

$$p_{bd} = p'_{bd} + p_p = 4708 + 3464 = 8172 \text{ psi} \tag{4.15}$$

The breakdown pressure is increased when the well is deviated (compare Example 4.1 with Example 4.2). Breakdown pressure is only for initiating fractures; once a fracture is created, fracture pressure is used to estimate the bottomhole pressure required to propagate the fracture. The injection pressure in the fracture is usually several hundred psi higher than the fracture pressure (or closure pressure). In the field, the closure pressure is approximated by the minimum horizontal stress. The difference between injection pressure in the fracture and closure pressure is defined as net pressure. This is discussed in the next section.

4.1.3 Fracture Geometry Prediction

Predicting fracture geometry (fracture width, length, and height) is the first step of fracture design. There are three different approaches to estimate fracture geometry: two-dimensional (2D) analytical models, pseudo-3D models, and fully 3D models. 2D geometry models assume that fracture height is a constant through fracture propagation, and only fracture width and length are growing during pumping with an assumed geometric shape. Pseudo-3D models assume that in addition to width and length growth, fracture height can grow in a stepwise fashion and correlations are used to estimate the height growth. Fully 3D models simulate rock failure and fracture propagation in length, width, and height directions under a true 3D stress field in a 3D coordinate system.

The commonly used 2D fracture geometry models are the Perkins-Kern-Nordgren (PKN) model, the Kristianovich-Geertsma-de Klerk (KGD)

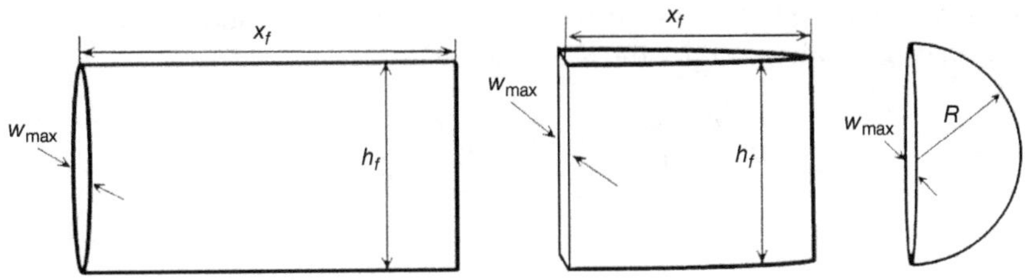

FIGURE 4.3 Geometry assumptions for PKN model, KGD model, and radial fracture model.

model, and the penny-shaped model, and each assumes an ideal fracture geometry such that an analytical solution can be obtained. Figure 4.3 shows the assumed geometries for the PKN model, the KGD model, and the radial fracture model (penny-shaped fracture). All models assume that fractures created are bi-wing, linear fractures.

The PKN model (Perkins and Kern, 1961; Nordgren, 1972) assumes that at the wellbore, a created fracture has an elliptical-shaped opening with the maximum width located at the middle of the fracture-height direction. Along the fracture length, the fracture also has an elliptical opening. Fracture height is a constant. With the assumed shape, at an injection rate of q_i, and for a rock with Young's modulus E the maximum fracture width at the wellbore w_{max} and fracture half-length x_f are related as

$$w_{max} = 3.27\left[\frac{q_i \mu x_f}{E'}\right]^{1/4} \tag{4.16}$$

where E' is the plane strain modulus and is related to Young's modulus E by

$$E' = \frac{E}{1-\upsilon^2} \tag{4.17}$$

In Eq. (4.16), μ is fluid viscosity and υ is Poisson's ratio. Equation (4.16) is in consistent unit. The maximum fracture width at the wellbore can be expressed in terms of average fracture width, which is more convenient for design purpose. Converting to the field units, q is in bbl/min, μ is in cp, x_f is in ft, and E' is in psi, the average width is

$$\bar{w} = 0.19\left[\frac{q_i \mu x_f}{E'}\right]^{1/4} \tag{4.18}$$

The average width calculated from Eq. (4.18) is in inches.

The KGD model (Khristianovic(h) and Zheltov, 1955; Geertsma and de Klerk, 1969) was developed by assuming that fracture openings at the wellbore are rectangular-shaped and width decreases along the fracture with an elliptical geometry. Similarly, following the unit system used in Eqs. (4.16) and (4.18), the maximum and average width can be calculated by the KGD model as:

$$w_{max} = 2.7\left(\frac{q_i \mu x_f^2}{E' h_f}\right)^{1/4}\left(\frac{\pi}{4}\right) = 2.12\left(\frac{q_i \mu x_f^2}{E' h_f}\right)^{1/4} \tag{4.19}$$

and

$$\bar{w} = 0.34\left(\pi\frac{q_i \mu x_f^2}{E' h_f}\right)^{1/4}\left(\frac{\pi}{4}\right) = 0.27\left(\pi\frac{q_i \mu x_f^2}{E' h_f}\right)^{1/4} \tag{4.20}$$

For penny-shaped fractures, the width equations depend on the assumptions used for the fracture opening at the wellbore, and it can be either PKN (as shown in Fig. 4.3) or KGD type of opening. For PKN-typed opening at the wellbore (Gidley et al., 1989), with consistent units,

$$w_{max} = 1.66\left(\frac{q_i \mu R}{E'}\right)^{1/4} \tag{4.21}$$

and

$$\bar{w} = 0.79\left(\frac{q_i \mu R}{E'}\right)^{1/4} \tag{4.22}$$

For the KGD-type opening,

$$w_{max} = 2.65\left(\frac{q_i \mu R}{E'}\right)^{1/4} \tag{4.23}$$

and

$$\bar{w} = 1.36\left(\frac{q_i \mu R}{E'}\right)^{1/4} \tag{4.24}$$

Example 4.3 Fracture width with PKN model.
Calculate the average fracture width during injection for a treatment designed with a pumping rate of 65 bpm and fracture fluid viscosity of 100 cp. The density of the fracture fluid is 64 lbm/ft^3. Young's modulus is 10^6 psi and Poisson's ratio is 0.2.

Solution Using Eq. (4.4), we can build a relationship between fracture width and half length.

$$\bar{w} = 0.19\left(\frac{65 \times 100 \times (1 - 0.2^2) x_f}{10^6}\right)^{1/4} = 0.053 x_f^{1/4} \tag{4.25}$$

Figure 4.4 shows the relationship between fracture half-length and fracture average width defined by PKN model. For a 700-ft-long fracture, the average width at the wellbore is 0.272 in.

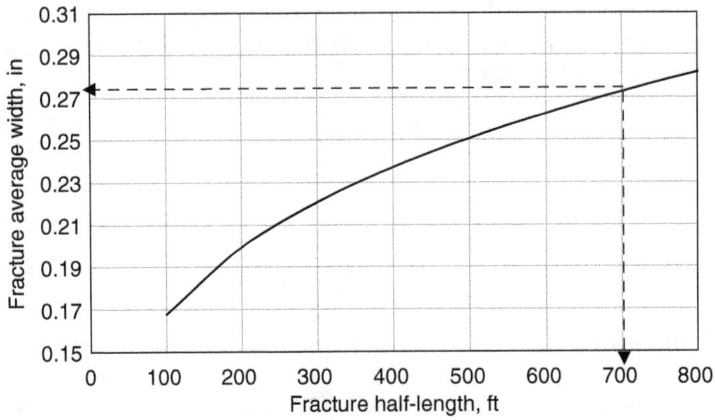

FIGURE 4.4 Fracture geometry calculation by PKN model for Example 4.2.

4.1.4 Treatment Design

Once fracture width and half-length are determined by a fracture geometry model, treatment designs can be carried out to determine how to pump the targeted fracture.

Pumping Time

Pumping time is estimated from the volumetric balance based on the geometry of fractures calculated by fracture geometry models. Since the total volume pumped, V_i, should be the sum of the created fracture volume, V_{frac}, and the leakoff volume of fluid into the formation, V_{leakoff}, we have

$$V_i = V_{\text{frac}} + V_{\text{leakoff}} \tag{4.26}$$

where the injection volume V_i is simply the product of injection rate and time

$$V_i = q_i t_i \tag{4.27}$$

and the created fracture volume V_{frac} is

$$V_{\text{frac}} = 2x_f \overline{w} h_f \tag{4.28}$$

V_{leakoff} can be estimated by Eq. (4.29)

$$V_{\text{leakoff}} = K_L C_L 2(x_f h_f) r_p t_i^{1/2} \tag{4.29}$$

In Eq. (4.29), C_L is the leakoff coefficient and can be obtained from a minifrac test before the fracture pumping, or from lab measurements with the similar rock, fracturing fluid, and pumping conditions. K_L is a fracture-efficiency-related parameter and defined as (Nolte, 1986)

$$K_L = \frac{1}{2}\left[\frac{8}{3}\eta + \pi(1-\eta)\right] \tag{4.30}$$

and fracture efficiency, η, is defined as

$$\eta = \frac{V_{\text{frac}}}{V_i} \tag{4.31}$$

Substituting Eqs. (4.27) to (4.31) into Eq. (4.26), we have

$$q_i t_i = 2x_f h_f \overline{w} + K_L C_L (2A_f) r_p \sqrt{t_i} \tag{4.32}$$

where r_p is the height ratio of payzone thickness h to fracture height h_f. Injection time t_i can be solved from Eq. (4.32). Notice that Eq. (4.32) should be solved in consistent units.

Example 4.4 For the fracture described in Example 4.2, how long does it take to pump this fracture (width is 0.272 in, half-length is 700 ft, and height is 100 ft)? Use injection rate of 65 bpm. Leakoff coefficient is 0.001 ft/min$^{0.5}$ and K_L is 1.47.

Solution From Eq. (4.32), using the given data, we have

$$\begin{aligned} 65 \times 5.615 t_i &= 2 \times 700 \times \left(\frac{0.272}{12}\right) \times 100 \\ &\quad + 1.47 \times 0.001 \times 2 \times 2 \times 700 \times 100 \sqrt{t_i} \end{aligned} \tag{4.33}$$

Rearranging Eq. (4.33),

$$365 t_i - 411.6\sqrt{t_i} - 3173.3 = 0 \tag{4.34}$$

Solving for t_i,

$$t_i = \left(\frac{-(-411.6) + \sqrt{(-411.6)^2 - 4 \times 365 \times (-3173.3)}}{2 \times 365}\right)^2 = 12.7 \text{ min} \tag{4.35}$$

Using 13 min as the total injection time, then

$$V_i = 65\left[\frac{\text{bbl}}{\text{min}}\right](5.615)\left[\frac{\text{ft}^3}{\text{bbl}}\right](13)[\text{min}] = 4745 \text{ ft}^3 \tag{4.36}$$

and

$$\eta = \frac{3173.3}{4745} = 0.67 \tag{4.37}$$

Pad Pumping

In order to create a fracture with proppant packed in it, pumping starts with clear fracture fluid without proppant, and this is called pad fluid. Proppant is only added after a portion of fracture is created with pad fluids. The volume of pad fluid depends on the leakoff velocity or fracture efficiency. Formations with higher leakoff velocity require higher pad volume. Nolte (1986) presented the following equation [Eq. (4.38)] that can be used to estimate the volume of pad fluid and its pumping time

$$V_{\text{pad}} \approx V_i\left(\frac{1-\eta}{1+\eta}\right) \tag{4.38}$$

Define

$$\varepsilon = \frac{1-\eta}{1+\eta} \tag{4.39}$$

then,

$$V_{\text{pad}} = V_i\varepsilon \tag{4.40}$$

and simply,

$$t_{\text{pad}} = \frac{V_{\text{pad}}}{q_i} \tag{4.41}$$

Example 4.5 For the fracture design in Examples 4.2 and 4.3, calculate pad volume and pad pumping time.

Solution First we calculate the parameter ε by Eq. (4.39)

$$\varepsilon = \frac{1-0.67}{1+0.67} = 0.2 \tag{4.42}$$

From Eqs. (4.40) and (4.41), we have

$$V_{\text{pad}} = 0.2 \times 4745 = 949 \text{ ft}^3 \tag{4.43}$$

and

$$t_{\text{pad}} = \frac{942}{65 \times 5.615} = 2.9 \text{ min} \tag{4.44}$$

The examples show that to create this 700-ft-long fracture with an average width of 0.27 in at the wellbore, the total pumping time is 13 min, and the first 3 min should be pad fluid pumping without proppant.

Proppant Schedule

After a fracture is initiated by pad fluid pumping, proppant is gradually added to the fluid to finish fracturing. Fracture fluid with proppant is referred as slurry. Proppant concentration is ramped up during slurry pumping with a schedule that satisfy two requirements; the concentration of proppant in slurry cannot increase too fast to avoid screen-out; and the maximum concentration of proppant has to be below some constrains at the end of the pumping time, which commonly set by the pump capacity of solid handling. Pumping schedule (proppant concentration as a function of pumping time) depends on the formation leakoff behavior. A guideline by an exponent function is

$$c_p(t) = c_f \left(\frac{t - t_{\text{pad}}}{t_i - t_{\text{pad}}} \right)^{\varepsilon} \tag{4.45}$$

where c_f is the pump solid-handling capacity, and is also referred as the end-of-job concentration. In field operation, the proppant concentration increases as a step-function following the curve defined by Eq. (4.45).

Example 4.6 Generate a pumping schedule for the fracture designed in Examples 4.3 to 4.5. The end-of-job concentration, c_f, for the pump selected is 6 ppg.

Solution Use Eq. (4.45), the proppant concentration as a function of injection time is listed in Table 4.1. The first 8 min of pumping is pad pumping, and the proppant concentration during this time is zero. Suggested proppant schedule uses a step change in proppant concentration, marching along the concentration curve. Figure 4.5 shows the theoretical concentration function and the suggested real-life concentration schedule for pumping.

Injection Time, min	Proppant Concentration, ppg
0–3	0
3–4	3.8
4–5	4.4
5–6	4.7
6–7	5.0
7–8	5.2
8–9	5.4
9–10	5.6
10–11	5.7
11–12	5.9
12–13	6.0

Table 4.1 Proppant Schedule during Fracture Pumping

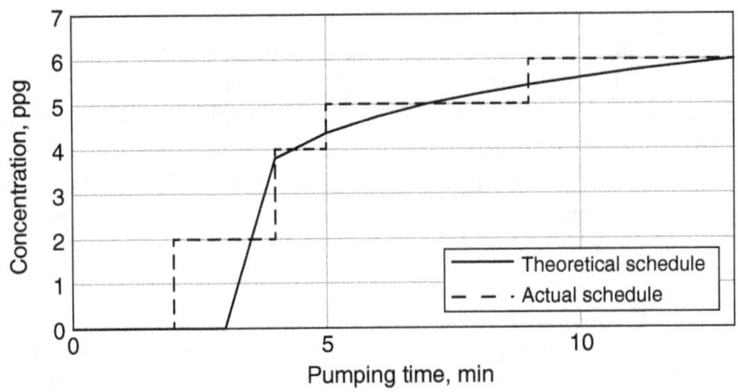

Figure 4.5 Pumping schedule.

Proppant Mass and Fracture Width after Closure

If assuming that all proppant is placed in a fracture uniformly (this is unlikely in reality), then the total proppant mass needed to create a fracture that has a geometry of x_f, h_f, and $\bar{w}$ is

$$M_p = \overline{c_p}(V_i - V_{\text{pad}}) \tag{4.46}$$

where $\overline{c_f}$ is the average concentration of proppant. For a uniformly distributed proppant pack,

$$\overline{c_p} = \frac{1}{t_i - t_{\text{pad}}} \int_{t_{\text{pad}}}^{t_i} c_f \left(\frac{t - t_{\text{pad}}}{t_i - t_{\text{pad}}} \right)^{\varepsilon} dt \tag{4.47}$$

with appropriate integration,

$$\overline{c_p} = \frac{c_f}{\varepsilon + 1}(1 - 0) = \frac{c_f}{\varepsilon + 1} \tag{4.48}$$

then

$$M_p = \frac{c_f}{\varepsilon + 1}(V_i - V_{\text{pad}}) \tag{4.49}$$

When fracturing is completed, after pumping is shut down, the created fracture will close upon the solid proppant pack. To estimate fracture width after closure, we assume that the length of the fracture does not change. With a fully, uniformly packed fracture, the fracture width reduces to the corresponding solid proppant volume,

$$w_p = \frac{M_p}{2x_f h_f \rho_p (1 - \phi_p)} \tag{4.50}$$

In Eq. (4.50), ρ_p is proppant density and ϕ_p is proppant porosity. Along with proppant permeability, size, and strength, this information is provided by the vendors when selecting proppant.

The fracture width calculated by Eq. (4.50) is the width after closure. It is smaller than the average fracture width calculated by the fracture geometry models [Eqs. (4.18), (4.20), and (4.24)]. This is the width used when calculating fracture conductivity for production.

Example 4.7 **Proppant mass and propped fracture width.**
Calculate the total proppant mass required to create the fracture described in Example 4.5. What is the fracture width after closure if the proppant selected has a density of 180 lbm/ft³ and a porosity of 0.38?

Solution The injection volume V_i for the case is 4745 ft³ (Example 4.3), and the pad volume V_{pad} is 942 ft³ (Example 4.4). Using the exponent parameter $\varepsilon = 0.2$ (Example 4.4), and the end-of-job concentration given in Example 4.5, $c_f = 6$ ppg, from Eq. (4.49), we have

$$M_p = \frac{6}{0.2+1}(4745-942)\left[\frac{\text{lbm}}{\text{gal}}\right](7.48)\left[\frac{\text{gal}}{\text{ft}^3}\right] = 142{,}232 \text{ lbm} \tag{4.51}$$

And from Eq. (4.50)

$$w_p = \frac{142{,}232}{2\times 700\times 100\times 180\times(1-0.38)} = 0.009 \text{ ft} = 0.11 \text{ in} \tag{4.52}$$

4.1.5 Fracture Conductivity and Fractured Well Productivity

Fracture conductivity measures the flow capacity of a fracture after closure. It is defined as the product of the propped fracture width and the permeability of the proppant,

$$C_f = wk_f \tag{4.53}$$

The common unit used for fracture conductivity is mD-ft. To connect fracture flow capacity to fractured well performance, we use dimensionless conductivity, C_{FD}, defined as

$$C_{FD} = \frac{wk_f}{x_f k} \tag{4.54}$$

In Eq. (4.54), x_f is fracture half-length and k is reservoir permeability. Equation (4.54) presents an important concept that the created fractures should measure up with the reservoir that the fractures are going to produce from. The numerator of Eq. (4.54) represents the ability of the fracture to flow, and the denominator represents the capacity of the reservoir. Obviously, the two should be balanced. Optimal dimensionless fracture conductivity that yields most efficient fracture design has been described by Economides et al. (2002). It is understood that for low-permeability reservoirs, fracture half-length x_f is more crucial; and for moderate- to high-permeability reservoirs, fracture width w is more important for better production performance. Fracture geometry is determined by the formation rock properties, and fracture design parameters (pumping rate, volume, and type of proppant and fluid selected) can control the geometry to a certain degree. Once a fracture is created in a reservoir, the four parameters in Eq. (4.54) are all known, and the well productivity can be predicted.

To predict fractured well production, simply analytical models, semi-analytical source solutions, or full reservoir numerical simulations all can perform the task. Here we use two analytical approaches to illustrate the procedure.

Pseudo Radial Flow Equation at Pseudo-Steady-State Condition for Fractured Well Production

Meyer and Jacot (2005) presented an analytical approach for pseudo steady state for fractures well productivity with finite conductive fractures under pseudo radial flow condition. It uses an equivalent wellbore radius to represent a fracture in a radial flow drainage. The flow pattern shown in Fig. 4.6. The equation should be used when the fracture half-length x_f is relatively smaller than the reservoir drainage radius, r_e because of the radial flow assumption.

Define equivalent wellbore radius for a fractured well as

$$r'_w = \frac{x_f}{\left(\frac{\pi}{C_{FD}} + 2\right)} \tag{4.55}$$

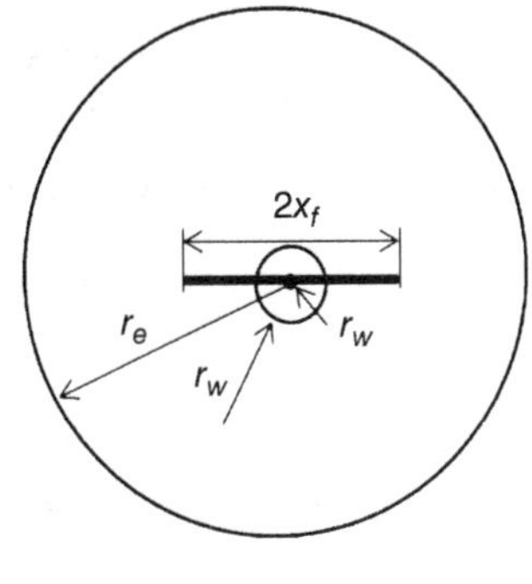

FIGURE 4.6 Radial flow assumption for the analytical equation of fractured well flow rate.

Then the skin factor after fracturing is

$$s_f = -\ln\left(\frac{r'_w}{r_w}\right) = -\ln\left(\frac{x_f}{r_w\left(\frac{\pi}{C_{FD}+2}\right)}\right) \tag{4.56}$$

Notice that the skin factor after fracturing is always a negative number. The dimensionless productivity index, J_D, is then calculated with the fracture skin factor,

$$J_D = \frac{1}{\ln\left(\frac{0.472r_e}{r_w}\right)+s_f} \tag{4.57}$$

Once we obtain J_D, the productivity index, *PI*, and flow rate, *q*, can be readily calculated from J_D. For single-phase liquid wells, the liquid flow rate is

$$q = (PI)(\Delta p) = J_D\left(\frac{kh}{141.2B\mu}\right)\Delta p \tag{4.58}$$

and for single-phase gas well, the gas flow rate is

$$q = (PI)(\Delta(p^2)) = J_D\left(\frac{kh}{1424TZ\mu}\right)\Delta(p^2) \tag{4.59}$$

The additional parameters in Eq. (4.59) are average reservoir temperature, *T*, in Rankin, and fluid compressibility factor, *Z*.

Example 4.8 Conductivity and production rate.
From Example 4.6, we calculated that the fracture width after closure equals to 0.018 ft. If the reservoir permeability is 0.1 mD, and the proppant used to create the fracture has porosity of 0.38, permeability of 16 Darcies, and density of 180 lb/ft^3, what is the conductivity of the fracture? What is the flow rate at 1000 psi drawdown? The reservoir fluid viscosity is 0.8 cp, the formation volume factor is 1.1, the wellbore radius is 0.25 ft, and the drainage radius is 879 ft.

Solution From Eq. (4.37), we have

$$C_{FD} = \frac{0.009\times 16000}{700\times 0.1} = 2.1 \tag{4.60}$$

The equivalent wellbore radius after fracturing is

$$r'_w = \frac{700}{\frac{\pi}{2.1}+2} = 200 \text{ ft} \tag{4.61}$$

and

$$s = -\ln\left(\frac{200}{0.25}\right) = -6.7 \tag{4.62}$$

therefore

$$J_D = \frac{1}{\ln\left(\frac{0.472 \times 879}{0.25}\right) + (-6.7)} = 1.4 \quad (4.63)$$

and

$$q = \frac{1.4 \times 0.1 \times 100 \times 1000}{141.2 \times 1.1 \times 0.8} = 108 \left[\frac{\text{bbl}}{\text{day}}\right] \quad (4.64)$$

For a given amount of proppant mass, when the combination of fracture geometry changes, conductivity changes, and therefore productivity changes too. Figure 4.7 shows a plot of dimensionless productivity as a function of fracture half-length for the total proppant of 150,000 lb. As fracture length increases, for the reservoir permeability of 0.1 mD, the productivity exhibits an optimal condition when the fracture half-length is about 720 ft. Beyond this length, longer fracture does not yield higher J_D. The dimensionless conductivity is around 2. This coincides with the optimal theory by Economides et al. (2002). When the reservoir permeability is smaller (0.01 mD), within a reasonable fracture half-length, we do not reach the optimal condition. When designing fracture treatments, optimal conditions should be used as a guideline; as long as field operation conditions allow. The limitations of pumping rate, pumping pressure, proppant concentration, and other parameters may prohibit to achieve the optimal conditions in the field.

Linear Flow Equation at Transient Flow Condition for Multistage Fractured Horizontal Wells

Apparently when reservoir permeability is very low, such as in unconventional reservoirs, the flow of a fractured well is dominated by the linear flow around the fracture area during the majority of producing time, and the flow is more likely under the transient condition instead of pseudo-steady-state condition. Bello and Wattenbarger (2010)

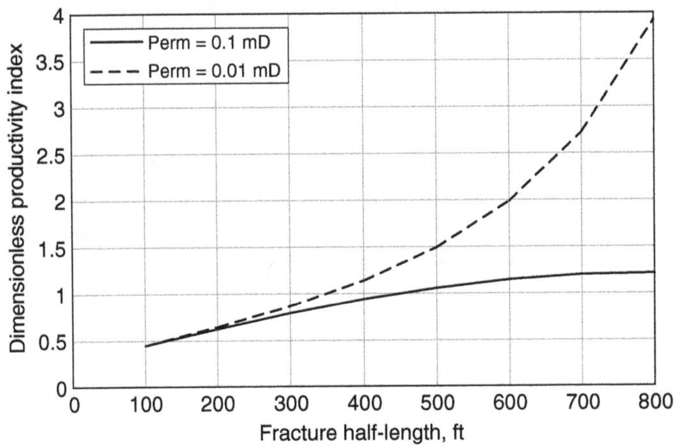

FIGURE 4.7 Dimensionless productivity J_D as a function of fracture half-length.

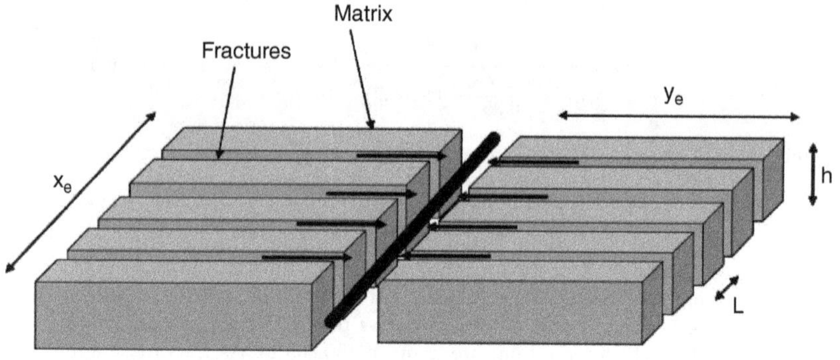

FIGURE 4.8 Flow pattern of linear flow in multistage fractured well in low-permeability unconventional reservoirs (Bello and Wattenbarger, 2010).

presented an analytical model for different flow periods of a multi-stage fractured horizontal well. We use the matrix-dominated linear flow equation under transient flow condition (Region 4 in Bello and Wattenbarger's work) to estimate well flow rate. This region is for long-term production. Figure 4.8 illustrates the simplified linear-flow pattern for such a case. In Fig. 4.8, x_e is the wellbore length, y_e is corresponding to the fracture half length, h is the height of the fractures, and L is the spacing between a pair of fractures.

The model assumes that the spacing between all fractures is a constant, and the reservoir is homogeneous. To calculate flow rate from all fractures, we first define the dimensionless time groups as,

$$t_{DAcm} = \frac{0.000633 k_m t}{(\varphi \mu c_t)_m A_{cm}} \tag{4.65}$$

where subscript m in Eq. 4.65 denotes matrix. The contact area of matrix, A_{cm}, is defined as

$$A_{cm} = \frac{(2x_e)h(2y_e)}{L} \tag{4.66}$$

The reservoir parameters involved in Eqs. (4.65) and (4.66) are permeability k_m, in mD, producing time t, in days, porosity, ϕ, fluid viscosity, μ, in cp, and total compressibility c_t in 1/psi. With the dimensionless time, t_{DAcm}, the dimensionless productivity for linear matrix flow is

$$q_{DL,m} = \frac{1}{2\pi\sqrt{\pi t_{DACm}}} \tag{4.67}$$

Similar to the radial flow equations, for liquid wells, liquid flow rate in bbl/day can be calculated by

$$q = \frac{k_m \sqrt{A_{cm}}(p_i - p_{wf})}{141.2B\mu} q_{DLm} \tag{4.68}$$

and for gas well,

$$q = \frac{k_m \sqrt{A_{cm}}(p_i^2 - p_{wf}^2)}{1424 T \mu Z} q_{DLm} \tag{4.69}$$

In Eq. (4.69), temperature is in °R, and flow rate is in Mcf/day.

Example 4.9 A 5000-ft long horizontal well in a gas reservoir is planned to be hydraulically fractured with 100 fractures along the wellbore (L = 50 ft). The fractures have 300-ft half-length (y_e) and 100-ft height. The reservoir permeability is 0.005 mD, porosity is 0.05, total compressibility is 10^{-4} 1/psi, and the reservoir fluid viscosity is 0.02 cp, the average gas compressibility factor Z is 0.9, and the average temperature is 640°R. Calculate gas flow rate at 7 days (one week) if the initial reservoir pressure is 4000 psi, and 1000 psi drawdown (p_i-p_{wf}) is applied to the well for production.

Solution We first calculate the dimensionless groups:

$$A_{cm} = \frac{2 \times 5000 \times 100 \times 2 \times 300}{50} = 1.2 \times 10^7 \tag{4.70}$$

and,

$$t_{DAcm} = \frac{0.00633 \times 0.005 \times 7}{0.05 \times 0.02 \times 10^{-4} \times 1.2 \times 10^7} = 0.000185 \tag{4.71}$$

The dimensionless productivity is

$$q_{DL,m} = \frac{1}{2\pi\sqrt{\pi \times 0.000185}} = 6.6 \tag{4.72}$$

Expanding the summation term in Eq. (4.53),
Finally, the production rate in MMscf/day (1000 Mscf/d) at the given condition is

$$q = \frac{0.005\sqrt{1.2 \times 10^7}(4000^2 - 3000^2)}{1424 \times 640 \times 0.02 \times 0.9} \times 6.6 = 49 \text{ MMcf/day} \tag{4.73}$$

4.2 Fracture Materials

Fracture materials include solid particles that keep the fractures open after closure (propping agents, or proppant), fluids that create fractures and carry the proppants, and chemicals that are added to the fracture fluids for various reasons.

4.2.1 Propping Agents for Fracturing (Proppant)

There are many factors that need to be considered when selecting proppant and fluids for fracture treatments. This includes technical evaluation and economic concern.

The main function of proppant is to support the fracture open after closure with a permeable path. Proppant can be natural sands or man-made ceramic particles. Proppants are measured by their properties such as strength, density, size distribution, porosity, and permeability. Two important factors to evaluate a type of proppant are mechanical strength of the material and permeability of the proppant pack. In addition, proppant should be easy to transport into the fracture with minimum settling, thus, proppant density is also an important factor.

Natural sand proppants have been used in the industry since the early times of fracturing practice. Until today, a large portion of the proppant pumped in the field is still natural sands. The biggest benefit of using natural sand is its economic value. Overall, natural sands have less quality compared with man-made proppant. While all other properties (density, strength, porosity, and permeability) are determined naturally, there are better quality natural sands which have more uniformed size distribution, more regular rounded shape and higher strength. Figure 4.9 (Brannon, 2011–2012) shows the commonly used proppants. The two common natural sands, the brown sand and the Ottawa white sand, are on the left column. As shown in the picture, Ottawa white sand is more uniformly sized and more regular shaped, and is a better proppant compared with the brown sand. On the other hand, the quality of man-made ceramic proppants can be designed to meet the specific requirements. For example, light-weighted proppant is easier to transport with less settling during pumping, and therefore creates more uniformly packed fractures (middle and right pictures in the top row in Fig. 4.9). Bauxite has high strength to resist being crushed after closure, yielding higher conductivity (right picture in the lower row in Fig. 4.9). Ceramic proppant in general has higher strength compared with the natural sands. Their size and shape can be controlled, giving higher porosity and permeability.

Both natural sand and ceramic proppant are sometimes resin-coated. The main purpose of resin-coating proppant is to prevent proppant flowing back during production, and therefore to maintain the conductivity of the created fracture.

FIGURE 4.9 Different types of proppants (Brannon, 2011–2012).

Proppant Mesh Size	Particle Size Range, μm
16/20	850–1180
16/30	600–1180
20/40	420–850
30/50	300–600
40/70	212–420
70/140 (100)	106–212

TABLE 4.2 Proppant Size

Proppant is specified by mesh size when designing a fracture treatment. The size is defined by two numbers; for example, 20/40 mesh; setting the range of the particle size that more than 90% of particles fall in between the sizes. Table 4.2 lists the commonly used proppant sizes and the corresponding particle diameters in μm.

Large-sized proppant yields high conductivity, but it is also more difficult to transport during fracturing because of its weight. On the contrast, small-sized proppant is easier to transport, but has low conductivity for production. From production performance point of view, large-sized proppant should be selected for moderate to high-permeability formations to ensure sufficient conductivity for production; and low-permeability formations can tolerate smaller sized proppant, especially with the designed fracture width being small. For conventional reservoirs, the norm of proppant size is around 20/40. Since the unconventional resource development, proppant size used in the field tends to become smaller. More commonly used proppant size in the unconventional fields is 40/70 mesh, sometimes even 100 mesh. This is driven by two reasons. First, to avoid gel damage in fractured wells, lower concentrations of polymer gels are used in low-permeability formations, which decreases the ability of carrying large-sized proppant. Small-sized proppant reduced the screen out rate. Also, most unconventional shale formations have natural fractures, and using smaller proppant enhances proppant transport into the natural fractures.

A hybrid approach is to combine small proppant and large proppant. In such a treatment, smaller-size proppant is pumped first and throughout most of the treatment, and then larger proppant is pumped at the end of the treatment to build a high-permeability proppant pack near the wellbore to enhance the production, as flow rate increases toward the wellbore.

4.2.2 Fracture Fluids

The functions of fracture fluids are to initiate fractures, transport proppant into the fracture, and distribute the proppant as evenly as possible. A required property for fracture fluid is that the viscosity used to carry the proppant has to diminish when the fracture treatment is finished, thus the fracture fluid flows back to the surface without difficulty once production starts. Fracture fluid viscosity is generated by mixing polymers with water to create gelled fluids, and breakers are added to break the gel when needed. Because of these functions, fracture fluid apparent viscosity is a key parameter when selecting/designing fluids.

The fracture fluids in the early days of hydraulic fracturing were mainly water or oil. As the fracture technology advanced, polymers, surfactants, and other additives are used today to aqueous phase fluids to provide fracture fluids that have stable rheology properties under reservoir condition and meet the requirement of damage prevention for fractured well performance. Other additives include viscous-elastic surfactant (VES), foaming agent, and chemically reactive fluid such as HCl and acetic acid. Three basic kinds of fracture fluids are used in fracture treatments today, linear gel, cross-linked gel, and slick water. In hybrid treatments, more than one type of fracture fluids are used alternately.

Linear Gel

Linear gel as fracture fluid has been used in the industry for fracturing since the 1950s (Jennings, 1996). The most common polymer for linear gel is guar gum from natural guar beans. Guar molecules in dry powder form have their long chain tightly coiled up. When added to water-based fluids, guar hydrates, the long chain relaxes to a straighten chain, and creates a non-Newtonian viscosity to the aqueous phase. Guar can be modified to hydroxypropyl guar (HPG), or carboxymethyl hydroxypropyl guar (CMPGH). Guar derivatives are more advanced linear gel fracture fluids, which exhibit stable rheology compared with guar gel. The chemical structures of guar and HPG are displayed in Fig. 4.10.

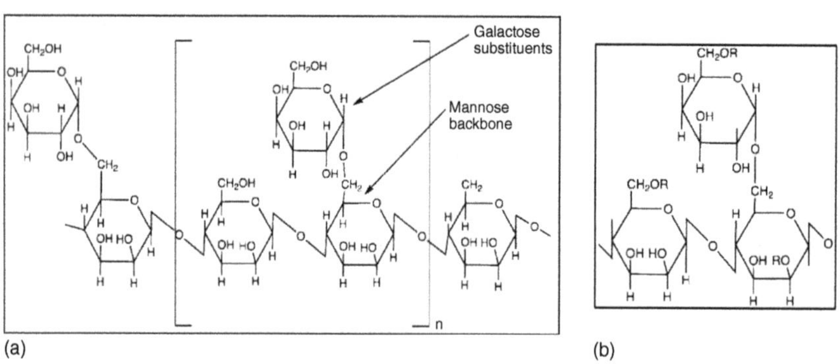

FIGURE 4.10 Chemical structure of guar (*a*) and HPG (*b*).

Crosslinked Gel

As hydraulic fracturing technology advanced, massive hydraulic fracture jobs with increased proppant concentration and large amount of proppant were pumped in the field starting from the 1970s. Linear gels no longer satisfied the requirements of the designed job. Crosslinked gels became popular because of their much higher and stable viscosity as fracture fluids. Crosslinked gels are guar-based linear gels with additives. Borate, for example as a crosslinked gel polymer, is a large number of boron-containing oxyanions and can form a complex polymeric structure in a low pH-value environment, as shown in Fig. 4.11. The apparent viscosity of crosslinked gel is in the range of hundreds to thousands cp, and it is stable at high temperature. The complex structure of the molecule is reversible when pH increases, and this feature is used to break the gel after fracturing.

Slick Water

Fracturing with low proppant concentration or even without proppant has been used in low-permeability formations. From the 1970s "river frac" in Oklahoma to "water frac" for horizontal wells in Austin Chalk development in 1980s (Meehan, 1995), the method showed success in low-permeability carbonate formations with heavy natural fracture systems. Slick water was introduced as fracture fluid as the activities of fracturing unconventional reservoirs increased. The trend of fracture geometry shifted to narrower width and longer contact through fracture network for unconventional reservoirs, proppant load was reduced to avoid damage in permeability-sensitive formations, but injection rate was increased significantly. Slick water is a

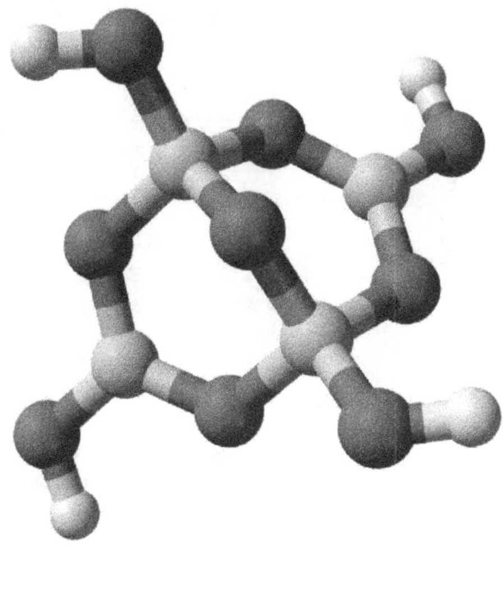

Figure 4.11 Chemical structure of Borate crosslinked gel. The dark color: oxygen, grey color: boron, and white color: hydrogen. (*Wikipedia*.)

nongelling, water-based fluid. Lower viscosity ensures higher pumping rate. To reduce frictional pressure drop, low concentration of acrylamide polymer is added as friction reducer.

In North America, the number of slick water fracturing treatments doubled from 21% in 1998 to more 40% in 2009 (Brannon, 2012) due to the great success in Barnett shale. The majority of fracture fluids used in Marcellus Shale, Woodford Shale, Utica Shale, and Fayetteville Shale is slick water (Ely et al., 2014). In the Bakken Shale and Eagle Ford Shale, a combination of slick water and gel fluid (hybrid fracture fluid) is also commonly used.

Friction reducers enhance linear flow, and thus can reduce frictional pressure drop up to 80%. Friction reducers are evaluated in the laboratory for their performance and stability at downhole conditions for the required time. Figure 4.12 shows a testing result of frictional pressure drop reduction as a function of polymer concentration for six different friction reducers (Baser et al., 2010). The frictional pressure drop with friction reducer is compared with the frictional pressure drop when pumping water without friction reducers. An optimal concentration behavior is observed for most of the tested samples (peak reduction on the plot), and the performance ranged from close to 80% friction reduction (samples A, C, and D) to about 50% reduction (sample F). Samples A, C, and D exhibit the peak performance at lower polymer concentration (less than 3 ppg). These three samples are more effective and have higher efficiency compared with other samples that were tested. When selecting friction reducers, another concern is the sustained stability of the chemicals as a function of time. The performance of friction reducers should be stable throughout the pumping.

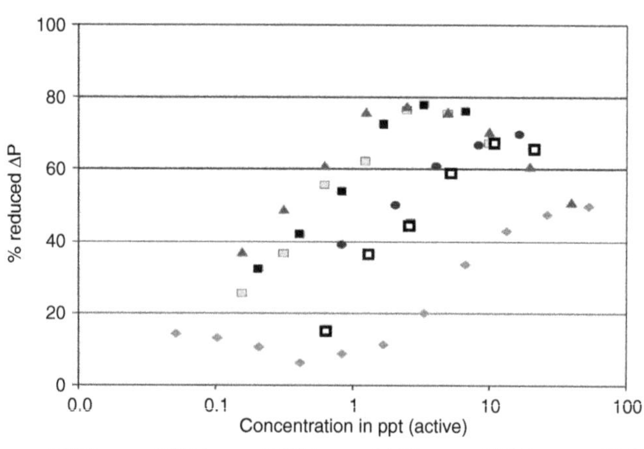

FIGURE 4.12 Sample of friction reducer tests (Baser et al., 2010).

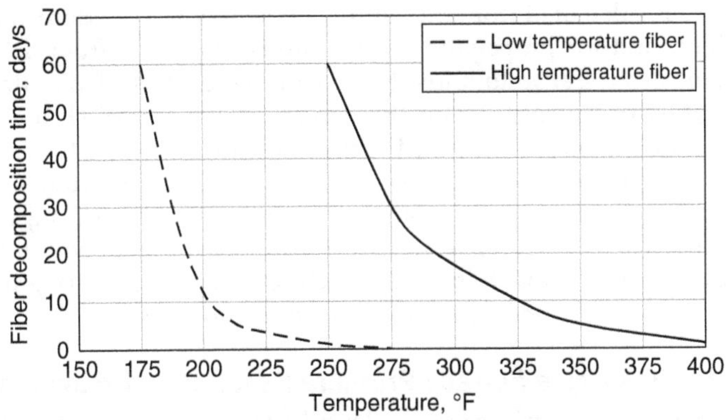

FIGURE 4.13 Fiber decomposing characteristics (Bustos et al., 2007).

Fiber-Based Fracture Fluids

Fibers-based fracture fluids enhance the fracture fluids in several aspects. The original application is mainly on proppant flowback prevention. In addition, reducing leakoff volume and control height growth are also observed benefits. Fibers used in fracture fluids can be made of glass, ceramic, metal, or carbon (Card et al., 1995).

One of the most obvious functions of fiber is to enhance proppant transport. Fiber latten can effectively carry the proppant with less settling, which results in a more evenly distributed fracture pack, and longer propped fracture length. Fiber decouples proppant transport from viscosity. Gel concentration of fracture fluid is greatly reduced. For gel-damage sensitive formations, fiber-based fracture fluids help improve well performance after fracturing (Bustos et al., 2007).

Fiber decomposes and is self-cleaned after fracturing. The decomposition characteristics are critical when designing fiber-based fracture fluids. If decomposition happens before the pumping job is completed, proppant may not be distributed as designed, lowering the efficiency of fracturing. On the other hand, if fiber does not decompose, fiber itself plugs the fracture pack and reduces the permeability of proppant pack. Fibers are tested in the lab for stability and decomposition as a function of time. Figure 4.13 shows an example of testing result (Bustos et al., 2007). Sample B can stand much higher temperature compared with sample A.

4.3 Bottomhole Treating Pressure and Surface Injection Pressure for Fracturing

Bottomhole treating pressure is an important information that indicates the progress of a fracture job in real time. The required surface injection pressure for breaking the formation and initiating the fracture is calculated from the estimated breakdown pressure at bottomhole; and for propagating a fracture is calculated from the net pressure desired at bottomhole. Surface pressure is recorded in almost all fracture treatments. For breakdown and pad injection, clear fluids (gel or slick water) are pumped. The calculations for clear fluids and slurry are presented separately here. The pressure calculations presented here have two applications: to determine surface injection pressure when designing a fracture job, and to estimate bottomhole pressure and therefore to monitor fracture propagation during treatments when bottomhole pressure gauges are not present.

4.3.1 Determine Surface Pressure Required to Breakdown

To generate breakdown pressure described by Eq. (4.1) at bottomhole conditions, surface injection pressure at the tubing head p_{ti} is calculated considering the pressure drop in the wellbore,

$$p_{ti} = p_{bd} - \Delta p_{PE} + \Delta p_f \tag{4.74}$$

Δp_{PE} in Eq. (4.74) is hydrostatic pressure drop created by the weight of the fluid column in the tubing, and Δp_f is frictional pressure drop caused by fluid flow in the tubing. For single-phase liquid injection without solids (such as pad fluid injection),

$$\Delta p_{PE} = \rho g h \tag{4.75}$$

and

$$\Delta p_f = \frac{2 f_f \rho q^2 L}{\left(\frac{\pi D^2}{4}\right)^2} \tag{4.76}$$

where ρ is fluid density, h is vertical depth of the well, L is tubing length, f_f is frictional factor, q is injection rate, and D is tubing diameter. f_f, the frictional factor, can be obtained either from Moody's diagram or by correlation equations (Economides et al., 2012), Eqs. (4.75) and (4.76) should be in consistent units. For the field units,

$$\Delta p_{PE} = \frac{\rho h}{144} \tag{4.77}$$

and

$$\Delta p_f = \frac{7.35 \times 10^{-7} f_f \rho q^2 L}{D^5} \tag{4.78}$$

In Eqs. (4.77) and (4.78), ρ is in lbm/ft^3, h and L are in ft, q is in bbl/day, and D is in inches.

Equation (4.74) shows that higher hydrostatic pressure drop helps to reduce surface pumping pressure, but higher frictional pressure drop requires higher surface pumping pressure. In terms of pumping, fluids with higher density help to reduce the surface injection pressure, but high injection rates can lead to high frictional pressure drop. For gelled fracture fluids, polymers added to the fluids to increase viscosity also serve as friction reducers. For slick water, low-concentration polymers are added as friction reducers to fracture fluids. Friction reducers reduce frictional factor, f_f, and therefore release the requirements of high surface pumping pressure.

Example 4.10 Surface injection pressure calculation.
The target formation that is planned to be fractured is located at a depth of 8000 ft. The tubing ID is 3½ in. The fracture fluid is water with friction reducer (slick water), and the friction reducer results in an average friction factor of 0.001. The density of the fracture fluid is 66.8 lb/ft^3, and the viscosity is 5 cp. Injection rate is 65 bbl/min (bpm). Calculate the surface injection pressure for the breakdown pressure estimated in Example 4.1.

Solution From Eqs. (4.60) and (4.61),

$$\Delta p_{PE} = \frac{66.8 \times 8000}{144} = 3711 \text{ psi} \tag{4.79}$$

and

$$\Delta p_f = \frac{7.35 \times 10^{-7} \times 0.001 \times 66.8 \times (65 \times 24 \times 60)^2 \times 8000}{3.5^2} = 6552 \text{ psi} \tag{4.80}$$

Thus, the surface injection pressure is

$$p_{ti} = 5424 - 3711 + 6552 = 8265 \text{ psi} \tag{4.81}$$

Surface injection pressure has to be safely lower than the pressure specifications of all equipment and manifold, commonly specified from 10,000 to 15,000 psi. The example here uses an injection rate of 65 bpm. If the injection rate is increased to 70 bpm, the frictional pressure drop increases to 7599 psi, and the injection pressure required at the surface is 9312 psi. Frictional pressure drop is proportional to injection rate to the second power. Increasing injection rate can quickly raise the requirement of surface injection pressure.

4.3.2 Estimate Bottomhole Pressure during Injection to Monitor Treatments

During fracture propagation after breakdown, the bottomhole pressure has to be higher than the closure pressure of the formation, and the difference between the bottomhole pressure and the closure pressure (also known as fracture pressure) is referred to as net pressure. Net pressure is one of the parameters that controls fracture propagation. During fracturing, surface pressure is kept high enough to provide the required bottomhole treating pressure, and therefore provide the designed net pressure for fracture growth. For slick water without proppant, the calculation presented in the above section can be used inversely to estimate bottomhole pressure from the recorded surface injection pressure. When pumping slurry (proppant in gelled fracture fluid), pressure drop calculation is different. Lord and McGowen (1986) presented a correlation that estimates friction pressure drop as a function of polymer gel concentration and proppant concentration (also called proppant loading) as,

$$\Delta p_{G,P} = \sigma(\Delta p_0) \tag{4.82}$$

where $\Delta p_{G,P}$ is frictional pressure drop for slurry and Δp_0 is frictional pressure drop for water. The drag ratio, σ, is calculated by the correlation,

$$\ln\left(\frac{1}{\sigma}\right) = 2.1505 - \frac{8.024}{v} - \frac{0.2365G}{v} - 0.1639\ln(G) - 0.05266C_p \exp\left(\frac{1}{G}\right) \tag{4.83}$$

In Eq. (4.83), G is cross-linked gelling agent concentration in the unit of lb/Mgal, C_p is proppant concentration in lb/gal (ppg), and v is velocity of injection fluid,

$$v = \frac{q}{A} \tag{4.84}$$

or in the field unit

$$v = \frac{17.15q}{D^2} \ \frac{\text{ft}}{\text{s}} \tag{4.85}$$

where, with the conversion factor of 17.15, q is injection rate in bpm, and D is pipe diameter in inches.

Once σ is estimated from Eq. (4.83), frictional pressure can be calculated by Eq. (4.82) using water pressure drop as a reference,

$$\Delta p_0 = 0.40429 D^{-4.8} q^{1.8} L \tag{4.86}$$

where again D is in inches, q is in bpm, and L is in feet.

Example 4.11 Surface injection pressure calculation for slurry injection. Calculate the surface injection pressure required when the fracture propagation pressure gradient is 0.8 psi/ft. The well is 8000 ft deep, the tubing ID is 3½ in, and the injection rate is 65 bpm. The proppant concentration is 2 ppg and the polymer gel concentration is 150 lb/Mgal. Use a slurry (fracture fluid plus proppant) density of 90 lb/ft³ and a net pressure of 400 psi.

Solution We first calculate pressure drop for water by using Eq. (4.86)

$$\Delta p_0 = 0.40429 \times (3.5^{-4.8}) \times (65^{1.8}) \times 8000 = 14{,}504 \text{ psi} \tag{4.87}$$

The fluid velocity v is

$$v = \frac{17.15 \times 65}{3.5^2} = 91 \ \frac{\text{ft}}{\text{s}} \tag{4.88}$$

Then we can calculate the drag ratio

$$\ln\left(\frac{1}{\sigma}\right) = 2.1505 - \frac{8.024}{91} - \frac{0.2365\times 150}{91} - 0.1639\ln(150) - 0.05266$$

$$\times 2\times\left(\exp\left(\frac{1}{150}\right)\right) = 0.7452 \tag{4.89}$$

$$\sigma = 0.475 \tag{4.90}$$

and

$$\Delta p_{G,P} = \sigma(\Delta p_0) = 0.475\times 14504 = 6890 \text{ psi} \tag{4.91}$$

The surface injection pressure is

$$p_{ti} = 6400 + 400 + 6890 - \frac{90\times 8000}{144} = 8690 \text{ psi} \tag{4.92}$$

Required surface pressure is less during fracture propagation compared with breaking down the formation to initiate a fracture. Once slurry enters the wellbore, surface pressure should decrease because the heavier slurry yields higher hydrostatic pressure drop [Eq. (4.92)].

4.3.3 Understanding Treatment Records

Surface pressure, injection rate, and proppant concentration are usually recorded during fracture treatment. This information is extremely helpful during pumping for treatment monitoring and post fracturing for treatment evaluation. In cases where bottomhole pressure is not measured, it can be calculated by the methods introduced in the above sections. Figure 4.14 shows a typical treatment record for a fracture job. On the plot, the curves with the square and triangle symbols are the proppant concentration in lbm/gal at surface and bottomhole condition; notice the time delay of the two curves. The circle-symbol curve is the surface injection rate, which is kept a roughly constant

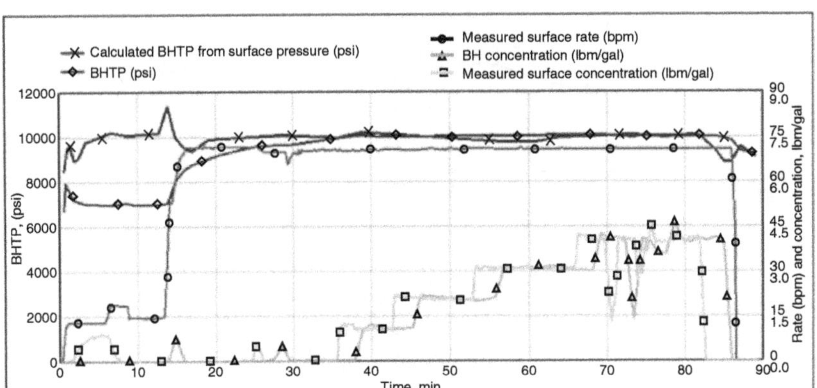

FIGURE 4.14 An example of fracture treatment record (Bazan and Meyer, 2015).

of 70 bpm after 14 min of pumping. The curve with the diamond symbols is the bottomhole pressure measured by the gauge and the cross-symbol curve is the calculated bottomhole pressure. The calculated bottomhole pressure matched gauge-measured pressure very well. The recorded surface pressure and calculated/recorded bottom pressure reflect breakdown of the formation (early-time sudden pressure drop after a continuous increasing pressure), and fracture propagation (slightly decreasing bottomhole pressure indicates a PKN fracture propagation and slightly increasing bottomhole pressure a KGD fracture). As proppant concentration increases, surface pressure decreases gradually when bottomhole pressure is kept constant because proppant weight increases the hydrostatic pressure drop. A sudden increase in surface treating pressure can be an indication of screen-out, and a rapid decrease in treating pressure can be caused by uncontrolled height growth. Without quantitative calculation, a simple Nolte-Smith plot can help to identify the pumping problems during a treatment. Figure 4.15 illustrates an example of a Nolte-Smith plot (Gidley et al., 1989).

In the Nolte-Smith plot, treating pressure is shown as a function of logarithm of injection time, the curve is divided into five regions, each representing an idealized phenomenon of fracture propagation. In region I, the curve has a slope ranging from ⅛ to ¼, indicating a normal fracture propagation (restricted height and unrestricted length growth). In region II, the curve has a flat line with a slope of 0, representing the stable height growth. Region III splits into two sections; III-a has a slope of one (unit slope) and III-b has a slope of two (double slope). With the slope suddenly increasing, region III-a suggests screen-out, occurring with restriction in length growth of the fracture. Region III-b is referred to as tip screen-out. In general tip

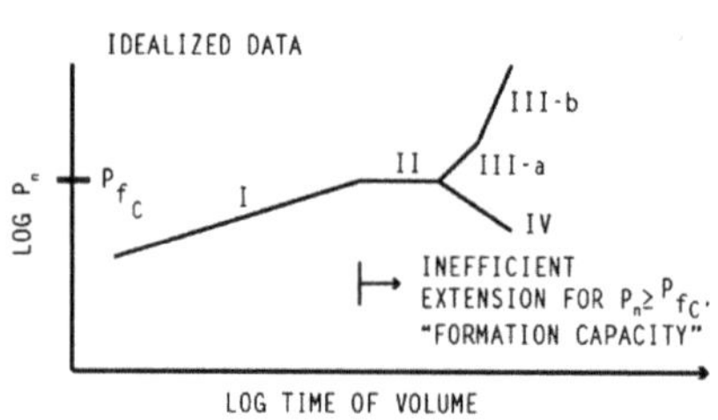

APPROXIMATE LOG-LOG SLOPE		INTERPRETATION
1/8-1/4	I	RESTRICTED HEIGHT AND UNRESTRICTED EXTENSION
0	II	a) STABLE HEIGHT GROWTH b) FISSURE OPENING
1-1 (UNIT)	III-a	RESTRICTED EXTENSION-two active wings
2-1 (DOUBLE)	III-b	RESTRICTED EXTENSION-one active wing
NEGATIVE	IV	UNSTABLE HEIGHT GROWTH (ie Run-away)

FIGURE 4.15 Nolte-Smith plot for qualitative diagnose of fracture treatment (Gidley et al., 1989).

screen-out ends the treatment with a forced shutdown of pumping. On the other hand, if the slope is decreasing (region IV), it indicates an unrestricted height growth. This is undesired because most likely the fracture is growing out of the payzone. The Nolte-Smith plot is a straightforward and useful tool for fracturing operation diagnosis.

4.3.4 Diagnostic Fracture Injection Test

The diagnostic fracture injection test (DFIT) is one of the most commonly used methods to determine fracture and reservoir parameters from prefracture injection pressure. The concept is modified and extended from the original minifrac test presented by Nolte (1979), and integrated with transient pressure analysis approach later (Craig and Blasingame, 2006; Barree et al., 2009). The main parameters from minifrac tests are leakoff coefficient and fracture closure pressure. In addition, DFIT can also be used for estimating other reservoir parameters such as pore pressure, permeability, and reservoir storage. Barree et al. summarized DFIT procedures using Fig. 4.16 (Barree et al., 2015). Fluid is injected with low injection rate until formation breakdown. Injection rate is increased after breakdown to a full pumping scale, and then held constant for 3 to 5 min to establish a stable pressure. The injection rate is reduced in two steps, first to 75% of the full scale, and then 50%. Each step lasts 10 to 15 seconds, and this provides step-injection pressure data. The injection is then shut down, and instantaneous shut-in pressure (ISIP) is recorded. From this moment on, leakoff pressure is recorded for estimating leakoff coefficient. When the fracture is completely closed, pressure decline is due to the fluid flow into the reservoir, and transient pressure analysis can be conducted to estimate permeability, reservoir initial pressure, and fracture geometry. Closure pressure is identified from the pressure derivative change, and it is approximated as the minimum closure stress of the formation.

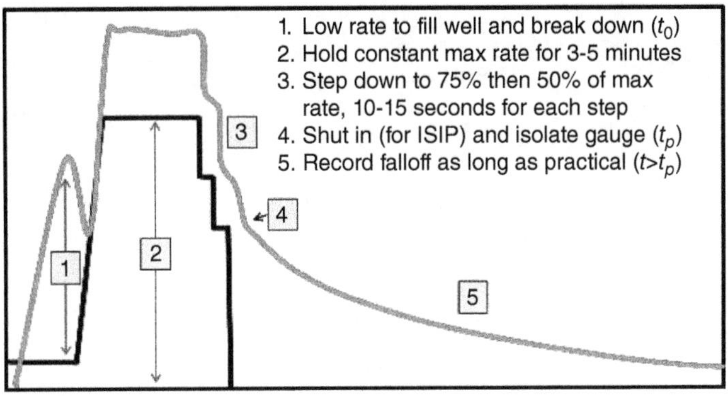

FIGURE 4.16 DFIT test procedure (Barree et al., 2015).

Here we only introduce the procedure to estimate leakoff coefficient and closure pressure. The G function was defined as (Nolte, 1979; Barree et al., 2009)

$$G(\Delta t_D) = \frac{4}{\pi}[g(\Delta t_D) - g_0] \tag{4.93}$$

where

$$g(\Delta t_D) = \frac{4}{3}\left[(1+\Delta t_D)^{\frac{2}{3}} - (\Delta t_D)^{\frac{2}{3}}\right] \tag{4.94}$$

and

$$\Delta t_D = \frac{t - t_0}{t_0} \tag{4.95}$$

In Eq. (4.95), t_0 is pumping time and t is total time (pumping and shut-in). With G function, the following relationship can be used to estimate leakoff coefficient C_L,

$$\Delta p(\Delta t_{D,1}, \Delta t_{D,0}) = \frac{C_L h_f E' \sqrt{t_0}}{h^2} G(\Delta t_{D,0}, \Delta t_{D,1}) \tag{4.96}$$

where h_f is fracture height and h is permeable formation thickness.

From the derivative curve, $Gd(p)/d(G)$ versus G, we can also identify the closure stress, as illustrated in Fig. 4.17. Barree et al. (2009) showed that when a tangent line from the origin deviates from the Gdp/dG curve, the closure pressure is the pressure at the deviating point as the dashed line [1] marked.

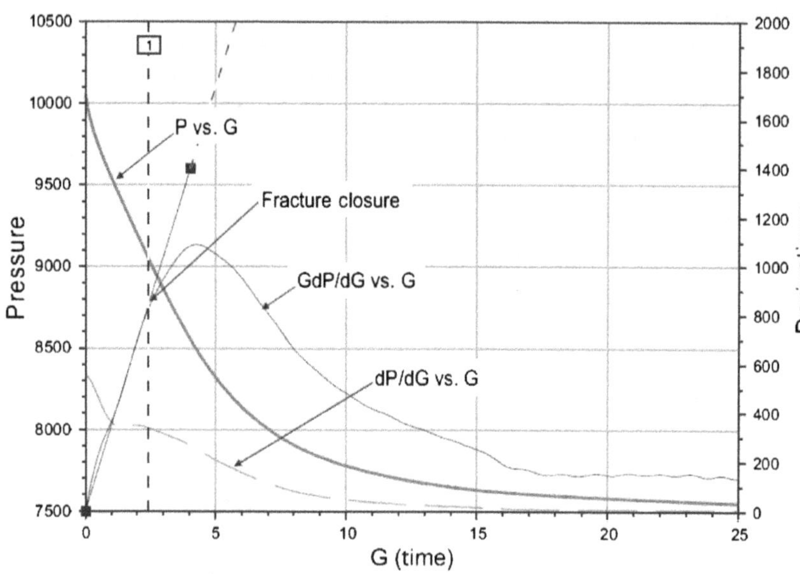

FIGURE 4.17 Using derivative plot to identify closure pressure (Barree et al., 2009).

4.4 Tasks of Completion for Successful Fracture Stimulation

Completion design for hydraulic fracturing has unique requirements in addition to the basic functions of completions mentioned in Chapter 2. Completion for fracturing should provide good communication between the wellbore and formation rock, reduce near-wellbore friction for easy breakdown and fracture initiation, and be able to place proppant with desired size and amount into the fracture without proppant bridging near the wellbore or screen-out anywhere in the fracture (Behrmann and Nolte, 1999; Ceccarelli et al., 2010). Limited entry completion for fracturing is designed to better distribute fracture fluids. Unlike completion for production where the dominant reservoir parameter is reservoir permeability; rock mechanical properties such as in-situ stress distribution are the most important parameters in designing successful completions for hydraulic fracturing. Starting from well structure design to completion selection and completion design, the effect of rock mechanical properties needs to be considered in each phase.

There are mainly two categories of completions for hydraulic fracturing: completion with cemented casing/liner and perforations, and completion without cementing (usually referred as openhole completion). For vertical wells in conventional reservoirs, a cased and cemented wellbore with perforations is the most commonly used completion method for fracturing for many years.

In the recent development of unconventional resources, horizontal wells are deployed to produce from these extremely low-permeability formations, presenting new challenges to completion design for hydraulic fracturing. Openhole completion with liners and sliding sleeves that have the ability of communicating with reservoir have been developed to meet the challenge. Today both cased/cemented/perforated completion and openhole/liner/sleeve completion are used in horizontal wells for fracturing. Commonly fracture treatments are pumped in multiple stages to place numerous fractures along a horizontal wellbore for better production.

4.5 Completion for Fracturing Vertical and Deviated Wells

The most commonly used completion method for fracturing vertical and deviated wells is perforation through cemented casing. As mentioned before, the biggest difference between perforation for production and perforation for hydraulic fracturing is the effect of in-situ stress. When perforating for production, in-situ stress has a minimum effect on production performance of the well. The effect of perforations on production is considered through a skin factor, which has been discussed in Chapter 3. When perforating for hydraulic fracturing because fractures preferably propagate in a direction that is perpendicular to the minimum stress direction; perforation orientation becomes critical for successful fracturing. Perforation design for hydraulic fracturing has some unique considerations compared with perforating for production.

4.5.1 Near-Wellbore Stress Impact and Perforation Phasing

The near-wellbore problems of perforating for fracturing are related to excess pressure in the near-wellbore region and fracture screenout because of tortuosity. These problems are related to perforation. Fractures initiated from perforations develop in alignment with the perforation orientation. When perforations are misaligned with preferred orientation for fractures to propagate, the initiated fractures will attempt to twist or turn once leaving the near-wellbore region to realign with the in-situ stress field in the formation. From a theoretical point of view, perforations should be aligned with the maximum stress direction, and that is the best direction for fracture to propagate. This means that perforation phasing should be 180°, aligned in the maximum stress direction, and any other orientations of perforation would add difficulty for fracture propagation. Figure 4.18*a* illustrates this ideal situation. In reality, this requires two conditions: accurate information about the in-situ stress field and oriented perforating. Oriented perforating today is not unusual, especially for horizontal well applications. On the other hand, the assessment of accurate stress fields is much harder to achieve. If the stress field information has uncertainty, which is almost always the case, oriented perforating

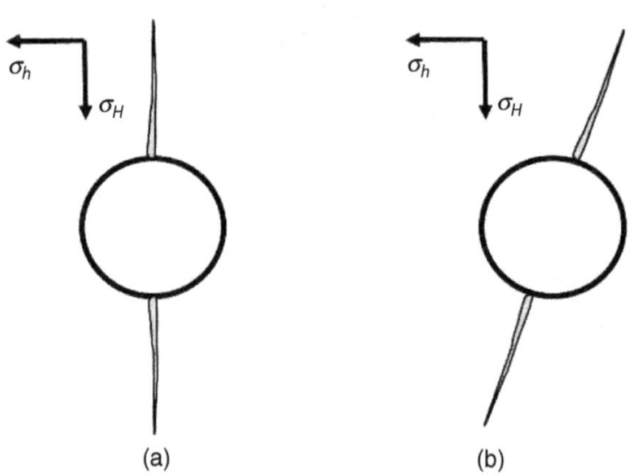

FIGURE 4.18 Illustration of perforation orientation regarding to in-situ stress field orientation: (*a*) aligned perforation, (*b*) unaligned perforation.

loses its value because the perforation can be placed in an accurate direction, but it may not be the preferred direction. Figure 4.18*b* shows a schematic of slightly unaligned perforations with preferred stress direction.

When perforations are not aligned with the principal stress direction, fractures can be initiated if the misaligning is within 15° to 30° (Behrmann and Nolte, 1998). For higher misaligned angle, near-wellbore tortuosity impacts fracture propagation. Near-wellbore tortuosity is a phenomenon that a fracture twists and/or turns because of the impact of in-situ stress during its propagation. When fractures tend to align with the in-situ stress field from the orientation associated with the perforations once in the far-field, it creates narrower gaps at the turning points. This can result in proppant bridging at the turning points, and screen-out. To avoid near-wellbore tortuosity problem, Abass et al. (1994) suggest that instead of 180° phasing, 60° phasing should be used based on the lab experimental observation. The argument is that in cases of uncertainty, at least some of the perforations will be in the preferred orientation and be able to initiate fractures.

Another common problem in perforating for fracturing is the creation of micro-annulus between the casing and cement during perforating or at the initial stage of fracture pumping (Romero et al., 2000). Figure 4.19 is a schematic of this phenomenon. If the perforation is misaligned, the fracture in such situation would possibly propagate around the wellbore first and break into the formation at the location of preferred orientation. The arrows in Fig. 4.19 indicate the flow path. The restriction areas, or the pinch-points (see Fig. 4.19) would likely result in a screen-out. As the number of perforations increases, multiple fractures are more likely to initiate instead of single or bi-wing fractures. The fractures that are not in the preferred orientation aligned with the principal stress will stop propagating once getting into the far-field from the wellbore; also, these undeveloped fracture are possible sources of screen-out when slurry pumping starts.

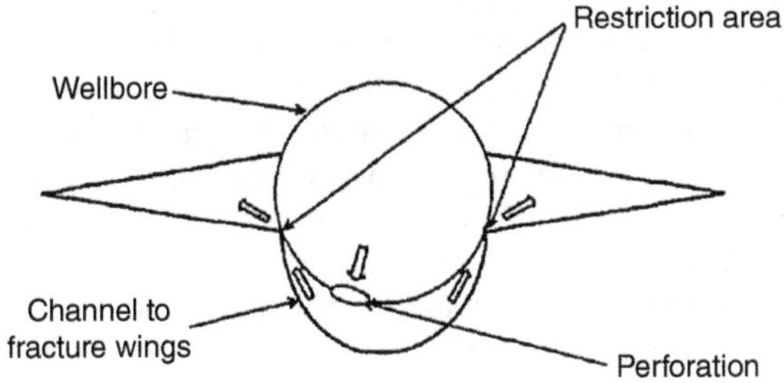

Figure 4.19 Microannulus created between casing and cement, causing restrictions for fracture propagation in the case of misaligned perforation with principal stress field (Romero, 2000).

Gun	Fracture Initiation Pressure	Micro-Annulus Pinch Point	Multiple Fractures
0°, 1 SPF	bad	worst	best
180°, 1 SPF	bad	bad	good
90°, 2 SPF	good	bad	bad
120°, 1.5 SPF	best	good	bad
60°, 3 SPF	best	best	worst

TABLE 4.3 Perforating Gun Trade-Off, Vertical Well, Not Oriented, Microannulus (*adopted from Behrmann and Nolte, 1998*)

Behrmann and Nolte compared the benefits and problems of different perforation phasing design for fracturing (Behrmann and Nolte, 1998). Table 4.3 is adopted from their study. In summary, 0°/180° phasing is beneficial to avoid multiple fractures initiating, but with higher possibility of creating a micro-annulus. When the number of perforations increases (120°/60° phasing), the micro-annulus problem is reduced, but the chance of creating multiple fractures increases. Oriented perforating helps to initiate and propagate fractures when the in-situ stress field is better described with less uncertainty.

4.5.2 Perforation Length, Diameter, and Interval Length

Unlike perforating for production, where the length of perforations is critical for the perforations to extend outside the damage zone (Chapter 3); when perforating for fracturing, the length of perforations is less critical since the fracture will extend into the formation much deeper. A rough guideline is that a 6-in perforation length should be enough for fracture initiation and propagation (Behrmman and Nolte, 1999). It is observed that deep penetration could be beneficial to bypass near-wellbore tortuosity (Ceccarelli et al., 2010), especially for deep formation (over 4000 m or 12,000 ft). From experimental observations, deep penetration also helps to reduce the breakdown pressure (Wutherich and Walker, 2012). On the other hand, perforation diameter becomes more important when perforating for fracturing. Perforation hole-size needs to be large enough to place the proppant into the fracture. Small hole-size generates high perforation friction, which is only desirable in limited entry design (discussed later in Section 4.5.4). When the perforation is too small, proppant can bridge at the perforation entry, resulting in high treating pressure at the surface or even a screen-out. It is suggested that perforation-hole diameter should be at least 6 times bigger than the proppant size. In case of mixed proppant size, the largest grain size should be used to design perforation diameter. For example, the perforation size should be 6 times larger than 20-mesh size for 20/40 mesh proppant.

Perforation interval design for fracturing is also different compared with perforation interval design for production. When perforating for production, the intervals should cover most of the payzone for higher reservoir contact and better production performance. When perforating for fracturing, large intervals have higher possibility to create more than one fractures, instead of single or bi-wing fractures. Multiple competing fractures usually result in some fractures being narrower and more difficult to transport proppant into the fractures, increasing the risk of failure (screen-out). It is recommended, practically, that the perforation interval should be around two to four times the wellbore diameter.

4.5.3 Perforation Friction

Perforation friction refers to the pressure drop across a perforation hole in the casing. This frictional pressure adds additional pressure drop to the surface pumping pressure. When frictional pressure drop through perforations becomes large, it can increase the surface pumping pressure significantly. The pressure drop is calculated by

$$\Delta p_{\text{perf}} = 0.2369\left(\frac{q}{n}\right)^2 \frac{\rho}{d_p^4 C_d^2} \tag{4.97}$$

where q is the pumping rate in bpm, ρ is the fluid density in lb/gal (ppg), n is the number of perforations, d_p is the diameter of perforation in inches, and C_d is the discharge coefficient. The constant in the equation, 0.2369, is a unit conversion factor for the mentioned field units. Discharge coefficient C_d is a function of the fluid rheology. Evaluation of discharge coefficients by experimental studies presented the numerical values of discharge coefficients for specific perforation sizes (Lord, 1994). El-Rabba et al. also presented empirical expressions for discharge coefficients (El-Rabba et al., 1999). For clean linear gel,

$$C_d = \left(1 - \exp\left(\frac{-2.2d_p}{\mu_a^{0.1}}\right)\right)^{0.4} \tag{4.98}$$

and for cross-linked gel,

$$C_d = \left(1 - \exp\left(\frac{-3.79d_p^{0.32}}{\mu_a^{0.25}}\right)\right)^{0.6} \tag{4.99}$$

In Eqs. (4.98) and (4.99), μ_a is apparent viscosity of the fluid. Notice that the correlations presented in Eqs. (4.98) and (4.99) are based on limited testing data and should be used with caution.

Perforation friction pressure drop should be added to Eq. (4.74) when calculating the surface injection pressure,

$$p_{ti} = p_{bd} - \Delta p_{PE} + \Delta p_f + \Delta p_{\text{perf}} \tag{4.100}$$

Example 4.12 Calculate perforation friction pressure drop and recalculate surface injection pressure for breakdown in Example 4.1. Using injection rate of 65 bpm, fluid density is 70 lbm/ft^3. The apparent viscosity is 100 cp for a linear gel, and 150 for a crosslinked gel. The perforations are 2 SPF for 20 ft and the perforation diameter is 0.375 in.

Solution For linear gel, from Eq. (4.98), we first calculate the discharge coefficient,

$$C_d = \left(1 - \exp\left(\frac{-2.20.37}{100^{0.1}}\right)\right)^{0.4} = 0.69 \tag{4.101}$$

There are 40 perforation holes over 20 ft of completion interval, and the perforation friction, from Eq. (4.77), is

$$\Delta p_{\text{perf}} = 0.2369\left(\frac{65}{40}\right)^2 \frac{70\left[\frac{\text{lbm}}{\text{ft}^3}\right]}{7.48\left[\frac{\text{gal}}{\text{ft}^3}\right](0.375^4)(0.69^2)} = 622 \text{ psi} \tag{4.102}$$

Adding the perforation friction to the breakdown pressure, from Examples 4.1 and 4.10, the surface injection pressure is

$$p_{ti} = 5424 - 3711 + 6652 + 622 = 8987 \text{ psi} \tag{4.103}$$

For crosslinked gel,

$$C_d = \left(1 - \exp\left(\frac{(-3.79)(0.37^{0.32})}{150^{0.25}}\right)\right)^{0.6} = 0.694 \tag{4.104}$$

The discharge coefficient is insensitive to the type of gel or the apparent viscosity of the fluid.

4.5.4 Limited Entry Concept

Limited entry refers to a technique that uses a controlled perforation number to balance fracture propagation in multiple layered formations. When a multiple layered formation has different properties in different layers, fractures initiate in the layers that are easier to breakdown and propagate. Limited entry uses small number of perforations and hole sizes to generate the large perforation friction at the layers that are propagating fractures, thus to slow down the fracture propagation, and redirect fracture fluid to other layers. This acts as a mechanical diversion method. The key to the success of limited entry is to correctly design the perforation parameters to generate the desired perforation friction (Cramer, 1987).

During slurry pumping, when proppant is added to fracture fluids, the perforation friction starts decreasing because of the combination of two facts. First, the edge of perforation holes become smooth by sand erosion; this increases the discharge coefficient. The second fact is that perforation hole-size becomes large due to the sand erosion. Proppant erosion is related to the total amount of proppant pumped, proppant size, concentration, and pumping rate. Based on 20/40 mesh sand experiments, Crump and Conway (1988) proposed a relationship between the proppant pumped and the discharge coefficient,

$$C_d = 0.56 + 1.65 \times 10^{-4} M_p \tag{4.105}$$

where M_p is proppant mass in pounds. Equation (4.105) assumes that the minimum discharge coefficient is 0.56, and it also limits C_d to a maximum value of 0.89. Cramer (1987) proposed a relationship of perforation diameter to the proppant pumped

$$d_p = d_{p,\text{initial}} + 4.29 \times 10^{-6} M_p \tag{4.106}$$

Equations (4.105) and (4.106) are empirical correlations. They should only be used as a guideline.

Example 4.13 Limited entry, perforation friction pressure calculation. Recalculate perforation frictions for Example 4.12 after 50,000 lb of proppant is pumped through a 20-ft perforation interval with 2 SPF.

Solution There are total of 40 perforations. The discharge coefficient by Eq. (4.85) gives a new discharge coefficient after pumping 50,000 lb of proppant,

$$C_d = 0.56 + 1.65 \times 10^{-4} \times \left(\frac{5000}{40}\right) = 0.766 \tag{4.107}$$

and

$$d_p = 0.375 + 4.29 \times 10^{-6} \times \left(\frac{50{,}000}{40}\right) = 0.38 \text{ in} \tag{4.108}$$

Using these two parameters to recalculate frictional pressure drop, we have

$$\Delta p_{\text{perf}} = 0.2369\left(\frac{65}{40}\right)^2 \frac{70\left[\frac{\text{lbm}}{\text{ft}^3}\right]}{7.48\left[\frac{\text{gal}}{\text{ft}^3}\right](0.38^4)(0.766^2)} = 478 \text{ psi} \tag{4.109}$$

Frictional pressure drop through the perforation is 24% lower than the originally designed friction (Example 4.12, 622 psi).

Realizing that the near-wellbore pressure drop is a combination of several effects, and the pressure drop through the perforation hole in the casing is only a part of it. Other pressure drop components include flow through the perforation tunnel, pressure drop caused by perforation compaction and formation damage, and near-wellbore tortuosity caused by stress change, perforation orientation, and fracture propagation. Limited entry only controls the pressure drop through the perforation holes on the casing, and all other components are formation property controlled. Once the fracture is created, all other pressure drops change much more rapidly and drastically compared with the erosion caused by proppant passing the perforation holes on the casing. The impact of limited entry and proppant erosion is reduced.

4.6 Completion for Fracturing Horizontal Wells

The completion technology for fracturing horizontal wells has advanced dramatically for the last three decades. The technology has made tremendous contribution in the Austin Chalk development in the 1980s, in North Sea offshore development from the 1990s until today, and it is one of the most critical components that made economic oil and gas production from the unconventional shale plays possible. In the current practice, there are two leading types of completions in horizontal well fracturing; openhole with liner and sliding sleeve fracture ports, and cemented casing with perforation and isolation plugs. The detailed designs vary broadly. The most common way is to start from the toe and consecutively fracture toward the heel along the horizontal wellbore.

One of the important parameters when designing multiple-stage fracturing of horizontal wells is the length treated by each fracture stage. Depending on the wellbore length, this determines the number of stages in a fracture treatment. The stage interval length is determined by production performance from the well, the reservoir mechanical and flow properties, and drilling and completion limitations.

Completions for fracturing horizontal wells need to provide the following functions: (1) isolation between each stage, (2) communication between the wellbore and reservoir for the targeted stage, (3) initiation and propagation of fractures, and (4) protection of the stages that have been fractured.

4.6.1 Plug and Perf

Plug and perf is a cased-hole completion for hydraulic fracturing. Perforating a cased hole yields better control of fracture initiation and propagation. The completion starts by setting a casing in the horizontal section of a well, and cementing the casing in place before fracturing. When the horizontal section is cemented and cleaned, the well has no communication with the reservoir, and is ready to be fractured.

Fractures are first created from the toe section of a horizontal wellbore, and move toward the heel of the wellbore, one stage at a time. In each stage, there are usually multiple clusters of perforations aimed at generating multiple fractures in one stage.

The fracturing operation uses a plug and perf assembly tool to perforate the cemented casing and isolate between stages. The tool usually has a bridge plug at the end and perforating guns following the plug setting tool section. Depending on the number of clusters, the perforation guns are stacked in series. Figure 4.20*a* shows a plug and perf assembling with two perforating clusters. The bridge plugs are made of composite drillable material which temporarily blocks the communication inside the casing between the section of the wellbore that has been fractured and the section that is going to be fractured. Figure 4.20*b* shows the schematic of a composite plug. When squeezed by the two slips, the packing material expands and sets the plug against the casing. These plugs should be easy to drill out after all stages of fracturing are completed.

The perforation assembling with wireline is lowered down the vertical part of the wellbore mainly by gravitational forces; and pumped down to the designed location in the horizontal borehole. Since there is no communication for the first stage between the cemented liner and the formation before perforating the first stage, a communication point needs to be established in the wellbore to assist pumping down the wireline for the first stage. This is usually done by open an annulus circulating valve with a designated pressure.

The main steps in a plug and perf fracturing procedure include the following steps:

1. Run the perf gun tool assembly on a wireline into the hole to the targeted location. The first stage is at the toe of the horizontal section. There is no plug for the first stage. The assembling is pumped down through a pressure-operated communication point.

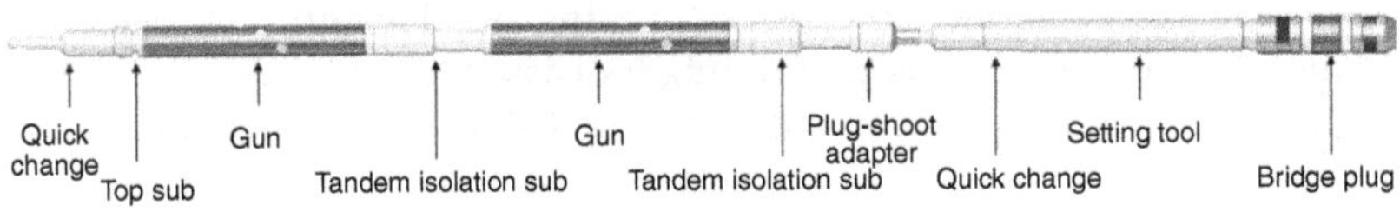

(a)

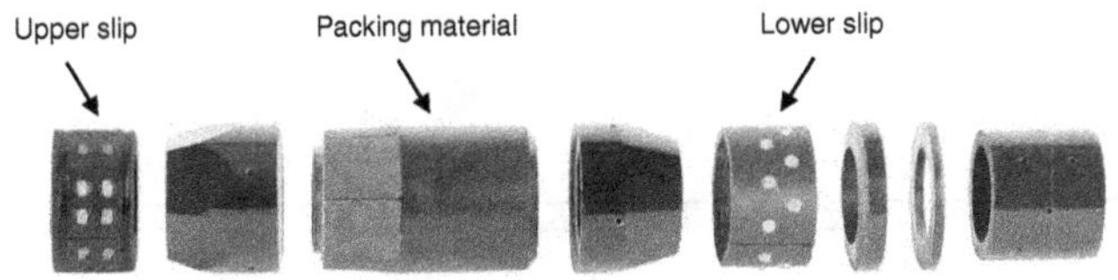

(b)

FIGURE 4.20 Plug and perf completion for fracturing: (*a*) tool assembling (McNelis et al., 2015); (*b*) composite/drillable plug (*courtesy of Weatherford*).

2. Perforate the first stage for multiple clusters by multiple perforating guns.
3. Pull the perf assembling out of the hole.
4. Pump the designed fracture job for the first stage and clean the hole after all proppant are pumped.
5. Run the plug and perf assembly into the hole to the next stage location.
6. Set the composite bridge plug.
7. Perforate the second stage and pull the assembling out of the hole.
8. Pump the second fracture stage.

These steps repeat until all stages are pumped. Once the last stage is completed, the plugs are drilled out and the well is ready to produce.

There are different means to run a plug and perf assembly into a wellbore. The most commonly used method is wireline deployed assembly with a wireline logging truck. The assembly is attached to the wireline and pumped down to the desired location. The assembly can also be set into the wellbore by coiled tubing or drill pipe. Figure 4.21 illustrates a wireline operated plug and perf fracturing procedure.

4.6.2 Jet Perforating and Fracturing

Jetting as a completion and stimulation technique is more commonly used in acid stimulation. In the early 2000s, hydraulic jetting procedure for perforating and fracturing was introduced to the industry and applied successfully in several different regions around the world (Surjaatmadja et al., 2005). The method is also referred to as "pinpoint fracturing." Jetting perforating and fracturing can be applied to both uncemented and cemented completions.

The completion uses a coiled tubing with a jetting head to perforate the casing or wellbore wall (in case of openhole completion). Depending on the perforation design (phasing and density), jet heads can be arranged around the perforating tool assembling.

Jetting delivers high velocity streams (up to 700 ft/min) with low-concentration solids added to it to erode casing and create a cavity in

Figure 4.21 Wireline operated plug and perf fracturing.

the formation. Once the perforation is finished, clean fracture fluid is pumped down the annulus between the coiled tubing and the casing to initiate and propagate fractures. Slurry is followed when the fracture is established to complete the fracture procedure. For multistage fracturing, borehole is not completely cleaned after fracturing in each stage, and the proppant in the borehole is used as isolation between the stages that has been fractured from the stage that will be fractured to prevent fracture fluid flowing to the stages that have been fractured.

For multiple staged fracturing in a horizontal well, the main steps of jetting perforating and fracturing are the following (Fig. 4.22):

1. Run a coiled tubing into the hole with jetting assembly to the location of the first-stage fracture.
2. Jet through the tubing a high-velocity stream with solids to perforate casing and create cavity in the formation.
3. Pump clean fracture fluid through the annulus between casing and tubing to initiate and propagate fracture.
4. Add proppant to the fracture fluid to create the designed fracture geometry.

FIGURE 4.22 Jetting fracturing procedure (East et al., 2008).

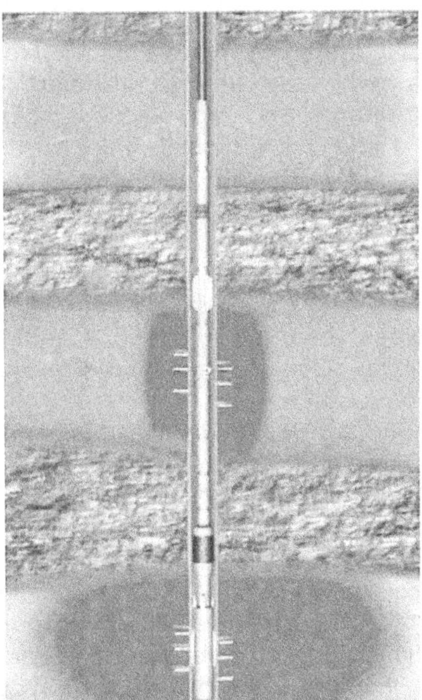

5. Circulate to partially clean the borehole when all proppant is pumped and fracturing is completed.
6. Pull tubing string to the second stage location.
7. Repeat steps 2 to 5 to fracture the second stage.

The most-noticeable advantage of this technology is that perforating and fracturing can be achieved in one trip, and there is no need to pull the stimulation string out of the hole at each stage, as required by the plug and perf procedure. Shorter operation time makes the technology economically attractive. Another possible beneficial feature of jetting fracturing is that jetting perforation, compared with perforating guns, causes less stress-related tortuosity during the course of perforating, enhancing fracture initiation, and propagation.

4.6.3 Just-in-Time Perforation for Fracturing (JITP)

Similar to the plug and perf technique, this completion can be applied in the wells with a cemented liner. The procedure uses a perforating assembly to create multiple stage fractures, and uses ball sealers to isolate the stages. For multiple-staged fracturing, the perforating guns are run into hole to the position of the first stage, where the first set of clusters are perforated. Instead of pulling the perforating assembling, as required by the plug and perf procedure, the perforating assembling is kept in hole, and the fractures are pumped. When the first-stage fracturing is finished, the perforating assembling is positioned to the second stage. Before perforating the second stage, a number of ball sealers are pumped down hole, with the balls enforced to seat on the open perforations and seal off the communication between the wellbore and formation. Once the isolation is confirmed, the perforation guns are fired for second-stage perforation. The second-stage fractures are pumped and the second batch of sealing balls are pumped down hole to seal off the perforations. This procedure repeats until all stages are pumped. Figure 4.23 illustrates the JITP procedure. The ball sealers have the same size, and the size is determined by the perforation size.

One of the main advantages of perforating and using ball sealers for fracturing is that the entire treatment can be pumped in one trip without pulling the perforation assembling out of the hole in each stage. The full bore size without casing packers also allows higher pumping rate for better fracture propagation. Ball sealers may not be the best isolation means, especially for horizontal and deviated wells. The wellbore has to be pressurized during the entire procedure so that the balls are pressed against perforation holes stably. One other critical problem is the risk of having live perforating guns stuck in the hole during the entire fracturing operation.

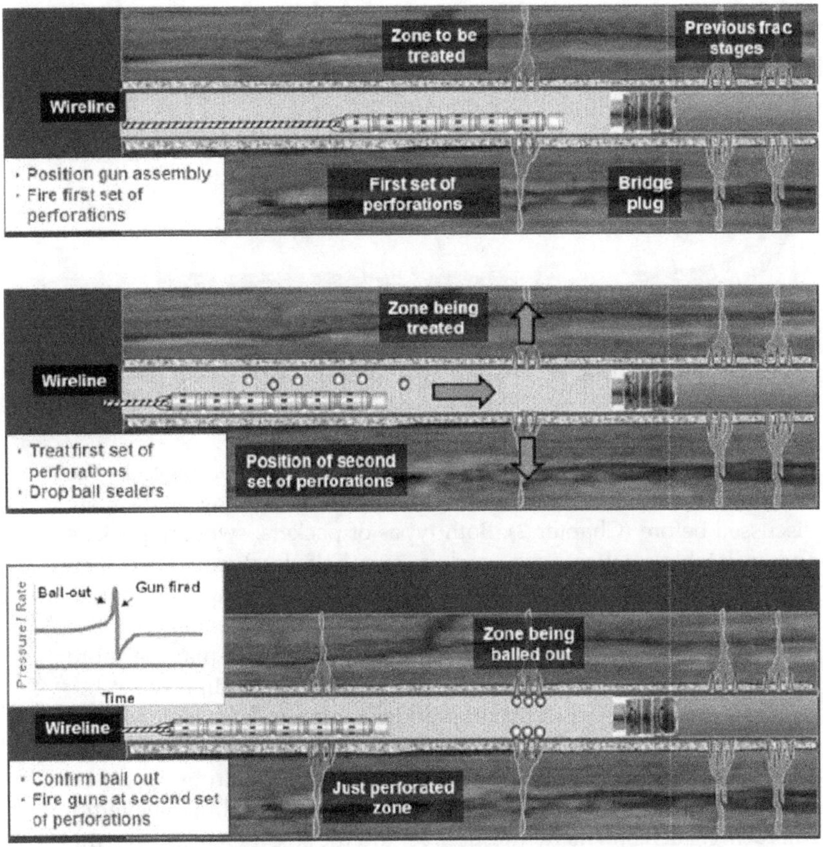

Figure 4.23 Procedure of just-in-time perforation for fracturing (Angeles et al., 2012).

4.6.4 Openhole Completion with Ball-Activated Sliding Sleeves

Openhole completion with uncemented liner for fracturing is one of the popularly used methods for multiple-stage fracturing horizontal wells. Initiated from unconventional reservoir fracturing, it is used in both unconventional and conventional reservoirs today. The completion uses external casing packers (ECPs) to isolate annulus communication between stages, and sliding sleeve valves (also referred to as frac ports) for the flow path between the wellbore and reservoir. ECPs are built as a part of the completion liner string. Sliding sleeves are activated by a series of frac balls that are dropped between fracture stages. Figure 4.24 shows a schematic of openhole uncemented liner completion for fracturing.

There are two main parts of an openhole sliding sleeve completion for fracturing, ECPs and sliding sleeve valves. The main function

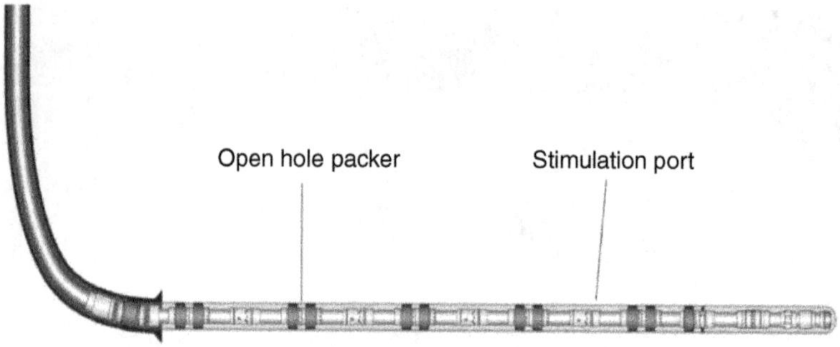

FIGURE 4.24 Completion schematic of uncemented liner with sliding sleeves for fracturing (Appleton et al., 2013).

of ECPs is to isolate communications in between the stages in the annulus between the wellbore and the completion string. ECPs were discussed before (Chapter 2). Both types of packers, swelling packers and inflatable packers, are used in openhole fracturing. As mentioned in Chapter 2, swelling packers are relatively simple to operate and more economical. Once they are set, the isolation is reliable. But because the setting time is not well-controlled, there may be more uncertainty involved. Sometimes it may take a long time to set the packers. This is a concern in completions for fracturing.

A series of balls are used in this fracturing procedure to activate the sliding sleeves. The balls have two functions, isolating upstream communication inside the liner and opening the sliding sleeve for fracturing. Each ball has a unique size, and the size increases from the first stage to the last stage. Inside the liner, there are corresponding-sized ball seats. At each stage, when a ball is pumped down to the ball seat, it will sit on the ball seat, shut off the communication, and isolate the liner from upstream. Further pressuring up the wellbore will activate the sliding sleeve valve to open. Each sleeve valve also has a unique pressure setting; from the toe to the heel, the pressure setting at each valve becomes higher. Because of the differences in ball size and sleeve pressure, the fractures are created one stages at a time, from the toe to the heel, consecutively. Figure 4.25 shows a set of fracture balls used in a multiple-stage fracturing job (the picture shows 24 different sizes). The ball sizes can be quite different when the stages of fracturing get higher. This limits the number of stages a fracture job can be pumped with this completion method. When the ball size gets too small, it has a higher uncertainty to be set correctly on the port to perform the functions. To ensure the sealing by the balls, the seat sometimes is modified from sharp edge to a slanted surface. This not only increases the sealing efficiency but also reduces the stress applied to the balls, thus reducing the failure of the balls.

FIGURE 4.25 Fracture balls for openhole sliding sleeve fracturing (Buffington et al., 2010).

The main steps for fracturing are the following:

1. Run the liner with the ECPs into the hole to the desired position.
2. Set the packers to establish the isolation in the annulus between the stages.
3. Drop the first ball for isolation the liner string and pressure test for isolation.
4. Drop the first-stage ball into the hole. The ball has two functions: isolate upstream communication inside the liner and open the sliding sleeve for fracturing. The first stage is always at the position toward the toe of the horizontal well.
5. Pump fracture fluid to initiate and propagate fractures at stage one. Add proppant to finish pumping the first stage.
6. Clean the hole by circulating the excess proppant out of the liner.
7. Drop the second ball into the hole. The ball has two functions; isolate stage one from the rest of the well inside the liner and open the second-stage frac port.
8. Pump the second-stage fracture. When finished, clean the hole.
9. Repeat steps 7 and 8 until all stages are pumped.
10. Put the well back on flow, and the balls will flow back with the reservoir fluid, and the well is ready for producing.

Perhaps the most significant feature of openhole sleeve completion for fracturing is its short operating time. During an entire fracture pumping, regardless the number of stages, as long as the operation is progressing as designed, there is no trip-out of the completion string.

4.6.5 Openhole with Surface-Controlled Frac Sleeves

For the completion of openhole with sliding sleeves discussed in Section 4.6.4, the balls used to open the sleeves and isolate the stages can cause operational problems such as balls breaking, or balls stuck in the borehole during flowback. As an improvement to compromise the problems with frac balls, hydraulically operated sliding sleeve valves controlled at the surface are used to replace frac balls. The valves are operated by hydraulic power controlled at the surface. Since there is no ball seats in the wellbore as it does in ball-activated sleeves, the additional benefit also includes full-access to the bore hole. Ideally, it does not require to start the first stage at the toe, and fracture does not have to be in location sequence. The sleeves can be used as flow control valves during production time for selective production from desired zones.

4.7 Completion Selection for Unconventional Reservoirs

It is well-understood by the industry that for conventional reservoirs, fractures with extended length are more likely to be created in hard-rocks (high Young's modulus), and for low-permeability reservoirs this is also preferred from production point-of-view. For soft rocks, fractures tend to develop in the width-direction more than they do in hard rocks, and if the reservoirs have moderate to high permeability, sufficient width is more important for production. For unconsolidated formation with high permeability, a frac-pack completion is effective for both sand control and production enhancement. The most common completion for fracturing in conventional reservoirs is through perforations in cased and cemented boreholes. Optimization is achieved through fracture pumping design and proppant/fluid selections.

Fracturing unconventional reservoirs is more challenging. The key to produce from these extremely low-permeability (on the order of hundreds of nano-Darcy) formations is to build extended contact with the reservoir by means of fracturing. This leads to the technique that is commonly used today in unconventional reservoirs: multiple-stage fractures along horizontal wells. Meanwhile because of the low production from low permeability, unconventional reservoirs make the drilling, completion, stimulation, and production operation extremely cost-sensitive. The outcomes of fracture treatments depend on a broad range of parameters, including but not limited to reservoir flow properties and rock mechanical properties, completion type, fracture fluids and proppant selection, and pumping conditions. The two dominant completion techniques, openhole with sliding sleeves and plug and perf, have been extensively applied in all shale formations, and are also successful, for unconventional shale plays.

From the early time Barnett Shale development, it was realized that for complex shale formations with natural fractures fabricated in a layered network pattern, formation rocks are easier to break into network patterns. Hydraulic fractures in such formations are unlikely bi-wing planar fractures. Fracture networks form along the weak joints of rock bedding, and extend to a large width band (Fisher et al., 2004; Warpinski et al., 2008). The understanding of fracture networks that developed over the combination of existing natural fracture and created hydraulic fractures leads to the concept of creating a complex fracture "fairway" (Fig. 4.26, Fisher et al., 2004). The breakthrough of the complex fracture network concept changed the conventional practice of fracturing. The complex formation structure adds difficulty to proppant transportation, especially at the corners of the bedding joints. Meanwhile, when a fracture network is developed during pumping, even with extremely low proppant concentration or no proppant at all, some conductivity of fracture maintains after pumping eases because of unsmooth fracture surfaces and possible shear displacements during fracturing. This conductivity is referred to as "unpropped conductivity," first introduced by Fredd et al. (2001). Barnett Shale samples were tested in the laboratory (Zhang et al., 2013). Unpropped fracture

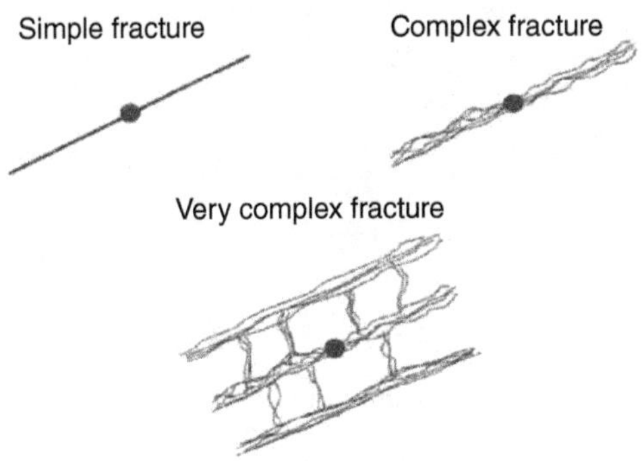

FIGURE 4.26 Fracture network in shale formations (Fisher et al., 2004).

conductivity contributes significantly to flow in unconventional shale formation. To promote fracture networks, much larger volume of liquids and smaller-sized proppant were used in the field. Fracturing with slick water without proppant or alternating between slick water and gelled slurry also become a common routine in Barnett Shale reservoirs. Openhole completion with sliding sleeves to isolate between stages and pumping fractures one stage at a time had a great success in this type of shale reservoirs. In each stage, the treatment is pumped over an open space of several hundreds of feet, and weak points over the stage interval can initiate a fracture. This completion can initiate multiple fractures over one stage and enhance fracture networks.

Some shale formations exhibit very different geological structures compared with the Barnett Shale. The liquid-rich formations in the Eagle Ford Shale and Bakken Shale both have high carbonate content (more than 50% in the payzone, sometimes even more than 80%). Natural fractures do exist in these shale formations, but they are in a more isolated manner, larger scales and lower density; when compared with the Barnett Shale. The formation rocks are also highly heterogeneous. Figure 4.27 shows an outcrop picture of Eagle Ford Shale. Eagle Ford Shale was classified into five facies, zone A to zone E; from the bottom upward (Donovan et al., 2012). The lower

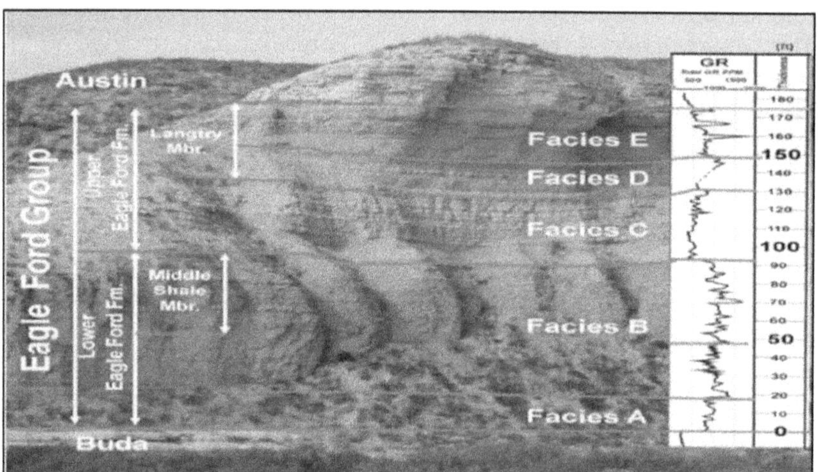

FIGURE 4.27 Eagle Ford outcrop: five different geological facies of the Eagle Ford Shale with their corresponding gamma ray response and the industry nomenclature for the formation—at Lozier Canyon, Southwest Texas (Donovan et al., 2012).

Eagle Ford zones consisting of Facies A and B have higher total organic content (TOC) and are better targets for oil and gas production. Notice that the lower Eagle Ford is darker-colored and less laminated when compared with the upper Eagle Ford (Facies C, D, and E). Hydraulic fractures in the lower Eagle Ford are possible with less fracture density in the fracture network compared with in the Barnett Shale. Both openhole with sliding sleeve completion and plug and perf completion are used in the Eagle Ford reservoir. The plug and perf completion has better defined initiation points for fracture propagation, and showed a great success in the Eagle Ford Shale. Because of better control on fracture initiation and propagation with the plug and perf completion, in the Eagle Ford Shale, about 85% of the liquid-rich wells are fractured with the plug and perf completion in 2015.

From field practices, it showed that the well production is closely related to the completion and fracture design. To create complex fracture networks, the number of fracture stages has been increasing and fracture spacing has been decreasing. Volumes of proppant and fluids pumped are also on the trend of increasing. Figure 4.27 shows a design trend of the fluid and proppant pumped in the Eagle Ford, and the statistical production history related to the stage spacing and volume of proppant and fluid pumped (Centurion et al., 2014). It is clear that a positive relationship exists between the 3-month production and the amount of proppant pumped, and an inverse relationship between well production and fracture spacing (Fig. 4.28). The field fracture design has been modified based on the observations from the production history. From 2011 to 2014, the amount of proppant and fluid pumped kept increasing. A similar study in Bakken formation presented by Lolon et al. (2016) is shown in Fig. 4.29. The result of the statistic study of Bakken fracture treatments has the same trend as with the Eagle Ford study. The well performance is positively correlated to both the fluid volume and the proppant mass, and inversely correlated to the cluster spacing. The cluster spacing has reduced from a couple of hundreds of feet in the early shale development time to 25 to 50 ft in both Eagle Ford and Bakken field in 2015.

Rassenfoss (2017) summarized completion evolution in fracturing practice in unconventional reservoir, using one of the largest producers in Permian Basin, Pioneer Resources, as example. As illustrated in Table 4.4, from 2013 to 2016, the proppant and fracturing fluid pumped were increased, and the cluster spacing and stage spacing were reduced, significantly. The changes are made based on field observation of better well performance. Even though currently we do not have fracture models, both for fracture propagation and for fractured well performance, that can explain the observation, the data-driven evolution is consistent in all shale formations—larger scale

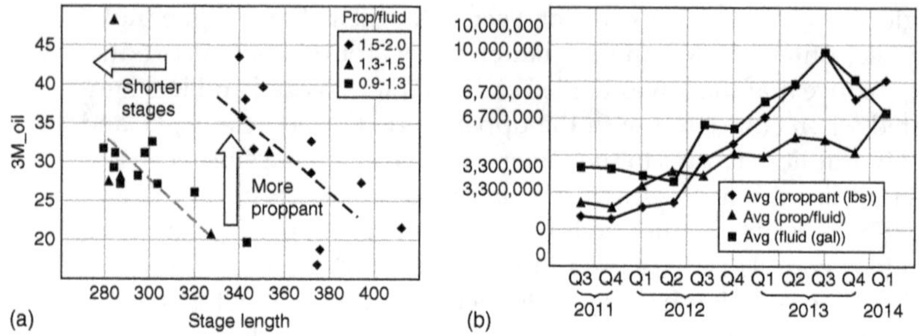

Figure 4.28 Statistic investigation of Eagle Ford Shale fracture design trend (Centurion et al., 2014).

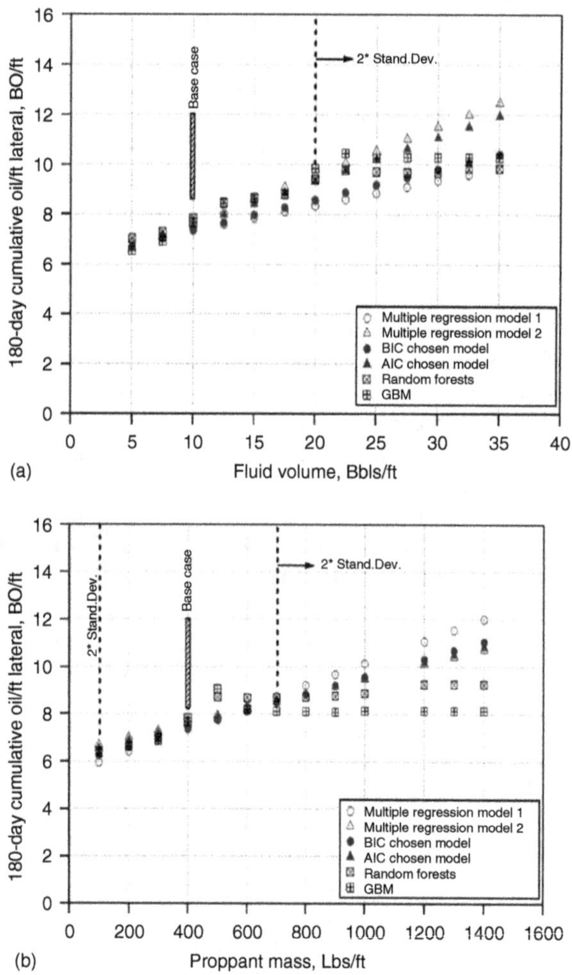

Figure 4.29 Fracture design parameters related to production, Bakken: (*a*) Fluid volume and (*b*) proppant mass (Lolon et al., 2016).

	Version 1.0 Initial Fracture Design (2013–2014)	Version 2.0 Current Fracture Design (Q3 2015–2017)	Version 3.0* Testing Underway (Q1–Q4 2016)
Proppant	1000 lb/ft	1400 lb/ft	Up to 1700 lb/ft
Frac fluid	30 bbl/ft	36 bbl/ft	Up to 50 bbl/ft
Cluster spacing	60 ft	30 ft	Down to 15 ft
Stage spacing	240 ft	150 ft	Down to 100 ft

Source: Adapted from Rassenfoss (2017).
*100 well program with 70 wells online.

Table 4.4 Summary of Pioneer Resources Fracture Practice during 2013 to 2017

treatment design with smaller cluster spacing yields better production performance.

In addition to completion technique advances, the procedure of fracturing in shale formations is also modified to meet the technical and economic challenges. Simultaneous fracturing is a procedure in which two or more nearby horizontal wells are fractured at the same time. The idea is that if two parallel horizontal wells are fracture at the same time, it may have higher possibility to create and build fracture networks between the wells.

Another example of utilizing completion and fracture procedure to increase the efficiency of multistage fracturing is a procedure called "zipper frac." In a zipper frac procedure, two wells are fractured alternately by the plug and perf completion method. When one well is perforating, the other well is fracturing. This procedure overcomes the biggest disadvantage of the plug and perf completion; namely longer operating time to pull the wireline tool out of the wellbore. The procedure is controlled by switching the manifold between perforating and fracturing of the two wells. When operated correctly, this procedure reduces the down time dramatically (on the order of minutes), improve the efficiency of the plug and perf procedure. Figure 4.30 depicts the procedure of zipper frac.

The benefit of zipper frac is not just in saving operation time and therefore, saving the cost to complete the well. Because the fractures from the two wells are created at the same time, it provides higher possibility to create fracture network. Fracture systems with complex fracture networks have higher capacity to transport hydrocarbon fluids.

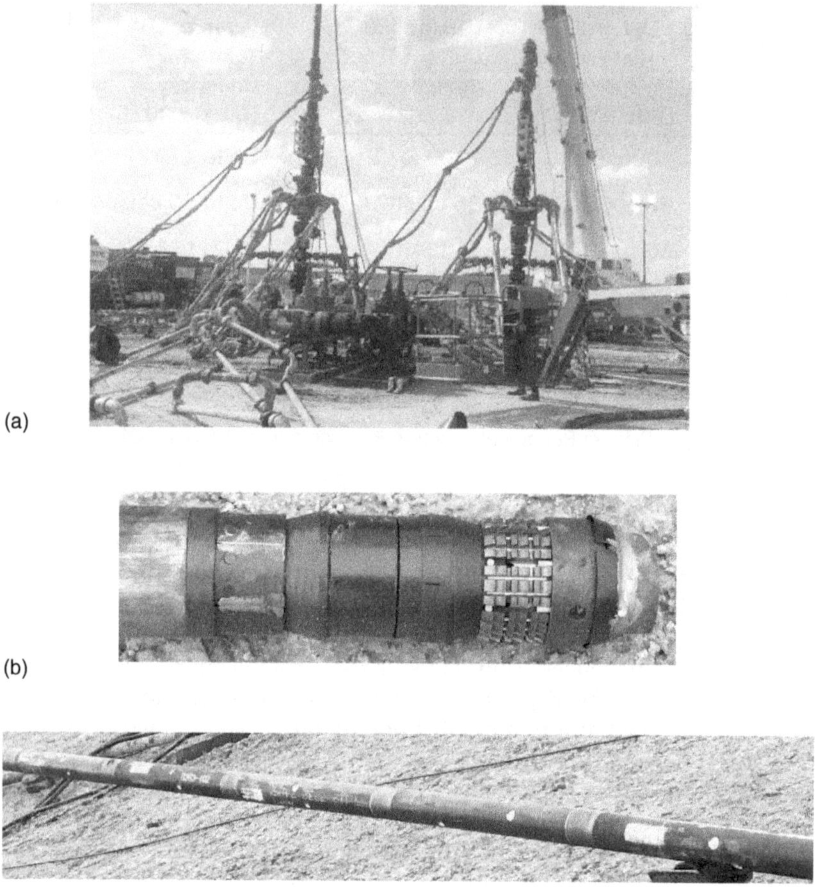

(a)

(b)

(c)

Figure 4.30 Illustration of procedure for zipper frac: (*a*) two wellheads, left one is perforating and right one is fracturing; (*b*) bridge plug; and (*c*) perforating tool assembling.

Nomenclature

A	= transformation matrix defined by Eq. (4.5)
A_f	= fracture face area, ft^2
B	= formation volume factor
C_d	= discharge coefficient
C_L	= leakoff coefficient, ft/min$^{1/2}$
C_f	= fracture conductivity, mD·ft
C_{FD}	= dimensionless fracture conductivity
c_f	= pump solid-handling capacity, lbm/ft^3
c_p	= proppant concentration, lbm/ft^3

$\overline{c_p}$	= average proppant concentration, lbm/ft^3
D	= pipe diameter, inches
d_p	= perforation diameter, inches
$d_{p,\text{initial}}$	= initial perforation diameter, inches
E	= Young's modulus, psi
E'	= plane strain modulus, psi
f_f	= friction factor
G	= cross-linked gelling agent concentration, lb/Mgal
$G(\Delta t_D)$	= G function defined by Eq. (4.93)
J_D	= dimensionless productivity index
h	= formation payzone thickness, ft
h_f	= fracture height, ft
K_L	= fracture efficiency parameter
k	= reservoir permeability, mD
k_f	= proppant permeability, mD
L	= wellbore length, ft
M_p	= proppant mass, lbm
n	= number of perforation
PI	= productivity index, bbl/day/psi
p_p	= pore pressure, psi
p_r	= reservoir pressure, psi
p_{bd}	= breakdown pressure, psi
p_{th}	= tubing head pressure, psi
Δp_{perf}	= perforation pressure drop, psi
Δp_{PE}	= hydrostatic pressure drop, psi
$\Delta p_{G,P}$	= frictional pressure drop for slurry, psi
Δp_0	= frictional pressure drop for water, psi
Δp_f	= frictional pressure drop, psi
q_i	= injection rate, bbl/min
R	= radius of penny-shaped fracture, ft
r_e	= reservoir drainage radius
r_p	= ratio of formation payzone thickness to fracture height
r_w	= wellbore radius, ft
r'_w	= equivalent wellbore radius, ft
s	= skin factor
s_f	= skin factor for fracture
T	= temperature, R
T_0	= tensile strength of the formation
t	= time
t_i	= injection time, minute

t_{Dxf}, t_{Dye} = dimensionless time parameters, Eqs. (4.65) and (4.66)
t_{pad} = pad fluid injection time
x_f = fracture half-length
V_i = injection fluid volume
V_{frac} = fracture volume
$V_{leakoff}$ = leakoff fluid volume
V_{pad} = pad fluid volume
v = velocity of injection fluid, ft/s
w = fracture width for production
$\overline{w}$ = average fracture width
w_{max} = maximum fracture width
w_p = propped fracture width
y_e = reservoir width, ft
Z = compressibility of gas
α = azimuth angel
β = deviation angle
ε = fracture efficiency parameter
ϕ_p = proppant pack porosity
γ = fracture orientation angle in deviated well
θ = angle along the well circumference
μ = fluid viscosity
ρ_p = proppant density
ν = Poisson's ratio
η = fracture efficiency
σ = drag ratio
σ_H = maximum horizontal in-situ stress
σ_h = minimum horizontal in-situ stress
σ_v = vertical in-situ stress

References

Abass, H. H., Brumley, J. L., and Venditto, J. J. (1994, January 1). "Oriented Perforations—A Rock Mechanics View," *Society of Petroleum Engineers*. doi:10.2118/28555-MS.

Angeles, R., Cole, S. W., Benish, T. G., Tolman, R. C., Gupta, J., and Ross, K. B. (2012, January 1). "One Year of Just-in-Time Perforating as Multi-Stage Fracturing Technique for Horizontal Wells," *Society of Petroleum Engineers*. doi:10.2118/160034-MS.

Barree, R. D., Barree, V. L., and Craig, D. (2009, August 1). "Holistic Fracture Diagnostics: Consistent Interpretation of Prefrac Injection Tests Using Multiple Analysis Methods," *Society of Petroleum Engineers*. doi:10.2118/107877-PA.

Barree, R. D., Miskimins, J., and Gilbert, J. (2015, May 1). "Diagnostic Fracture Injection Tests: Common Mistakes, Misfires, and Misdiagnoses," *Society of Petroleum Engineers*. doi:10.2118/169539-PA.

Baser, B., Shenoy, S., Gadiyar, B., Jain, S., and Parlar, M. (2010, September 1). "An Alternative Method of Dealing with Pressure: Friction Reducer for Water Packing of Long Horizontal Open Holes in Low-Fracturing-Gradient Environments," *Society of Petroleum Engineers*. doi:10.2118/123155-PA.

Bazan, L. W., and Meyer, B. R. (2015, November 9). "Fracture Complexity: Analysis Methodology and Signature Pressure Behavior of Hydraulic Fracture Propagation from Horizontal Wellbores," *Society of Petroleum Engineers*. doi:10.2118/176919-MS.

Behrmann, L. A., and Nolte, K. G. (1999, December 1). "Perforating Requirements for Fracture Stimulations. *Society of Petroleum Engineers*," *SPE Drilling and Completion Journal*, Vol. 14, No. 4. doi:10.2118/59480-PA.

Bello, R. O., and Wattenbarger, R. A. (2010, January 1). "Multi-stage Hydraulically Fractured Horizontal Shale Gas Well Rate Transient Analysis," *Society of Petroleum Engineers*. doi:10.2118/126754-MS.

Bocaneala, B., Barrett, C., Holland, B., Langford, M. E., McIntosh, K., Nitters, G., Norris, M. R., and Orski, K. (2015, September 8). "The Evolution of Completion Practices and Reservoir Stimulation Techniques in the North Sea," *Society of Petroleum Engineers*. doi:10.2118/175443-MS.

Brannon, H. D. (2011–2012). Hydraulic Fracturing Materials: Trends and Considerations. SPE Distinguished Lecturer Program.

Bustos, O. A., Powell, A. R., Olsen, T. N., Kordziel, W. R., and Sobernheim, D. W. (2007, January 1). "Fiber-Laden Fracturing Fluid Improves Production in the Bakken Shale Multi-Lateral Play," *Society of Petroleum Engineers*. doi:10.2118/107979-MS.

Card, R. J., Howard, P. R., and Feraud, J.-P. (1995, November 1). "A Novel Technology To Control Proppant Backproduction," *Society of Petroleum Engineers*. doi:10.2118/31007-PA.

Craig, D. P., and Blasingame, T. A. (2006, January 1). "Application of a New Fracture-Injection/Falloff Model Accounting for Propagating, Dilated, and Closing Hydraulic Fractures," *Society of Petroleum Engineers*. doi:10.2118/100578-MS.

Ceccarelli, R. L., Pace, G., Casero, A., Ciuca, A., and Tambini, M. (2010, January 1). "Perforating for Fracturing: Theory vs. Field Experiences," *Society of Petroleum Engineers*. SPE doi:10.2118/128270-MS.

Centurion, S., Junca-Laplace, J.-P., Cade, R., and Presley, G. (2014, October 27). "Lessons Learned from an Eagle Ford Shale Completion Evaluation," *Society of Petroleum Engineers*. doi:10.2118/170827-MS.

Cramer, D. D. (1987, January 1). "The Application of Limited-Entry Techniques in Massive Hydraulic Fracturing Treatments," *Society of Petroleum Engineers*. doi:10.2118/16189-MS.

Crump, J. B., and Conway, M. W. (1988, August 1). "Effects of Perforation-Entry Friction on Bottomhole Treating Analysis," *Society of Petroleum Engineers*. doi:10.2118/15474-PA.

Donovan, A. D., Staerker, T. S., Pramudito, A., Li, W., Corbett, M. J., Lowery, C. M., Romero, A. M., et al. (2012). "The Eagle Ford Outcrops of West Texas: A Laboratory for Understanding Heterogeneties within Unconventional Mudstone Reservoirs," *CGAGS Journal*, Vol. 1, pp. 162–185.

East, L. E., Bailey, M. B., and McDaniel, B. W. (2008, January 1). "Hydra-Jet Perforating and Proppant Plug Diversion in Multi-Interval Horizontal Well Fracture Stimulation: Case Histories," *Society of Petroleum Engineers*. doi:10.2118/114881-MS.

Economides, M. J., Hill, A. D., Ehlig-Economides, C., and Zhu D. (2012). *Petroleum Production Systems*, 2nd ed., Prentice Hall. ISBN: 0-13-703158-0.

Economides, M. J., Martin, T.; BJ Services Company. (2007). *Modern Fracturing: Enhancing Natural Gas Production*. ET Publishing, Houston, TX.

Economides, M. J., Oligney, R., and Valko, P. (2002). *Unified Fracture Design*. Orsa Press. ISBN-10:0971042705.

El-Rabba, A. M., Shah, S. N., and Lord, D. L. (1999, February 1). "New Perforation Pressure-Loss Correlations for Limited-Entry Fracturing Treatments," *Society of Petroleum Engineers*. doi:10.2118/54533-PA.

Ely, J. W., Fowler, S. L., Tiner, R. L., Aro, D. J., Sicard, Jr., G. R., and Sigman, T. A. (2014, October 27). "Slick Water Fracturing and Small Proppant: The Future of Stimulation or a Slippery Slope?" *Society of Petroleum Engineers*. doi:10.2118/170784-MS.

Geertsma, J., and de Klerk, F. (1969, December). "A Rapid Method of Predicting Width and Extent of Hydraulically Induced Fractures," *Journal of Petroleum Technology*, pp. 1571–1581.

Gidley, J. L., Holditch, S. A., Nierode, D. E. and Veatch, R. W. Jr. (1989). "Recent Advances in Hydraulic Fracturing," *SPE Monograph Series*, Vol. 12. *Society of Petroleum Engineers*. ISBN: 978-1-55563-021-1.

Gulbis, J. (1987). "Fracturing Fluid Chemistry," *Reservoir Stimulation*, M. J. Economides and K. G. Nolte (eds.), Schlumberger Educational Services.

Jennings, A. R. (1996, July 1). "Fracturing Fluids—Then and Now," *Society of Petroleum Engineers*. doi:10.2118/36166-JPT.

Khristianovic(h), S. A., and Zheltov, Y. P. (1955). "Formation of Vertical Fractures by Means of Highly Viscous Liquid," *Proceedings of Fourth World Petroleum Congress*, Sec. II, pp. 579–586.

Koloy, T. R., Sorheim, T., Braekke, K., and Lonning, P. (2014, October 27). "The Evolution, Optimization and Experience of Multistage Frac Completions in a North Sea Environment," *Society of Petroleum Engineers*. doi:10.2118/170641-MS.

Lolon, E., Hamidieh, K., Weijers, L., Mayerhofer, M., Melcher, H., and Oduba, O. (2016, February 1). "Evaluating the Relationship between Well Parameters and Production Using Multivariate Statistical Models: A Middle Bakken and Three Forks Case History," *Society of Petroleum Engineers*. doi:10.2118/179171-MS.

Lord, D. L., and McGowen, J. M. (1986, January 1). "Real-Time Treating Pressure Analysis Aided by New Correlation," *Society of Petroleum Engineers*. doi:10.2118/15367-MS.

Lord, D. L. (1994, January 1). "Study of Perforation Friction Pressure Employing a Large-Scale Fracturing Flow Simulator," *Society of Petroleum Engineers*. doi:10.2118/28508-MS.

Fredd, C. N., McConnell, S. B., Boney, C. L. and England, K. W. (2001). "Experimental Study of Fracture Conductivity for Water-Fracturing and Conventional Fracturing Applications," *SPE Journal*, Vol. 6, No. 3, pp. 288–298.

"Fracturing Flow Simulator," *Society of Petroleum Engineers*. doi:10.2118/28508-MS

Romero, J., Mack, M. G., and Elbel, J. L. (2000, May 1). "Theoretical Model and Numerical Investigation of Near-Wellbore Effects in Hydraulic Fracturing," *Society of Petroleum Engineers*. doi:10.2118/63009-PA.

Meehan, D. N. (1995, May 1). "Stimulation Results in the Giddings (Austin Chalk) Field," *Society of Petroleum Engineers*. doi:10.2118/24783-PA.

McNelis, L., Salt, W., and Scharf, T. S. (2015, September 28). "High-Performance Plug-and-Perf Completions in Unconventional Wells," *Society of Petroleum Engineers*. doi:10.2118/174922-MS.

Meyer, B. R., and Jacot, R. H. (2005, January 1). "Pseudosteady-State Analysis of Finite Conductivity Vertical Fractures," *Society of Petroleum Engineers*. doi:10.2118/95941-MS.

Nolte, K. G. (1979, January 1). "Determination of Fracture Parameters from Fracturing Pressure Decline," *Society of Petroleum Engineers*. doi:10.2118/8341-MS.

Nolte, K. G. (1986, July). "Determination of Proppant and Fluid Schedules from Fracturing Pressure Decline," *SPEPE*, pp. 255–265.

Nordgren, R. P. (1972, August). "Propagation of Vertical Hydraulic Fracture," *SPEJ*, pp. 306–314.

Palmer, I. D., and Veatch, R. W., Jr. (1990, August). "Abnormally High Fracturing Pressures in Step-Rate Tests," *SPEFE*, pp. 315–323.

Perkins, T. K., and Kern, L. R. (September 1961). "Widths of Hydraulic Fracture," *JPT*, pp. 937–949.

Prats, M. (1961, June). "Effect of Vertical Fractures on Reservoir Behavior—Incompressible Fluid Case," *SPEJ*, pp. 105–118.

Rassenfoss, S. (2017, February 1). "Rebound to Test If Cost Cuts Will Last," *Society of Petroleum Engineers*. doi:10.2118/0217-0035-JPT.

Smith, B. M., and Montgomery, C. (2015). *Hydraulic Fracturing*. CRC Press 2015. Print ISBN: 978-1-4665-6685-9. eBook ISBN: 978-1-4665-6692-7.

Surjaatmadja, J. B., East, L. E., Luna, J. B., and Hernandez, J. O. E. (2005, January 1). "An Effective Hydrajet-Fracturing Implementation Using Coiled Tubing and Annular Stimulation Fluid Delivery," *Society of Petroleum Engineers*. doi:10.2118/94098-MS.

Surjaatmadja, J. B., Willett, R. M., McDaniel, B. W., Rosolen, M. A., Franco, M., dos Santos, F. R., Cortes, M. (2007, June 1). "Selective Placement of Fractures in Horizontal Wells in Offshore Brazil Demonstrates Effectiveness of HydraJet Stimulation Process," *Society of Petroleum Engineers*. doi:10.2118/90056-PA.

Yew, C. H., Schmidt, J. H., and Li, Y. (1989, January 1). "On Fracture Design of Deviated Wells," *Society of Petroleum Engineers*. doi:10.2118/19722-MS.

Yew, C. and Wong, X. (2015). *Mechanics of Hydraulic Fracturing, Second Edition*. Gulf Professional Publishing. Waltham, MA. ISBN 978-0-12-420003-6.

Veatch, R. W., King, G. E., and Holditch, S. A. (2017). *Essentials of Hydraulic Fracturing: Vertical and Horizontal Wellbores*. PennWell. Tulsa, Oklahoma. ISBN-13: 978-1593703578, ISBN-10: 1593703570.

Wutherich, K., and Walker, K. J. (2012, January 1). "Designing Completions in Horizontal Shale Gas Wells: Perforation Strategies," *Society of Petroleum Engineers*. doi:10.2118/155485-MS.

Zhang, J., Kamenov, A., Zhu, D., and Hill, D. (2013, March 26). "Laboratory Measurement of Hydraulic Fracture Conductivities in the Barnett Shale," *International Petroleum Technology Conference*. doi:10.2523/IPTC-16444-MS.

CHAPTER 5

Completions for Acid Stimulation

Acid stimulation is a procedure in which reactive chemicals are injected into the formation to improve well productivity or injectivity. The goal of completion design for acid stimulation is to achieve the best stimulation outcomes. To understand completion design, we first review the fundamentals of acid stimulation, and then discuss how completion design can help improve stimulation outcomes and enhance well productivity. In the extended-reach environments where long reservoir intervals are stimulated, we discuss the critical issues in acidizing, including zonal flow control and isolation, fluid diversion, and acid injection velocity in the formation. The completion design for successful acid treatments will be addressed in detail for the aforementioned issues.

5.1 Fundamentals of Acid Stimulation

In this section, we introduce chemical reaction and reaction stoichiometry in sandstone matrix acidizing and carbonate acid stimulation. The mechanism of acidizing for sandstone and carbonate is significantly different. The treatment designs, including acid systems and additives selection, pumping conditions, and completion for acid stimulation, should take into account of these differences.

5.1.1 Types of Acid Treatments

Acidizing involves a wide range of stimulation treatments that can yield impressive production increases in many wells, if properly applied. Acidizing treatments are divided into two categories based on formation types: sandstone acidizing and carbonate acidizing. Carbonate acidizing can be further divided into two different treatment techniques: matrix acidizing and acid fracturing. Figure 5.1 illustrates sandstone matrix acidizing, carbonate matrix acidizing, and acid fracturing treatments. The word "matrix" here infers that acid injection is kept below the fracture pressure of the formation, and injected fluid flows into the formation through the porous medium.

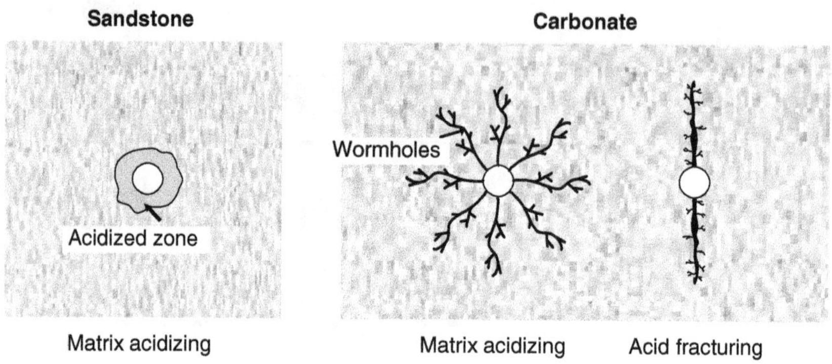

FIGURE 5.1 Acid treatments in sandstone and carbonate formations.

As discussed in Chapter 2, during the lifetime of a well, many activities can result in formation damage around the wellbore; for example, drilling and completion. Sandstone acidizing involves the dissolution of some minerals in the formation in the near-wellbore region to remove formation damage that is blocking or bridging the pore throats in the formation matrix, thus ideally recovering the original reservoir permeability. In most cases, the intention of acidizing in sandstone formations is to improve or to restore the permeability in the near-wellbore region without fracturing the formation. The change in the pore structure caused by dissolution of some of the minerals by acid is the mechanism by which matrix acidizing increases permeability. The invasion of filtrates caused by drilling muds and completion fluids can introduce an entirely different chemical environment. Acid dissolves a variety of damaging materials along with the formation materials. In an undamaged formation, the production increase from a matrix acidizing job is very low. This is because sandstone matrix acidizing is a near-wellbore treatment, with all of the acid reacting within about a foot around the wellbore.

In carbonate formations, injected acid moves into the formation and dissolves a portion of the rock to enhance the natural permeability of rock. Similarly to sandstone acidizing, matrix acidizing can be applied in carbonate formations. However, the manner in which the pore structure changes is fundamentally different in sandstones and carbonates. In sandstones, acid follows the permeability distribution of the formation, and more likely forms an injected fluid bank mimicking the permeability contour; while in carbonates, acid creates highly permeable channels that are referred to as wormholes. This phenomenon is illustrated in Fig. 5.1. The difference between acidizing in sandstone and carbonate leads to radically different approaches when modeling the acidizing process in these formations. Carbonate matrix acidizing creates a radially extending, highly permeable region immediately surrounding the wellbore, lowering pressure drop due to flow and improving well productivity significantly. This technique can be successfully applied in all carbonate rock types, for both high- and low-permeability reservoirs. It is, however, more successful in the reservoirs with moderate to high permeability. Unlike sandstone acidizing, the production performance of carbonate reservoirs can be improved by acid stimulation treatments even for undamaged wells. Field observations showed that in high-rate matrix acidizing applied in carbonate formations, the acid can penetrate distances much longer than a couple of feet beyond the wellbore, resulting in skin factors of about –4 (Furui et al., 2012a). This means that the well productivity can be doubled after high-rate matrix acidizing treatments even if there is no damage. This observation applies to both vertical and horizontal wells whether completed with openhole or cased hole techniques. Acid stimulation in carbonate reservoirs almost always improves well performance.

In acid fracturing, fluids (reactive or nonreactive) are pumped at pressures above the formation fracture pressure. The acid moves along a hydraulically created fracture plane, etching the face of the fracture to create an uneven surface because of the heterogeneous natural of carbonate formations, or creates permeable flow channels along the fracture walls. The uneven surfaces do not completely close up after pumping is stopped, providing conductivity for the created fractures. If channels are created along the fracture faces during acid pumping, the fracture conductivity is usually higher and more sustainable compared with an etched uneven surface. The flow path changes resulting from acid fracturing reduces the drawdown required to produce hydrocarbons, and improves well inflow performance. Acid fracturing is most successful in lower permeability carbonate reservoirs where it is easier to create long hydraulic fractures with minimum fluid losses. Stronger carbonate rock types provide better post-fracture conductivity than seen in softer rocks, where the acid etched fracture can close off under the closure stress over time. Both matrix acidizing and acid fracturing in carbonate formations offer potential for significant levels of production improvements. However, the mechanisms of these stimulation techniques are different.

5.1.2 Acidizing: Chemistry and Acid-Mineral Reaction Stoichiometry

Acidizing chemistry can be divided into two categories with regards to formation mineralogy. Carbonate reservoirs, including limestone, chalk, and dolomite, are stimulated with simple acid systems, such as HCl. Sandstone reservoirs are treated with more complex acid systems, containing HF or HF/HCl mixture to dissolve silicate-containing minerals such as clays and feldspars.

The amount of acid needed to dissolve a given amount of mineral is determined by the stoichiometry of the chemical reaction, which describes the number of moles of each species involved in the reaction. For example, the simple reactions between hydrochloric acid (HCl) and calcite and dolomite can be written as (Schechter, 1992).

$$\text{Calcite: } 2HCl + CaCO_3 \rightarrow CaCl_2 + CO_2 + H_2O \tag{5.1}$$

$$\text{Dolomite: } 4HCl + CaMg(CO_3)_2 \rightarrow CaCl_2 + MgCl_2 + 2CO_2 + 2H_2O \tag{5.2}$$

For sandstone acidizing, when hydrofluoric acid reacts with silicate minerals, numerous secondary reactions may occur that influence the overall stoichiometry of the reaction. For example, when HF reacts with quartz, the primary and secondary reactions are

$$4HF + SiO_2 \leftrightarrow SiF_4 + 2H_2O \qquad \text{(silicon tetrafluoride)} \tag{5.3}$$

$$SiF_4 + 2HF \leftrightarrow H_2SiF_6 \qquad \text{(fluosilicic acid)} \tag{5.4}$$

Silicon terafluoride and water are produced in the primary reaction. The SiF_4 produced may also react with the HF to form fluosilicic acid in a secondary reaction. A complication is that the fluosilicates may exist in various forms, so that the total amount of HF required to dissolve a given amount of quartz depends on the solution concentration.

The stoichiometry of the reactions presented above can be used to calculate a theoretical dissolution of mineral by various acids over a range of concentrations. The dissolving power of an acid with common carbonate minerals can be expressed as either the weight or volume of carbonate mineral dissolved. The gravimetric dissolving power, β, in terms of the mass of mineral consumed by a given mass of acid, is defined as

$$\beta = \frac{v_{\text{mineral}} MW_{\text{mineral}}}{v_{\text{acid}} MW_{\text{acid}}} \tag{5.5}$$

where v is the stoichiometric coefficient and MW is molecular weight. Often, the volume of carbonate mineral that can be dissolved per volume of treating acid is of more interest than the weight ratio. The dissolving power β calculated above can be converted to a volume ratio, χ, by multiplying the β value by the density ratio of the mineral and acid:

$$\chi = \beta \frac{\rho_{\text{acid solution}}}{\rho_{\text{mineral}}} \tag{5.6}$$

Example 5.1 A 28% HCl solution has a specific gravity of about 1.14, and $CaCO_3$ has a density of 169 lbm/ft³. For the reaction of these species, calculate the volumetric dissolving power.

Solution The gravimetric dissolving power between 100% HCl and $CaCO_3$ is

$$\beta_{100} = \frac{1 \times 100.1}{2 \times 36.5} = 1.37 \left[\frac{\text{lb}_m \text{CaCO}_3}{\text{lb}_m \text{HCl}}\right] \tag{5.7}$$

The subscript, 100, denotes 100% HCl. The dissolving power of any other concentration of acid is β_{100} times the weight fraction of acid in the acid solution. For 28% of HCl, the volumetric dissolving power is calculated by

$$\chi = (0.28 \times 1.37)\left[\frac{\text{lb}_m\text{CaCO}_3}{\text{lb}_m 28\%\text{HCl}}\right]\left(\frac{(1.14)(62.4)\left[\frac{\text{lb}_m 28\%\text{HCl}}{\text{ft}^3 28\%\text{HCl}}\right]}{169\left[\frac{\text{lb}_m\text{CaCO}_3}{\text{ft}^3\text{CaCO}_3}\right]}\right) \tag{5.8}$$

$$= 0.161\left[\frac{\text{ft}^3\text{CaCO}_3}{\text{ft}^3 28\%\text{HCl}}\right]$$

For a carbonate matrix acidizing treatment, if the injection volume is 75 gal of 28% HCl per foot thickness 1.6 ft^3 of $CaCO_3$ per foot have been dissolved after the acid injection. The dissolving power calculations are based on the assumption that all of the acid completely reacts in the presence of carbonate minerals. For weak organic acids, that may not be the case.

5.1.3 Reaction Rates of Acids with Carbonates and Sandstone Minerals

The rate at which minerals are dissolved has a strong impact on the penetration pattern of the acid treating fluid into the reservoir. The dissolution rate of the carbonate mineral is controlled by the reaction rate of the acid treating fluid with the carbonate compound and the rate of transfer of the acid to the mineral surface. The reaction rate of acid with carbonate minerals is very fast and in most cases, the limiting step is the acid transport to the mineral surface. It is this sensitivity to mass transfer that results in the creation of wormholes at the wellbore or fracture face.

Hydrochloric acid is a strong acid, meaning that when HCl is dissolved in water, the molecules almost completely dissociate to form hydrogen ions, H^+, and chloride ions, Cl^-. The reaction between HCl and carbonate minerals is actually a reaction of the H^+ with the mineral. With weak acids, such as acetic or formic acid, the reaction is also between H^+ and the mineral, with the added complication that the acid is not completely dissociated, thus limiting the supply of H^+ available for reaction. Because H^+ is the reactive species, the kinetics for the HCl reaction can also be used for weak acids by considering the acid dissociation equilibrium.

Schechter (1992) summarizes the reaction rate of HCl with carbonates, r_{HCl}, as follows:

$$-r_{\text{HCl}} = E_f C_{\text{HCl}}^{\alpha} \tag{5.9}$$

$$E_f = E_f^0 \exp\left(-\frac{\Delta E}{RT}\right) \tag{5.10}$$

where T is temperature, E_f is the reaction rate constant and C_{HCl} is the acid concentration. The constants α, E_f^0, and $\Delta E/R$ are given in Table 5.1. SI units are used in these expressions, so C_{HCl} has units of $\text{kg}\cdot\text{mol/m}^3$ and T is in K.

Mineral	α	$E_f^0\left[\dfrac{\text{kg moles HCl}}{\text{m}^2\text{-s-}\left(\dfrac{\text{kg moles HCl}}{\text{m}^3\text{acid solution}}\right)^{\alpha}}\right]$	$\dfrac{\Delta E}{R}(K)$
Calcite	0.63	7.314×10^7	7.55×10^3
Dolomite	$\dfrac{6.32 \times 10^{-4}T}{1-1.92 \times 10^{-3}T}$	4.48×10^5	7.9×10^3

Source: From Economides et al. (2012).

Table 5.1 Constants in HCl-Mineral Reaction Kinetics Models

The kinetics of a weak-acid carbonate mineral reaction can be obtained from the HCl reaction kinetics by (Schechter, 1992)

$$-r_{\text{weak acid}} = E_f K_d^{\alpha/2} C_{\text{weak acid}}^{\alpha/2} \tag{5.11}$$

where K_d is the dissociation constant of the weak acid and E_f is the reaction rate constant for the HCl-mineral reaction.

For sandstone formations hydrofluoric acid reacts with virtually all of the many mineral constituents of sandstone. Reaction kinetics for silicate-containing minerals can all be represented by

$$-r_{\text{mineral}} = E_f[1 + K(C_{\text{HCl}})^{\beta}]C_{HF}^{\alpha} \tag{5.12}$$

$$E_f = E_f^0 \exp\left(-\frac{\Delta E}{RT}\right) \tag{5.13}$$

and the constants α, β, E_f^0, and $\Delta E/R$ are given in Table 5.2.

Example 5.2 Calculate the reaction rate for 28% HCl with 100% calcite rock at 100°F (310.8 K).

Solution The reaction rate is calculated by

$$E_f = 7.314 \times 10^7 \exp\left(-\frac{7.55 \times 10^3}{310.8}\right) = 2.062 \times 10^{-3} \frac{\text{kg} \cdot \text{mol calcite}}{\text{m}^2 - \text{s} - \left(\dfrac{\text{kg} \cdot \text{mol HCl}}{\text{m}^3 \text{ solution}}\right)} \tag{5.14}$$

28 w% HCl acid needs to be converted to the correct unit for the above equation (molar concentration in $\text{kg} \cdot \text{mol/m}^3$),

$$\begin{aligned} \text{molar concentration} &= \frac{\text{(weight concentration)(density)}}{\text{molecular weight}} \\ &= \frac{0.28 \times 1076}{36.5} = 8.25 \left[\frac{\text{kg} \cdot \text{mol}}{\text{m}^3}\right] \end{aligned} \tag{5.15}$$

$$r_{\text{HCl}} = 2.062 \times 10^{-3} \times (0.8.25^{0.63}) = 0.0076 \frac{\text{kg} \cdot \text{mol calcite}}{\text{m}^2 - \text{s}} \tag{5.16}$$

Mineral	α	β	K [(kg · mol HCl/m^3)$^{-\beta}$]	$E_f^0\left[\frac{\text{kg}\cdot\text{mol mineral}}{\text{m}^2\cdot\text{s}\left(\frac{\text{kg}\cdot\text{mol HF}}{\text{m}^3\text{ acid}}\right)^{\alpha}}\right]$	$\frac{\Delta E}{R}(K)$
Quartz, SiO_2	1.0	–	0	2.32×10^{-8}	1150
Orthoclase, K-Feldspar, $KAlSi_3O_8$	1.2	0.4	$5.66 \times 10^{-2}\exp(956/T)$	1.27×10^{-1}	4680
Albite, Na-Feldspar, $NaAlSi_3O_8$	1.0	1.0	$6.24 \times 10^{-2}\exp(554/T)$	9.50×10^{-3}	3930
Kaolinite, $Al_4Si_4O_{10}(OH)_8$	1.0	–	0	0.33	6540
Sodium montmorillonite, $Al_4Si_8O_{20}(OH)_4–nH_2O$	1.0	–	0	0.88	6540
Illite, $K_{0\text{-}2}Al_4(Al, Si)_8O_{20}(OH)_4$	1.0	–	0	2.75×10^{-2}	6540
Muscovite, $KAl_2Si_3(OH)_2$	1.0	–	0	0.49	6540

Source: From Economides et al. (2012).

Table 5.2 Constants in HF-Mineral Reaction Kinetics Models

5.1.4 Additives in Acidizing Fluids

In addition to the acid, a wide variety of additives are included in most acid treatments. Additives are used to prevent excessive corrosion, sludge, emulsions, and iron precipitation. They are also used to improve cleanup, reduce drag force, divert the fluids to improve zonal coverage, and prevent precipitation of reaction products. The following section presents a brief description of various additives commonly used in carbonate and sandstone reservoirs. More detailed information is available in Ali et al. (2016) and Economides and Nolte (2000).

Corrosion Inhibitors

The most important acid additives are corrosion inhibitors. All acid systems require the addition of corrosion inhibitors to minimize the impact of acid attack on drillpipe, tubing, or any other metal parts of the well system that the acid contacts during treatments. Acid corrosion of casing or tubing can lead to well failure. This is of particular concern when using high concentrations of HCl or HF, especially for acidizing wells with high bottomhole temperatures. A variety of chemical compounds can be used to prevent acid corrosion. Commonly used chemical compounds that have been used as corrosion inhibitors are inorganic and organic corrosion inhibitors. Typical inorganic compounds include arsenic, copper, zinc, nickel, and antimony. Organic compounds include acetylenic alcohols, amines, thioureas, and thiophenols.

Standard test methods have been developed to evaluate the performance of corrosion inhibitors using representative metal samples (coupons) in specific acid formulations. Also it is recommended to conduct these tests at the conditions of bottomhole temperature and a pressure of at least 1000 to 1500 psi (Smith et al., 1978). In general, effective corrosion inhibition can be obtained in low-concentration HCl (≤15%) at temperatures up to 250°F for exposure times less than 24 hours. At temperatures greater than 250°F and/or longer exposure times, it may be necessary to reduce the HCl concentration or use a weaker, organic acid (acetic acid or formic acid) to adequately control the corrosion rate.

Common additives such as mutual solvent, short-chain alcohols, antisludge agents (anionic), and anionic surfactants used in acid systems can adversely affect corrosion inhibition (Ali et al., 2016). Solvents or surfactants may impede the adsorption of the inhibitor onto the metal surface. Therefore, all corrosion tests should be performed with the acid system containing all additives.

Mutual Solvents

Mutual solvents are chemical compounds that are soluble in both water and hydrocarbons. Mutual solvents are commonly added to acid systems (Hall, 1975; King and Lee, 1988) to reduce the surface tension of the acid, water-wet the formation surface, and solubilize hydrocarbons that could interfere with acid reaction on the surface of rock minerals. In addition to improving the contact of acid with mineral surfaces, mutual solvents also reduce the stability of emulsions that may be formed during acid treatments. As mentioned earlier, mutual solvents can impact the effectiveness of corrosion inhibitors and should be tested for corrosion inhibitor performance before being used in the field.

Surfactants

Surfactants are chemicals that possess both oil-soluble (lipophilic) and water-soluble (hydrophilic) functional groups. Typically, surfactants are classified by the ionic nature of the hydrophilic group. For example, many surfactants are salts of quaternary amines having a positive charge and are called cationic surfactants. Surfactants prepared from neutralized sulfonic acid have a negative charge and are called anionic surfactants. Some surfactants do not have a charge, but are polar, such as ethoxylated nonylphenol. These surfactants are referred to as nonionic surfactant. Finally, certain surfactants may contain either positive or negative charge depending on the pH environment. Typically, at low pH, these surfactants are cationic and at high pH the surfactant is anionic.

Surfactants reduce the surface tension between the acid and oil or gas and improve contact of the acid with the carbonate mineral surface. Surfactants also improve the well cleanup after acidizing. Additionally, surfactants may be used to prevent or treat sludge (Houchin et al., 1992; Ali et al., 1994). Sludge is an agglomeration of naturally occurring components in hydrocarbons and acid reaction products. Surfactants can promote the dispersion of these particles and/or prevent the particles from agglomerating.

Surfactants can change the wettability of mineral surfaces with the charge on the mineral surface and the charge on the hydrophilic group of the surfactant dictating whether the rock will be water wet or oil wet. The charge on a mineral surface is determined by the pH environment. Common minerals have a defined point of zero charge which is the pH value that results in a neutral surface. At pH values below the point of zero charge, the surface of the mineral is positively charged and at pH values above the point of zero charge, the surface is negatively charged.

It is important to remember that surfactants can increase the emulsion tendency of acid with certain crude oils. Therefore, compatibility testing of the acid containing surfactant and crude oil must be performed.

Alcohol

Alcohol (typically, methanol or isopropanol) can be used in acid treatments to lower the surface tension of the acid and enhance flowback after the treatment (Ali and Hinkel, 2000). This is most commonly done in dry gas wells and low-permeability formations where water block may occur and impede production. To be effective, alcohol is added at concentrations of up to 25% by volume. Alcohol in acid treating fluids can decrease the effectiveness of corrosion inhibitors.

Deemulsifiers/Nonemulsifiers

Deemulsifiers are usually oil-soluble surfactants that are added to emulsions to break an oil-water emulsion formed upon mixing of the acid and crude oil. The effectiveness of the deemulsifier and rate of breaking will depend on how quickly it can concentrate at the oil/water interface (Ali and Hinkel, 2000). This is often difficult to accomplish if the emulsion is formed in the reservoir.

Nonemulsifiers are surfactants added to acid systems to prevent emulsions. These surfactants are usually mixtures of surfactants and solvents that are blended to obtain a final composition with broader applications (Ali and Hinkel, 2000). Since nonemulsifiers are added to acid systems to prevent the formation of emulsions, the performance is generally better than deemulsifiers.

Iron Control

During acidizing treatments, iron can be dissolved by the acidic fluids. The sources of iron are from tubulars and/or iron containing minerals (hematite, pyrite, chlorite, magnetite, etc.) in the reservoir. Initially, the iron is in the ferrous (+2) oxidation state. As the acid spends, the ferrous ions convert to ferric ions (+3) and begin to precipitate as $Fe(OH)_3$ at a pH of about 1 to 2 (Taylor et al., 1999). However, Ferrous hydroxide $Fe(OH)_2$ begins to precipitate as the pH becomes greater than 6. Since the pH of the spent acid following acidization of the carbonate formation is around 4.5 (Chang et al., 2008), only the precipitation of ferric hydroxide in the spent acid is of concern. To prevent reduction of permeability due to deposition of $Fe(OH)_3$ within the reservoir, iron control additives are often included in acid treating fluids.

Iron control additives are usually chelants or reducing agents that prevent iron precipitation. Reducing agents convert the ferric ions to ferrous ions to prevent precipitation as the acid spends. Common reducing agents are citric acid, ascorbic acid, and erythorbic acid. Chelants bond with ferrous or ferric ions and form a stable, soluble compound that will remain in solution as the acid spends and the pH rises. Common chelants used in acidizing treatments are citric acid, ethylene diamine tetraacetic acid (EDTA) and nitrilotriacetic acid (NTA).

Friction Reducers

Often, acid treatments need to be conducted at pump rates that create turbulent flow conditions in the tubing. This can result in increased frictional pressure loss. This increased fractional pressure drop results in an increase in surface injection pressure. When the surface injection pressure reaches the equipment limit (maximum allowed pressure by the equipment specification), the injection rate has to be compromised. By adding a small amount of certain polymer, the frictional pressure loss can be reduced. Common friction reducers in acid include xanthan, guar, and synthetic polymers such as polyacrylamide.

Clay Stabilizers

Clay dispersion and/or swelling can cause a significant permeability reduction and associated loss of well productivity. Clay dispersion can be triggered by ionic imbalance in the aqueous environment in the reservoir caused by workover, treating, or water injection fluids; or by extreme fluctuation in fluid velocity in the reservoir. Clay swelling is generally caused by changes in ionic environment and/or the introduction of aqueous fluids. A variety of chemical products are available to minimize clay sensitivity to changing reservoir conditions. Clay stabilization can be achieved by specific cations (salts) in the aqueous fluids, such as K^+, NH_4^+, Ca_2^+, Mg_2^+, Al_3^+, and Zr_4^+ or by cationic, organic compounds, such as amines, polyamines, or quaternary amines in the acid treating fluids. Zhou et al. (1995) discussed the advantages and disadvantages of various clay stabilizers.

5.1.5 Acidizing Treatment Operations

The following guidelines should be followed when conducting an acidizing treatment:

- All solutions should be tested to ensure that they conform to the design formulations
- All necessary precautions should be taken to minimize the damage caused by the acidizing process itself
- The acidizing process should be monitored by measuring the rate and pressure (surface and/or downhole)

For carbonate acid treatments, the objective is to improve flow conditions in the reservoirs in addition to remove the near-wellbore formation damage. Therefore, even for undamaged wells acid treatment still can enhance the well productivity. On the other hand, for the wells in sandstone formations, there should be a clear indication that low well productivity is due, at least in part, to acid soluble formation damage. This can be determined through a pretreatment pressure transient test to measure skin effect, an analysis of other possible sources of skin effects, and an assessment of the source of formation damage. A pre-stimulation production log is also helpful in some cases in planning the treatment for multi-zone wells and long horizontal wells to identify damaged sections. Production logging is also used as a baseline for posttreatment analysis.

An important step in an acidizing treatment is cleaning all surface tanks, surface flow lines, and the tubing used to inject the acid. Hydrochloric acid will partially dissolve and loosen rust, pipe dope, and other contaminants from pipe walls, so any source of such material should be cleaned before the acid is introduced to the formation. Surface equipment can be cleaned prior to transport to the location or on site with the acid solution itself. If the production tubing is used for acid injection, an HCl solution or other cleaning fluids can be circulated down the tubing and back to the surface through the annulus before injection into the formation begins. This process is called tubing picking.

During the treatment, injection rates (including friction reducer pump rate), injection fluids density, and surface (and bottomhole if available) pressure should be monitored. As discussed in the later section, the use of surface and bottomhole pressures can provide the assessment of tubing friction changes and friction reducer performance giving more accurate prediction of the reservoir-face injection pressures, which minimizes uncertainties in post-stimulation skin evaluation. Occasional sampling of the injected solutions is also helpful. If any problems occur in the course of a treatment, the samples of injected solutions are an important diagnostic aid.

When acid injection is completed, the well should be flowed back immediately to minimize damage by the reaction product precipitation. If precipitations are not anticipated, the remaining tubing fluids can be completely displaced to the formation during overflush. A posttreatment pressure transient test run shortly after a stabilized rate has been attained is the most positive means of assessing the effect of the acid treatment. A post-stimulation production log also helps in diagnosing the treatment outcome in some wells.

5.2 Tasks and Designs of Completion for Successful Acid Stimulation

Completions for acid stimulation have been advanced as today more and more horizontal wells are drilled to produce oil and gas. Horizontal completions present unique challenges with respect to effective acid placement to treat the entire reservoir interval. As horizontal drilling technology has continuously improved in recent years, the industry has pursued longer horizontal wells, especially in low-permeability reservoirs. Horizontal completion lengths of 10,000 ft and longer are common today.

When treating long stimulation intervals, significant variations in reservoir properties along the wellbore frequently occur. This includes porosity, permeability, pore pressure, mineralogy, water saturation, natural fractures, and conductive faults. Many of these variations are especially pronounced in carbonate reservoirs.

With long completion intervals and large variations in reservoir properties as described above, it is a challenge to achieve an effective overall acid stimulation. A large variety of mechanical and chemical diversion techniques exist for improving acid treatment coverage in horizontal wells and are also reviewed in this section.

5.2.1 Sandstone Acidizing Design

The goal of a sandstone acidizing treatment is to remove near-wellbore formation damage while minimizing the damage caused by the acidizing process itself.

The first step in planning a matrix acidizing treatment in a sandstone formation should be a careful analysis to determine the cause(s) of impaired well performance. This analysis should begin with a measurement of the total well skin effect. This can be accomplished by a transient pressure test, or by analyzing the production history. When a large positive skin effect is present, sources of mechanical skin (partial completion, perforation skin, etc.) should be first identified based on the well structure and completion. If mechanical effects do not explain the flow impairment, this is indicative of the existence of formation damage. In general, damage due to drilling mud and completion fluids invasion or fines migration can be successfully treated with acid.

Once acid-treatable formation damage is identified as the cause of poor well productivity, the acidizing treatment design process can begin. A typical acid treatment in sandstone consists of the injection of HCl preflush, with a 50-gal/ft treatment interval length being a common preflush volume, followed by the injection of 50 to 200 gal/ft of HCl/HF mixture. A post-flush of diesel, brine, or HCl then displaces the HF/HCl from the tubing or wellbore. Once the treatment is completed, the spent acid should be immediately produced back to minimize damage by the precipitation of reaction products.

For years, the standard treatments consisted of 3 wt% HF and 12 wt% HCl mixture preceded by a 15 wt% HCl preflush for sandstones. In fact, 3% to 12% of HF/HCl mixture has been so commonly used that it is referred to generally as "full-strength mud acid." In recent years, however, the trend has been toward the use of lower strength HF solutions for the reduction in damaging precipitates from the spent acid and lessened risk of unconsolidation of the formation around the wellbore.

Based on extensive field experience, McLeod and Norman (2000) presented guidelines for acid selection, as shown in Tables 5.3 and 5.4. These guidelines should not be taken as hard-and-fast rules, but rather as starting points in a treating design. Because over-designed acidizing treatment in sandstone formations can easily result in damage such as precipitation and unconsolidation, all treatments should

Mineralogy	Permeability		
	>100 mD	20–100 mD	<20 mD
< 10% silt and <10% clay	15% HCl	10% HCl	7.5% HCl
>10% silt and >10% clay	10% HCl	7.5% HCl	5% HCl
>10% silt and <10% clay	10% HCl	7.5% HCl	5% HCl
<10% silt and >10% clay	10% HCl	7.5% HCl	5% HCl

Source: McLeod and Norman, 2000.

TABLE 5.3 Fluid Selection Guidelines for Pre-Flush Fluids

Mineralogy	Permeability		
	>100 mD	20–100 mD	<20 mD
<10% silt and <10% clay	12% HCl–3% HF	8% HCl–2% HF	6% HCl–1.5% HF
>10% silt and >10% clay	13.5% HCl–1.5% HF	9% HCl–1% HF	4.5% HCl–0.5% HF
>10% silt and <10% clay	12% HCl–2% HF	9% HCl–1.5% HF	6% HCl–1% HF
<10% silt and >10% clay	12% HCl–2% HF	9% HCl–1.5% HF	6% HCl–1% HF

Source: McLeod and Norman, 2000.

TABLE 5.4 Fluid Selection Guidelines for Mud Acid Fluids

be monitored closely during injection from injection pressure and rate. The monitoring procedure is straightforward. During acid injection, when keeping a relatively constant injection rate, a decreased injection pressure indicates a reducing skin factor; and vice versa; an increased injection pressure indicates an increased skin factor. The details of monitoring technique have been discussed by several authors (Paccaloni and Tambini, 1993; Zhu and Hill, 1996).

5.2.2 Carbonate Matrix Acidizing Design

The matrix acidizing process in carbonate formations is fundamentally different from that in sandstones. The standard treatments consist of 15–28 wt% HCl for carbonate formations. Because of simple reaction and high-reaction rate of carbonate formation with HCl, the acid systems are relatively simpler compared with the acid systems used in sandstone formations.

In carbonate matrix acidizing, the acid does not dissolve the rock uniformly around the wellbore as it is injected into the reservoir. Rather, the acid finds higher permeability regions in the reservoir pore

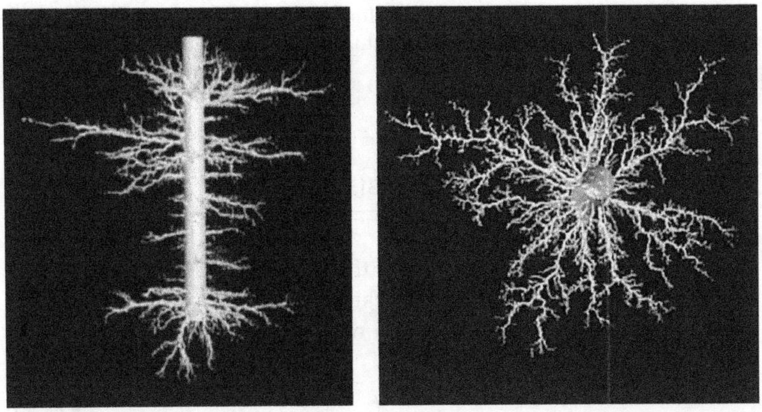

FIGURE 5.2 Radially growing wormholes created in a large block experiment (McDuff et al., 2010).

system and channels outward from the wellbore in a spherical manner, forming a network of small tunnels called *wormholes*. McDuff et al. (2010) presented CT scan images of a wormhole network created from a simulated borehole in a large limestone block experiment (Fig. 5.2).

As shown in the CT images, the acid treatment has dissolved a portion of the carbonate rock to create a wormhole network around the borehole. Fredd and Fogler (1999) and Fredd et al. (1997) have reported that at high injection rate, wormholes tend to branch more; and at low injection rate, acid does not penetrate into the formation, rather it dissolves the carbonate rock immediately once in contact, resulting in face dissolution.

The efficiency of matrix acidizing is measured by the extension of wormholes under a given acid injection volume. A relative concept, pore volume to breakthrough, is introduced to measure the efficiency. Pore volume to breakthrough is defined as the pore volumes of acid that are needed for wormholes to breaks through a give volume of the rock,

$$PV_{\mathrm{BT}} = \frac{\text{acid volume injected}}{\text{pore volume of rock}} \tag{5.17}$$

Equation (5.17) states that the smaller the pore volume to breakthrough, the less acid volume is needed for the wormholes to propagate through a given rock volume, and therefore, the higher the wormhole propagation efficiency.

Example 5.3 ***PV_{BT}* measurement in the large limestone block experiment.**
A large block flood test uses approximately 1 m^3 block of carbonate rock to test wormhole breakthrough efficiency. For a rock sample that has porosity of 0.2, it takes 2.5 liters injection for 15% HCl to break through from 27.25-in × 27.25-in × 32-in (~14 ft^3) block when acid is injected from a centered borehole at a high rate. Calculate the pore volume to breakthrough.

Solution We first calculate the pore volume of rock. Ignoring the volume of the borehole, the pore volume of the limestone block is calculated to be

$$PV = \phi V = 0.2 \times 14 = 2.8 \text{ ft}^3 \tag{5.18}$$

The pore volume to breakthrough, PV_{BT}, is then calculated by

$$PV_{BT} = \frac{(2.5)[\text{L}](0.034315)\left[\frac{\text{ft}^3}{\text{L}}\right]}{2.8[\text{ft}^3]} = 0.03 \tag{5.19}$$

An optimal injection rate yields the highest efficiency of matrix acidizing (lowest PV_{BT}). Injection rate is expressed in terms of interstitial velocity, defined as injection rate divided by the porous cross area of injection,

$$v_i = \frac{q}{A\phi} \tag{5.20}$$

Realized that PV_{BT} is a global parameter that includes many influential parameters in both reservoir properties and treatment conditions, detailed modeling for pore volume to breakthrough is extremely complex and tedious. Commonly, empirical correlations based on linear acid flooding experiments are used for treatment design. In a typical core flooding experiment, acid is injected through a cylindrical core samples with known porosity and permeability, and pressure drop across the core is measured as injection proceeds. Certain pressure drop to a minimum level is identified as the breakthrough point, and PV_{BT} is calculated from the volume of acid injected. This procedure repeats several times at different injection rate (interstitial velocity). From the lab experimental observations, a plot of pore volume to breakthrough versus interstitial velocity is generated. This type of plot is also referred to as a wormhole efficiency plot. Figure 5.3 shows such a plot (Buijse and Glasbergen, 2005). Figure 5.3 collected the data from several different experimental works. Regardless the experimental conditions, the curves exhibit the same features. All curves have a minimum point, indicating the optimal condition. At the optimal condition, wormholes propagate the furthest distance for a given acid volume. As injection rate increases (along the x-axis moving right), the pore volume to breakthrough decreases (better efficiency) until reaching the optimal condition. After the optimal point, we lose efficiency as injection rate increases. Notice that all curves have two well-fitted straight lines on a log-log plot, and the curves have much steeper slopes to the left of the optimum compared with the right part. This means that if injection rate misses the optimal condition, it is better to have a higher injection rate than the optimal. When injection rate is lower than optimal, we lose efficiency much faster.

Example 5.4 **PV_{BT} measurement from a liner acid flooding core test.**
For a rock sample with a porosity of 0.2, it takes 30 milliliter (mL) injection for the acid to breakthrough from a 1.5-in diameter and 6-in-long core at the injection rate of 10 mL/min. Calculate the pore volume to breakthrough and interstitial velocity for this case.

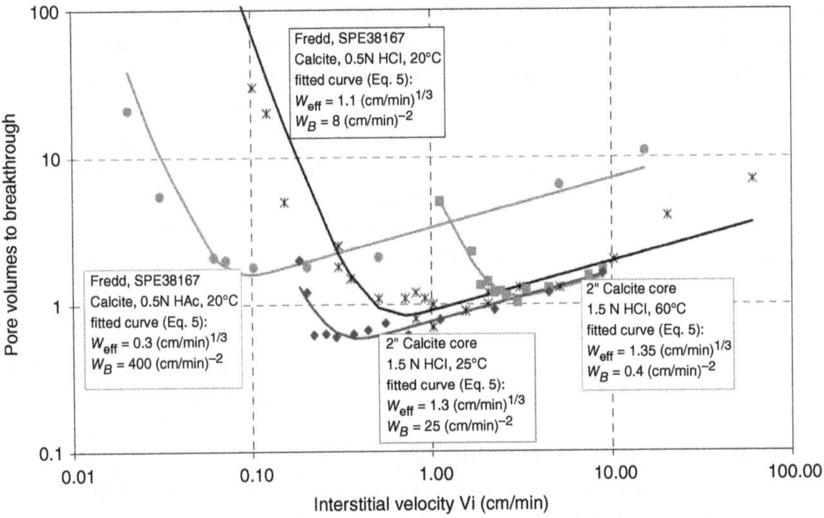

FIGURE 5.3 Wormhole efficiency curves for acidizing design (Buijse and Glasbergen, 2005).

Solution We first calculate the pore volumes to breakthrough. The pore volume of the core is

$$PV = \pi r^2 L\phi = 3.14 \times 0.75^2 \times 6 \times 0.2 = 2.12 \text{ in}^3 \tag{5.21}$$

and

$$PV_{BT} = \frac{30[\text{mL}] \times 0.061\left[\frac{\text{in}^3}{\text{mL}}\right]}{2.12[\text{in}^3]} = 0.86 \tag{5.22}$$

Notice that the PV_{BT} calculated above is much higher than those measured from the large block test (Example 5.3). The interstitial velocity in this case is

$$v_i = \frac{q}{A\phi} = \frac{10\left[\frac{\text{mL}}{\text{min}}\right] \times 0.061\left[\frac{\text{in}^3}{\text{mL}}\right]}{3.14 \times 0.75^2[\text{in}^2] \times 0.2} = 1.73\left[\frac{\text{in}}{\text{min}}\right] = 4.39\left[\frac{\text{cm}}{\text{min}}\right] \tag{5.23}$$

The wormhole efficiency curve is a good tool to design matrix acidizing treatment. The curves are constructed from laboratory experiments with correlation equations. The experiments have to use the acid system that will be pumped in the field, and rock samples that are as close to the formation to be treated as possible. If all conditions mimic the field treatment condition, injection rate is the only parameter that changes during a set of experiments. Once the minimum is identified, the curve can be created. The field injection rate is determined based on the observation of the optimal condition. It is also important to note that PV_{BT} values measured from linear acid-flooding experiments could be an order of magnitude larger than those measured from radial flow acid injection experiments. Thus, applications of linear acid-flooding experiment results to field conditions require some upscaling.

Buijse and Glasbergen Correlation for Wormhole Efficiency

Buijse and Glasbergen (2005) developed an empirical model of wormhole propagation based on the observation of pore volumes to breakthrough from acid corefloods related to interstitial velocity. They recognized that the wormhole propagation velocity, which is inversely related to the pore volumes to breakthrough, has a consistent functional dependence on interstitial velocity for different rocks and different acid systems. Using the two straight-line feature of each curve, they derived a function that captures this relationship. The wormhole propagation velocity can be expressed as

$$v_{wh} = \frac{dr_{wh}}{dt} = \left(\frac{v_i}{PV_{\mathrm{BT,opt}}}\right)\left(\frac{v_i}{v_{i,\mathrm{opt}}}\right)^{-\gamma}\left\{1-\exp\left[-4\left(\frac{v_i}{v_{i,\mathrm{opt}}}\right)^2\right]\right\}^2 \tag{5.24}$$

where v_i is the interstitial velocity. $PV_{\mathrm{BT,opt}}$ and $v_{i,\mathrm{opt}}$ are optimum pore volumes to breakthrough and optimum interstitial velocity identified from the efficiency curves based on a series of linear coreflooding experiments. Note that when γ equals to 1/3, then Eq. (5.24) is equivalent to the linear-coreflooding experiment results as shown in the original Buijse-Glasbergen model. In this model, the wormhole velocity depends on the interstitial velocity at the wormhole front, r_{wh}, and this velocity decreases as the wormholes extend away from the wellbore. For a given injection rate (or corresponding interstitial velocity), wormhole length can be estimated for a designed treatment.

Example 5.5 Using the middle curve on Fig. 5.3, calculate a wormhole length at 20 minutes injection time if the injection rate is 10 bbl/min. Use 0.2 for porosity, 100 ft for formation thickness, and 0.354 ft wellbore radius. Also calculate the post-stimulation skin factor.

Solution Reading from Fig. 5.3, the optimal interstitial velocity is 1.0 cm/min, and the optimal pore volume to breakthrough is about 0.9. The interstitial velocity at the wellbore is

$$v_i = \frac{q_i}{2\pi rh\phi} = \frac{10\left[\frac{\text{bbl}}{\text{min}}\right]\times 5.615\left[\frac{\text{ft}^3}{\text{bbl}}\right]}{2\times 3.14\times 0.354\ [\text{ft}]\times 100\ [\text{ft}]\times 0.2} = 1.27\ [\text{ft/min}] = 38.5\ [\text{cm/min}] \tag{5.25}$$

$$v_{wh} = \left(\frac{38.5}{0.9}\right)\left(\frac{38.5}{1.0}\right)^{-1/3}\left\{1-\exp\left[-4\left(\frac{38.5}{1.0}\right)^2\right]\right\}^2 = 12.7\ (\text{cm/min}) = 0.417\ [\text{ft/min}] \tag{5.26}$$

At 1 min of injection, r_{wh} is calculated to be

$$r_{wh} = r_w + v_{wh}\Delta t = 0.354 + 0.415\times 1.0 = 0.771\ (\text{ft}) \tag{5.27}$$

At the next time increment, the interstitial velocity and the wormhole growth rate at $r_{wh} = 0.771$ ft are calculated to be 17.7 cm/min and 0.248 ft/min. The wormhole penetration length is then calculated to be $r_{wh} = 0.769 + 0.248 \times 1.0 = 1.017$ (ft) at 2 min.

As shown in the above calculations, as wormhole extends, v_i becomes smaller very quickly. Finally, at 20 min, the wormhole penetration length is 3.3 ft long.

Since wormholes are much larger than the pores in non-vuggy carbonates, the pressure drop through the region penetrated by wormholes is small and can often be neglected. Thus the high permeability of the wormhole network effectively reduces near-wellbore pressure loss during production. With this assumption, the skin effect due to carbonate matrix acidizing can be calculated from the depth of penetration of wormholes, r_{wh},

$$s = -\ln\left(\frac{r_{wh}}{r_w}\right) \tag{5.28}$$

Hence, in the example calculations,

$$s = -\ln\left(\frac{3.325}{0.354}\right) = -2.2 \tag{5.29}$$

Furui Correlation for Wormhole Efficiency

Furui et al. (2012a) postulated that the velocity at the tip of the propagating wormholes drives the wormhole propagation rate, and this interstitial velocity at the tip of the wormholes, $v_{i,\text{tip}}$, is considerably higher than the average interstitial flux, v_i calculated from linear acid-flooding experiments:

$$v_{i,\text{tip}} = \frac{d_{\text{core}}}{d_{wh}} v_i \tag{5.30}$$

where d_{core} is the diameter of core samples used in the laboratory to generate the curve, and d_{wh} is the diameter of the dominant wormhole from the lab experiments. Note that Eq. (5.30) is an empirical relationship based on findings obtained from laboratory and numerical study results (Furui et al., 2012a).

From this observation, they derived a modified Buijse and Glasbergen wormhole propagation model as

$$v_{wh} = v_{i,\text{tip}} N_{Ac} \left(\frac{v_{i,\text{tip}} PV_{\text{BT,opt}} N_{Ac}}{v_{i,\text{opt}}}\right)^{-\gamma} \left\{1 - \exp\left[-4\left(\frac{v_{i,\text{tip}} PV_{\text{BT,opt}} N_{Ac} L_{\text{core}}}{v_{i,\text{opt}} r_{wh}}\right)^2\right]\right\}^2 \tag{5.31}$$

where N_{Ac} and $d_{e,wh}$ are the acid capacity number and the effective diameter of the wormhole cluster defined by

$$N_{Ac} = \frac{\phi \beta C \rho_{\text{acid}}}{(1-\phi)\rho_F} \tag{5.32}$$

$$d_{e,wh} = d_{\text{core}} N_{Ac} PV_{\text{BT,opt}} \tag{5.33}$$

For radial flow, the tip velocity is approximated as

$$v_{i,\text{tip}} = \frac{q}{\phi h \sqrt{\pi m_{wh}}} \left[(1-\alpha_z)\frac{1}{\sqrt{d_{e,wh} r_{wh}}} + \alpha_z \left(\frac{1}{d_{e,wh}}\right)\right] \tag{5.34}$$

where m_{wh} and α_z denote the number of the dominant wormholes along the angular direction (normally m_{wh} is thought to be 4 to 6) and the wormhole axial spacing, ranging from 0 to 1. When $\alpha_z = 0$, the dominant wormholes are closely spaced along the axial direction (2D radial flow geometry). For this case, the injection velocity at the wormhole tip declines proportionally to $1/r_{wh}^{0.5}$. This extreme case represents formations where the permeability in the axial direction is significantly lower than that in the radial direction. When $\alpha_z = 1$, the dominant wormholes are sparsely spaced along the axial direction with very little effect of low-length/stalled wormholes. In this case, the injection velocity at the wormhole tip does not decline as r_{wh} increases. For typical acid stimulation design, it is recommended to use $\alpha_z = 0.25$ to 0.5 for vertical wells ($k_a << k_r$) and $\alpha_z = 0.5$ to 0.75 for horizontal wells ($k_a \approx k_r$). Once again, r is 1/3.

Example 5.6 Repeat calculations for Example 5.5 using the Furui et al. model. Assume $\beta = 0.383$, $\rho_{acid} = 1.14$ g/cm^3, $\rho_F = 2.71$ g/cm^3, $d_{core} = 1$ in (2.54 cm), and $L_{core} =$ 6 in (15.24 cm). Also for wormhole geometry parameters, use $m_{wh} = 6$ and $\alpha_z = 0.5$.

Solution The acid capacity number is calculated by

$$N_{Ac} = \frac{0.2 \times 0.383 \times 1.14}{(1-0.2) \times 2.71} = 0.0403 \tag{5.35}$$

The effective diameter of the wormhole cluster, $d_{e,wh}$ is calculated by

$$d_{e,wh} = 2.54 \times 0.0403 \times 0.9 = 0.092 \text{ [cm]} = 3.02 \times 10^{-3} \text{ [ft]} \tag{5.36}$$

The interstitial velocity at the tip of the dominant wormhole is calculated by

$$\begin{aligned} v_{i,\text{tip}} &= \frac{10 \times 5.615}{0.2 \times 10\sqrt{3.14 \times 6}} \left[(1-0.5) \frac{1}{\sqrt{3.02 \times 10^{-3} \times 0.354}} + 0.5 \left(\frac{1}{3.02 \times 10^{-3}} \right) \right] \\ &= 116.7 \text{ (ft/min)} = 3558.3 \text{ [cm/min]} \end{aligned} \tag{5.37}$$

The wormhole growth rate is calculated by

$$\begin{aligned} v_{wh} &= 3558.3 \times 0.0403 \times \left(\frac{3558.3 \times 0.9 \times 0.0403}{1.0} \right)^{-(1/3)} \left\{ 1 - \exp\left[-4 \left(\frac{3558.3 \times 0.9 \times 0.0403 \times 15.24}{1.0 \times 0.354} \right)^2 \right] \right\}^2 \\ &= 28.4 \left[\frac{\text{cm}}{\text{min}} \right] = 0.932 \left[\frac{\text{ft}}{\text{min}} \right] \end{aligned} \tag{5.38}$$

At 1 min of injection, r_{wh} is calculated to be

$$r_{wh} = r_w + v_{wh}\Delta t = 0.354 + 0.932 \times 1.0 = 1.286 \text{ [ft]} \tag{5.39}$$

At the next time increment, the tip interstitial velocity and the wormhole growth rate at $r_{wh} = 1.286$ ft are calculated to be 3415.2 cm/min and 0.906 ft/min. The wormhole penetration length is then calculated to be $r_{wh} = 1.286 + 0.906 \times 1.0 =$ 2.192 (ft) at 2 min. Finally, at 20 min, the wormhole penetration length becomes 18.21 ft long.

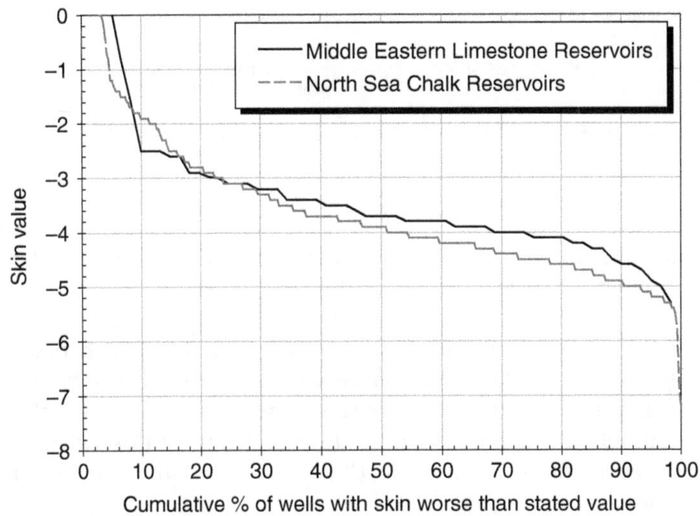

FIGURE 5.4 Field poststimulation buildup test data for carbonate matrix acidizing (Furui et al., 2010).

The resultant skin factor is calculated by

$$s = -\ln\left(\frac{18.21}{0.354}\right) = -3.9 \tag{5.40}$$

A field survey carried out over more than 400 wells shows that matrix acidizing techniques can effectively stimulate carbonate reservoirs with superb outcomes (Furui et al., 2010). The skin factor for all wells is in the range of –3.5 to –4.0 (P50 skin values), as shown in Fig. 5.4.

Example 5.7 PV_{BT} **measurement from field data**

Figure 5.4 shows that the P50 skin value for North Sea chalk reservoir is estimated to be –3.9 at the average acid injection volume of 75 gal per pay length. Assuming the chalk porosity is 30%, calculate the wormhole penetration length and PV_{BT} estimated from the calculated wormhole length. Assume that the wellbore radius is 0.3542 ft.

Solution Assuming that the permeability in the wormhole region is significantly larger than the original rock permeability, then the skin factor can be simplified as

$$s = -\ln\left(\frac{r_{wh}}{r_w}\right) \tag{5.41}$$

where r_{wh} is wormhole penetration length, and can be calculated by

$$r_{wh} = r_w \exp(-s) = 0.3542\exp(3.9) = 17.5 \text{ [ft]} \tag{5.42}$$

The pore volume of the rock is simply,

$$PV = \pi(r_{wh}^2 - r_w^2)\phi = 3.14(17.5^2 - 0.3542^2) \times 0.3 = 288.4 \text{ [ft}^3\text{/ft]} \tag{5.43}$$

The volume of the acid injected into the formation is

$$V_{\text{acid}} = 75\left[\frac{\text{gal}}{\text{ft}}\right] \times 0.13369\left[\frac{\text{ft}^3}{\text{gal}}\right] = 10\left[\frac{\text{ft}^3}{\text{ft}}\right] \tag{5.44}$$

and the pore volume to breakthrough measured from the field data can be estimated by

$$PV_{\text{BT}} = \frac{10}{288.4} = 0.035 \tag{5.45}$$

Figure 5.5 summarizes various PV_{BT} measurements and calculation results from the large block radial flow test, small linear core (1.5-in × 6-in) and large linear core (4-in × 20-in) tests, comparing with the field observations. As shown in the figure, the large block acid-flooding experiment (radial flow) matches very well to the field-observed PV_{BT} value. This indicates that the large block test is a good representation to wormholing observed in the field while the linear flow acid-flooding experiments underestimate PV_{BT} values and thus wormholing efficiencies.

In summary, carbonate reservoirs are routinely acid stimulated successfully, and can often achieve even better results if the reservoir is naturally fractured. These skin values represent productivity index ratios before and after stimulation (q_a/q_o) in the 1.5 to 2 range for undamaged wells. Matrix acid treatments can be pumped in all types of carbonates and are equally effective in low- or high-permeability reservoirs. They are generally cheaper and easier to implement than more complex acid fracturing treatments that are often pumped in lower permeability carbonate reservoirs. When reservoir permeability is too low, injection rate may be limited by the pumping pressure required to reach the optimal conditions. In such a case, acid fracturing is an option for well stimulation for carbonate formations.

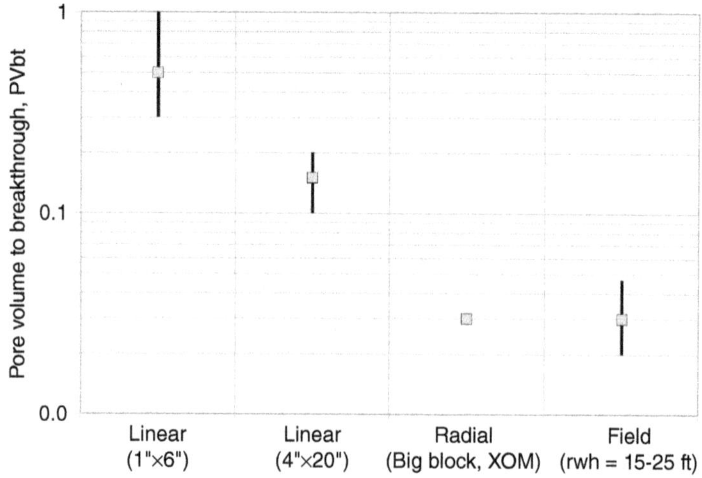

FIGURE 5.5 Comparison of various PVBT measurements.

5.2.3 Carbonate Acid Fracturing Design

Hydraulic fracturing treatments with acid, typically referred to as acid fracturing or acid frac, are also commonly applied in carbonate reservoirs to improve well flow performance. Acid fracturing treatments are generally designed for lower permeability carbonate reservoirs.

Hydraulic fractures are created by pumping viscous fluids into the reservoir above formation fracturing pressure and by propagating these hydraulic fractures deep into the reservoir. Hydraulic fracturing promotes better well productivity by allowing reservoir fluid to flow from the reservoir matrix to the fracture and then to the well, as opposed to radial flow through the matrix over the entire distance to the wellbore. Once inside the fracture, fluids reach the wellbore via flow down the high-permeability fracture where pressure loss is much lower than in the surrounding formation. Well productivity improvement via hydraulic fracturing is therefore mainly a function of fracture half-length and fracture conductivity. As can be expected it is easier to achieve high fracture length and high relative conductivity in low-permeability reservoirs than high-permeability reservoirs, and as a result acid fracturing is generally most successfully applied in lower permeability reservoirs.

Skin values and associated well productivity improvements for acid fractured wells are correlated to fracture half-length, x_f and fracture conductivity, $k_f w$. Calculations from dozens of buildup tests show that typical acid fracture lengths range from 30 to 200 ft (King, 1986). The longer fractures are usually in cooler, lower permeability formations.

The conductivity of an acid fracture is difficult to predict because it depends inherently on a stochastic process: If the walls of the fracture are not etched heterogeneously, very little fracture conductivity will result after closure. Thus the approach taken to predict acid fracture conductivity has been empirical one. Based on the acid distribution in the fracture, the amount of rock dissolved as a function of position along the fracture is calculated. Then an empirical correlation is used to calculate fracture conductivity based on the amount of rock dissolved. The methods for predicting fracture conductivity of an acid fracture should not be expected to be very accurate.

Nierode and Kruk Model

The amount of rock dissolved in an acid fracture is represented by a parameter called the ideal width w_i, defined as the fracture width created by acid dissolution before fracture closure. If all the acid injected into a fracture dissolves rock on the fracture face (i.e., no live acid penetrates into the matrix or forms wormholes in the fracture walls), the average ideal width is simply the total volume of rock dissolved divided by the fracture area, or

$$w_i = \frac{\chi V}{2(1-\phi)h_f x_f} \tag{5.46}$$

where χ is the volumetric dissolving power of the acid, V is the total volume of acid injected, h_f is the fracture height, and x_f is the fracture half length. Nierode and Kruk (1973) presented the correlation for the conductivity of an acid fracture based on extensive laboratory measurements of acid fracture conductivity and correlates conductivity with the ideal width, the closure stress, σ_c, and the rock embedment strength, S_{rock}. The Nierode-Kruk correlation is

$$k_f w = C_1 e^{-C_2 \sigma_c} \tag{5.47}$$

where

$$C_1 = 1.47 \times 10^7 w_i^{2.4} \tag{5.48}$$

$$C_2 = (13.9 - 1.3 \ln S_{rock}) \times 10^{-3} \text{ for } S_{rock} < 20000 \text{ psi} \tag{5.49}$$

$$C_2 = (3.8 - 0.28 \ln S_{rock}) \times 10^{-3} \text{ for } S_{rock} > 20000 \text{ psi} \tag{5.50}$$

For the constants given in these equations, $k_f w$ is in mD-ft, w_i is in inches, and σ_c and S_{rock} are in psi. The rock embedment strength is the force required to push a metal sphere to a certain distance into the surface of a rock sample. Laboratory-measured values of embedment strength for several common carbonate formations can be found in Table 5.5.

Formation	Rock Embedment Strength, psi
Desert Creek B Limestone	42,000
San Andres Dolomite	50,000–175,000
Austin Chalk-Buda Limestone	20,000
Bloomberg Limestone	93,000
Caddo Limestone	38,000
Canyon Limestone	50,000–90,000
Capps Limestone	50,000–85,000
Cisco Limestone	40,000
Edwards Limestone	53,000
Indiana Limestone	45,000
Novi Limestone	106,000
Penn Limestone	48,000
Wolfcamp Limestone	63,000
Clearfork Dolomite	49,000–200,000
Greyburg Dolomite	75,000–145,000
Rodessa Hill Laminate	170,000
San Angelo Dolomite	100,000–160,000

Source: Adopted from Nierode and Kruk (1973).

Table 5.5 Measured Rock Embedment Strength of Various Dry Carbonate Rocks

Example 5.8 Calculate fracture conductivity by Nierode-Kruk correlation for Indiana limestone. The ideal fracture width is 0.1 in.

Solution We first calculate the constants for the correlation. From Table 5.5, the embedment strength for Indiana limestone is 45,000 psi. Thus from Eqs. (5.48) and (5.49), we have

$$C_1 = 1.47\times 10^7(0.1^{2.47}) = 4.98\times 10^4 \tag{5.51}$$

$$C_2 = (3.8 - 0.28\ln(45000))\times 10^{-3} = 0.00145 \tag{5.52}$$

and

$$k_f w = C_1 e^{-C_2\sigma_c} = 4.98\times 10^4 e^{-0.00145\sigma_c} \tag{5.53}$$

Figure 5.6 plots the fracture conductivity as a function of closure stress. Conductivity reduced from 49,810 mD-ft at zero closure (0 psi) to about 2 mD-ft at 7000 psi.

Mou and Deng Model

To consider the effect of the geostatistic nature of carbonate formations, and leakoff volume, on acid fracture conductivity, Deng et al. (2012) presented a correlation that explicitly depends upon statistical variations in formation properties. The correlation uses three statistical parameters of the formation rocks, the correlation length in the horizontal flow direction, the correlation length in the vertical direction, and the standard deviation of the influential parameter. The parameter can be permeability or mineralogy distribution of the formation rock. The correlation was developed for three different types of rocks: one with conductivity dominated by permeability distribution effect, one dominated by mineralogy distribution, and one with combined effect of permeability and mineralogy. Here we show the correlation with permeability dominated effect.

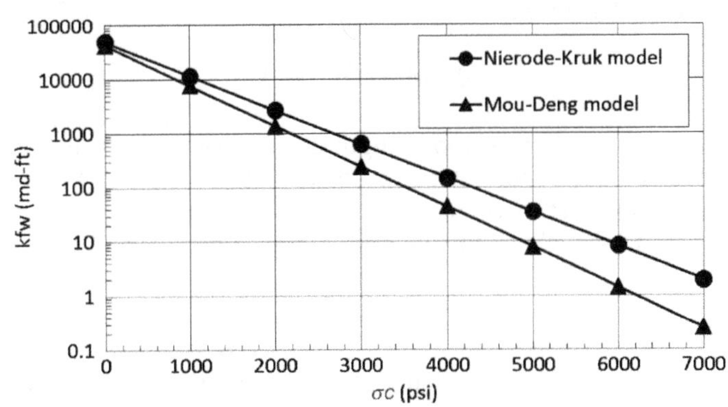

FIGURE 5.6 Example fracture conductivity calculations.

The conductivity correlation has the similar format as the Nierode-Kruk correlation,

$$k_f w = \alpha \exp[-\beta \sigma_c] \tag{5.54}$$

where

$$\alpha = (k_f w)_0 \{0.22(\lambda_{D,x}\sigma_D)^{2.8} + 0.01[(1-\lambda_{D,z})\sigma_D]^{0.4}\}^{0.52} \tag{5.55}$$

and

$$\beta = [14.9 - 3.78\ln(\sigma_D) + 6.81\ln(E)] \times 10^{-4} \tag{5.56}$$

In the above equations, σ_c is closure stress in psi, and E is Young's modulus in Mpsi (million psi). $\lambda_{D,x}$, $\lambda_{D,z}$, and σ_D are correlation length in horizontal direction along the fracture (x-direction), correlation length in vertical direction (z-direction), and dimensionless standard deviation, of permeability. The methods of obtaining these parameters were discussed by Beatty et al. (2011). The conductivity at zero closure stress $(k_f w)_0$ is

$$(k_f w)_0 = 4.48 \times 10^9 \bar{w}^3 [1 + \{a_1 \text{erf}[a_2(\lambda_{D,x} - a_3)] - a_4 \text{erf}[a_5(\lambda_{D,z} - a_6)]\}\sqrt{e^{\sigma D} - 1}] \tag{5.57}$$

where the constants are given by $a_1 = 1.82$, $a_2 = 3.25$, $a_3 = 0.12$, $a_4 = 1.31$, $a_5 = 6.71$, $a_6 = 0.03$, and $\bar{w}$ is average fracture width at zero closure stress in inches. In practice, ideal fracture width w_i, which is defined as dissolved rock volume divided by fracture surface area, is easier to obtain than average fracture width. For the high leakoff coefficient [>0.004 ft/(min)$^{0.5}$],

$$\bar{w} = 0.56\text{erf}(0.8\sigma_D) w_i^{0.83} \tag{5.58}$$

For the medium leakoff coefficient [~0.001 ft/(min)$^{0.5}$] with uniform mineralogy distributions,

$$\bar{w} = 0.2\text{erf}(0.78\sigma_D) w_i^{0.81} \tag{5.59}$$

Often, the vertical correlation length of permeability distribution is low, because the sedimentary carbonates are laminated. When the dimensionless vertical correlation length is low enough, for example, $\lambda_{D,z} < 0.02$, Eqs. (5.55) and (5.56) can be simplified to

$$\alpha = 0.12(k_f w)_0 (\lambda_{D,x}\sigma_D)^{0.1} \tag{5.60}$$

$$\beta = [15.6 - 4.5\ln(\sigma_D) - 7.8\ln(E)] \times 10^{-4} \tag{5.61}$$

Example 5.9 Calculate fracture conductivity using the Mou and Deng correlation. A highly laminated carbonate formation has a leakoff coefficient of 0.001 ft/min$^{0.5}$. The permeability correlation length in the horizontal direction is 50, the vertical permeability correlation length is 0.02, and the standard deviation of permeability is 0.7. The ideal fracture width is 0.05 in, and the Young's modulus for the rack is 1 Mpsi. Generate a plot of fracture conductivity versus closure stress using Mou and Deng model.

Solution We first calculate the average fracture width,

$$\bar{w} = 0.2\text{erf}(0.78 \times 0.7)0.05^{0.81} = 0.027 \tag{5.62}$$

At zero closure stress, the conductivity is

$$(k_f w)_0 = 4.48 \times 10^9 \times 0.027^3 \{1 + [1.82\text{erf}(3.25[50 - 0.12]) - 1.31\text{erf}(6.71[0.2 - 0.03])]\}\sqrt{e^{0.7} - 1}$$
$$= 251883 \text{ md-ft} \tag{5.63}$$

$$\alpha = 0.12 \times 251883(50 \times 0.7)^{0.1} = 43131 \tag{5.64}$$

$$\beta = [15.6 - 4.5\ln(0.7) - 7.8\ln(1)] \times 10^{-4} = 0.00172 \tag{5.65}$$

Thus, the fracture conductivity as a function of closure stress σ_c is

$$k_f w = 43131\exp[-0.00172\sigma_c] \tag{5.66}$$

Figure 5.6 plots the calculated fracture conductivity as a function of closure stress for this example to compare with the Nierode-Kruk model results. The conductivity by the Mou and Deng model declines from 43,131 mD·ft at zero closure to 0.3 mD·ft at 7000 psi closure. From Examples 5.8 and 5.9, when the closure stress is higher than 5000 psi, the conductivity because very small, 36 mD·ft by Nierode-Kruk model, and 8 mD·ft by the Mou and Deng model.

The skin effect model for hydraulic fractured wells can be adopted here to estimate the stimulation results.

Example 5.10 Calculate skin factor for Example 5.9 for a closure stress of 4000 psi. The conductivity at 4000 psi closure is 44 mD·ft. Using a fracture half-length of 200 ft long, and reservoir permeability is 0.1 and 1 mD. $r_w = 0.354$ ft, and $r_e = 1000$ ft.

Solution The dimensionless fracture conductivity for 0.1 mD permeability is [Eq. (4.54)]

$$C_{fD} = \frac{k_f w}{k x_f} = \frac{44}{0.1 \times 200} = 2.2 \tag{5.67}$$

And therefore, from Eq. (4.55), the equivalent wellbore radius is obtained by

$$r_w' = \frac{x_f}{\dfrac{\pi}{C_{fD}} + 2} = \frac{200}{\dfrac{\pi}{2.2} + 2} = 58 \text{ ft} \tag{5.68}$$

and from Eq. (4.56) the skin factor is calculated to be

$$s = -\ln\left(\frac{r'_w}{r_w}\right) = -\ln\left(\frac{58}{0.354}\right) = -5.1 \tag{5.69}$$

The folds of increase under steady state can be calculated as

$$\frac{Q}{Q_0} = \frac{\ln(r_e/r_w)}{\ln(r_e/r_w)+s} = \frac{\ln(1000/0.354)}{\ln(1000/0.354)-5.1} = 2.8 \tag{5.70}$$

For permeability of 1 mD, C_{fd} is 0.22, and r'_w is 12.3 ft, this leads to a skin factor of −3.5 after acid fracturing and a production increase of only 1.8 times.

In summary, low-permeability carbonate reservoirs can be effectively acid fractured to have a skin factor in the range of −4 to −5 if the rock is sufficiently strong to sustain the fracture conductivity after closure and during production. These skin values correspond to well inflow improvement ratios (after stimulation versus before stimulation) in the range of 2 to 3. In higher permeability reservoirs or carbonate reservoirs composed of weaker rock acid fracturing is less effective.

5.2.4 Acid Placement and Diversion

A critical factor to the success of an acidizing treatment is proper placement of the acid so that all of the productive intervals are contacted by sufficient volumes of acid. If there are significant variations in reservoir permeability, the acid will tend to flow primarily into the highest permeability zones. Once acidizing improves the flow condition, flow through the acidized location is easier, and more acid tends to flow to the already treated zones, leaving lower permeability zones virtually untreated. Even in relatively homogeneous formations, the formation damage may not be distributed uniformly; without the use of techniques to improve the acid placement, much of the damage may be left untreated. This problem is even more vital in horizontal well stimulation because of large contact of the reservoir. When it is determined that sufficient acid coverage cannot be obtained by allowing the acid to choose its own path, diversion becomes necessary during acid pumping for desired acid placement.

Mechanical Diversion Methods

The most common mechanical methods used to divert stimulation fluids are bridge plugs, packers, ball sealers, and limited entry perforating.

Bridge plugs that are run on tubulars are retrievable and can be moved and reset several times. Wireline bridge plugs cannot be moved once set, and their removal typically requires milling after the treatment. Wireline bridge plugs are used when several treatments are attempted in one day or when a rig is not over the hole during the treatment. They can be run in the hole quickly, and cleanout trips are

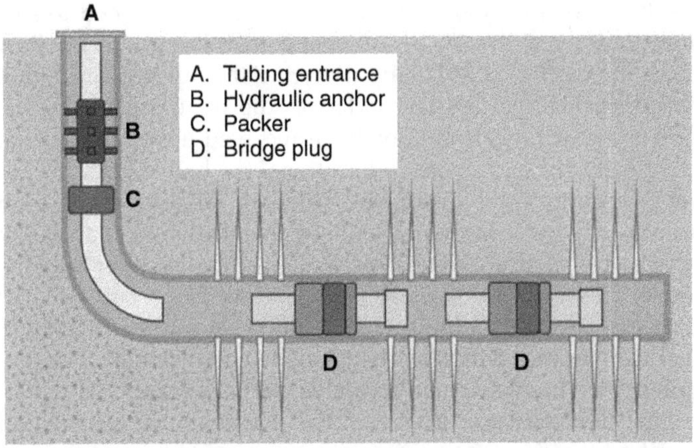

Figure 5.7 Multi-zone stimulation using bridge plugs (Jinzhou et al., 2013).

not required between stages. A typical treatment involves perforation of the bottom zone, stimulation treatment, and zonal isolation by setting a bridge plug immediately above the perforated interval. The next zone is then perforated and stimulated (Fig. 5.7).

When multiple-zone stimulation does not follow the consecutive routine to proceed from the lowest zone of interest up to higher intervals, a straddle packer system (Fig. 5.8) can be used. Straddle pachers provides a reliable method of isolation. These retrievable tools can easily be moved to cover any interval, as long as an unperforated casing is sufficiently long to provide a packer seat. However, if there are open perforations above the packer debris may enter the annular area. Small quantities of debris on top of a retrievable packer can stick the tool string, and this should be prevented.

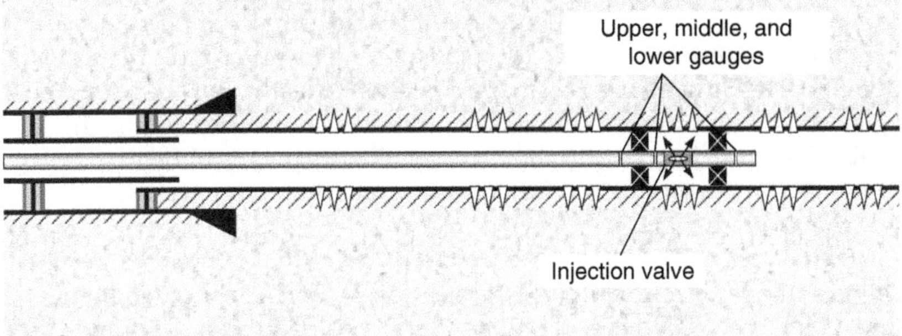

Figure 5.8 A straddle packer system.

Ball sealers are rubber-coated balls that are designed to seat on perforations from the inside of casing to block the fluid flowing through the casing, and thereby diverting the injected fluid to other perforations. Ball sealers are added to the treating fluid in stages. After a number of perforations have received acid and had higher fluid velocity through these perforations, the ball sealers are driven to the perforations to block further flow and divert the acid to other perforations. Most commonly, ball sealers are denser than the treating fluid, so that after the treatment, the ball sealers will fall into the rathhole. However, Erbstoesser (1980) showed that ball sealers that are slightly buoyant in the carrying fluid seat more efficiently than denser ball sealers. Balls may be produced back to the surface after completion of the treatment, so a ball catcher must be added to the flow line. The seating efficiency of ball sealers increases as injection rate and perforation friction increase.

Limited entry perforation is one of the mechanical diversion methods. The technique distributes fluid over a perforated interval by controlling frictional pressure drop across perforations. As discussed in Chapter 4, the frictional pressure drop is proportional to the second power of flow rate. The perforations that receive more fluid will have higher pressure drop through the perforation than the ones receiving less fluid. By limiting the number of perforations for multiple intervals, successful application of the limited entry technique increases the number of intervals that can be stimulated by acid. Generally, a 500- to 1000-psi pressure differential is considered necessary to provide adequate control of fluid placement. The total flow of fluid entering a given zone is restricted by controlling the size and number of perforations in that interval. The high pressure drop at the perforations forces fluid to go to another zone. This diversion technique has proved popular because of its simplicity and economics. The diversion does not require expensive tools; and it also does not require running and retrieving tools or making cleanup trips. The pressure drop calculation is detailed in Chapter 4, Section 4.4.

Chemical Diverters

Chemical diverters range from products that provide mechanical blocking of portions of the reservoir (suspended solids) to viscous diversion (gelling agents) to foam diversion. Ideally, the diverting agent should effectively limit acid penetration in the desired section(s) of the reservoir in a predictable, controllable manner and be easily removed during production (i.e., insoluble in acid but highly soluble in water or hydrocarbons).

Suspended solids that are effective as diverting agents include salt granules, benzoic acid flakes, waxes, oil-soluble resins, gilsonite, and polylactic acid. These products can be suspended in acid and form a low-permeability cake across the higher permeability zone(s) of the reservoir. By blocking access to higher permeability zone(s), acid is diverted to the lower-permeability zone(s). Once the treatment is completed, the particles are either produced off of the formation face or dissolved by the produced fluids.

Gelling agents can be added to the early stages of an acid treatment to increase the viscosity of the acid in the higher permeability zone(s). As the later stages of unviscosified acid are pumped, the low-viscosity acid will penetrate the lower-permeability zone(s). Typically, gelling agents are selected that will degrade after the acid treatment is finished, promoting cleanup in the higher permeability zone(s). Chemicals used to viscosify acid treatment fluids for diversion include guar, Xanthan, HEC, viscoelastic surfactants, and synthetic polymers.

In addition to simple linear polymer solutions, crosslinked systems can be used to divert acid. The crosslinked systems develop very high viscosity at low shear rates. More recently, crosslinked systems have been developed with pH-sensitive crosslinkers. The crosslinkers produce high viscosity as the acid starts to spend (pH ~3), but as the acid spends and the pH increases, the crosslinking is disrupted. Thus, the gelled fluid is easier to flow back after the acidizing treatment is completed.

Foamed acid is another method for acid placement. Foam can be pumped in stages during an acid treatment alternating with single-phase acid. Foam has higher viscosity to promote diversion and the foam structure also blocks pores, forcing the single-phase acid stages into the lower-permeability zones. The foam stages can be formulated to give limited foam stability, so that after acid treatments, the flow capacity is restored.

5.3 Completion Techniques for Acidizing Horizontal Wells

Acid treatment designs are different for horizontal wells compared with vertical wells. The major difference is related to the length of treatment intervals. While most vertical wells have reservoir contact between tens to hundreds feet, the acid stimulation length can be on an order of thousands of feet for a long horizontal well. Also due to difficulties of cementing and running casing, openhole completions are more frequently used in extended-reach and horizontal wells. These issues result in significant challenges in conveying and distributing acid along the treatment length. Even for vertical wells, the target reservoir interval may have significant variations in reservoir and geological properties, so that zonal isolation may be required for successful treatments. In these situations, engineers have to decide whether a well is completed by single or multiple zones. Both matrix acidizing and acid fracturing treatments can be performed in wells completing multiple zones. This section will review various completion techniques for single and multiple-zone completions, and address completion and stimulation designs for acid-stimulating wells.

5.3.1 Single-Zone Stimulation Completions (No Zonal Isolation)

Barefoot Acid Wash Completion

With the "barefoot" completion technique, there is no liner into the openhole after drilling process. Thus, there is no mechanical isolation involved with this completion. In most barefoot or uncemented liner applications in carbonate reservoirs, the reservoir section is typically drilled with a water-based mud that will generate a highly acid-soluble filter cake. Compared to using oil-based mud, water-based mud allows for better filter cake cleanup and uniform acid stimulation of the openhole sections.

After running the upper completion, acid stimulation is commonly executed using a short acid spray tool run on the end of a large-diameter coiled tubing or drill pipe. The acid spray tool normally rotates and/or pulsates to improve agitation and filter cake removal while the coiled tubing is moved up and down the wellbore (Haldar et al., 2008). In bullhead treatments, stimulation upstream of tubing tail is possible, but this would likely result in most of the acid being placed near the heel of the well, which is not desirable. Figure 5.9 illustrates acid stimulation treatment using a coiled tubing.

The overall stimulation with drill pipe will likely be better than with the coiled tubing method because drill pipe strings have larger diameter than coiled tubing's, allowing higher pump rates.

A barefoot acid wash completion is simple, fast, generally low cost, but depends on rig or coiled tubing unit requirements. Also the treatment is limited to matrix stimulation. The acid injection rate per foot is usually lower than desired optimal interstitial velocity, resulting in shallow wormhole penetration in carbonate stimulation.

Openhole with Predrilled Liner

Predrilled liners are sometimes used in openhole horizontal completion applications, mainly to ensure wellbore access (Fig. 5.10). The advantages of using a predrilled liner in an openhole completion

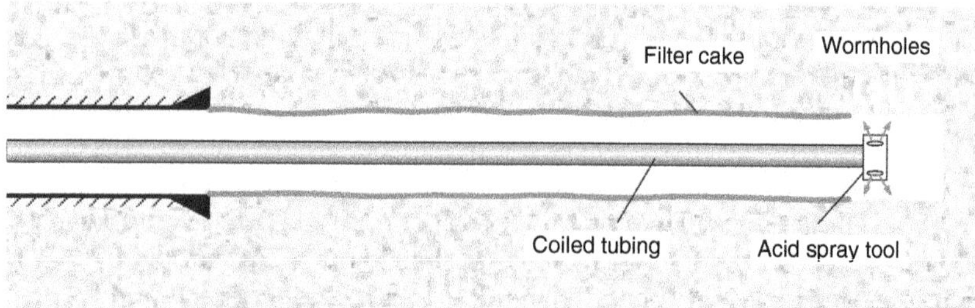

FIGURE 5.9 Barefoot completion for acid wash treatment.

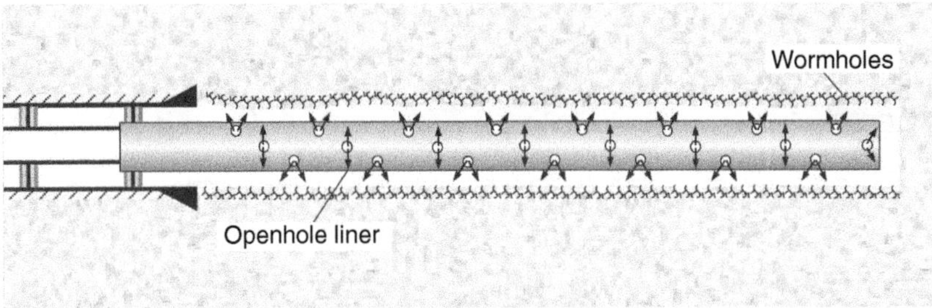

FIGURE 5.10 Openhole predrilled liner completion for limited entry acid stimulation.

include reducing the risk of wellbore blockage, releasing the restriction to production and providing a well reentry path.

Another significant advantage of this completion is that it is one of the simplest ways to use a limited entry design for effective acid diversion through the liner. The predrilled perforations distribute acid over the entire contact with the reservoir. When acid exits the liner and contact the reservoir, it dissolves filter cake and allows acid to penetrate into the reservoir and propagate wormholes. At the locations where wormholes are established, it takes more acid since the injectivity is increased. As flow rate through perforation increases, frictional pressure drop increases also, regulating the flow rate at the location.

Effective limited entry designs in horizontal wells typically involve drilling small holes and spacing them further apart, compared to vertical wells. Note that frictional pressure drop across the predrilled holes is highly sensitive to hole diameter. Thus the size of the perforation holes should be carefully determined.

Controlled acid jet (CAJ) liner completion is an example of openhole with predrilled liner completion for acid stimulation. It utilizes a nonuniform geometric distribution of the perforated holes, compensating for frictional pressure drops along the liner section. Hansen and Nederveen (2002) reported horizontal wells with 10,000- to 20,000 ft reservoir-contact length have been successfully completed and stimulated in the Dan/Halfdan oil field in the North Sea by the CAJ technique.

An openhole predrilled liner completion with limited entry perforations is simple, fast, and generally low cost. It works better for matrix stimulation. Because the flow area of the wellbore is too large for openhole predrilled liners, for fracturing treatments to generate enough velocity for the purpose of limited entry may be difficult. Even for matrix acidizing, to achieve moderate wormhole penetration in carbonate formations, liner length should be limited. Similar to barefoot acid wash completion long liner length with this stimulation technique results in shallow wormhole penetration and low-stimulation efficiency.

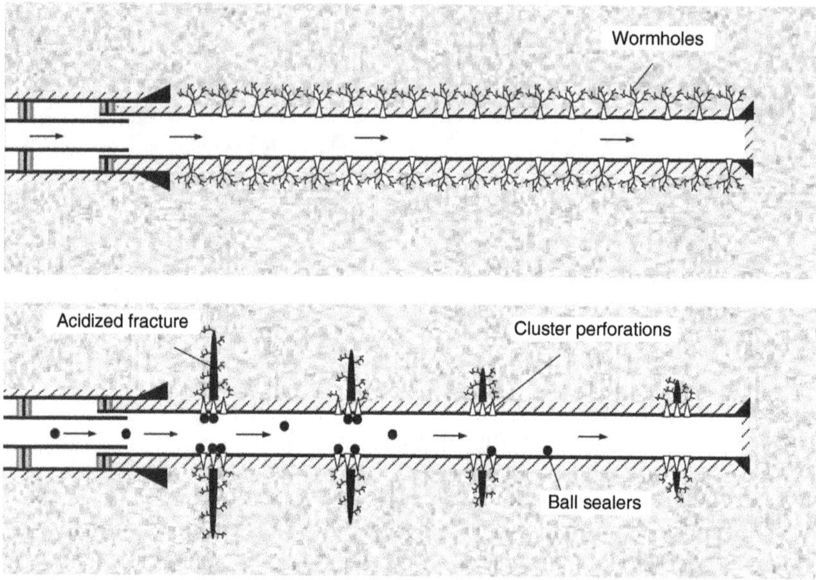

Figure 5.11 Cemented liner completion for matrix acidizing and acid fracturing.

Cemented Liner with Limited Entry Perforations and Ball Sealers

In this type of completion, a liner is cemented and perforated in the borehole (Fig. 5.11). For the purpose of effective acid stimulation, appropriate perforation design is important. The limited entry concept is commonly used to achieve the best efficiency of stimulation. For matrix acidizing, uniform perforation spacing patterns are typically used. In this case, the distance between perforations (i.e., perforation spacing) should be kept to less than 20 ft to ensure adequate stimulation coverage (10-ft perforation spacing is a good rule of thumb for cemented liners). Cluster perforations are typically used when acid fracturing a well. Cluster spacing depends on the optimum distance required between fractures. With cemented liners, it does not matter whether oil-based or water-based mud is used from acid stimulation perspective because perforations penetrate through the filter cakes.

As with most bullhead stimulation methods, pumping the acid at high rates is crucial to achieving acid diversion with limited entry. Ball sealers are often used in this application to enhance diversion. Ball sealer diversion becomes more effective with limited entry designs.

The cost of a cemented liner completion is higher than openhole liner completion because of the requirement of cementing and perforating. Both matrix acidizing and acid fracturing are possible with this completion technique. When ball sealers are used for diversion, it may require to clean out the holes after treatment unless degradable balls are used.

5.3.2 Multiple-Zone Stimulation Completions

Most conventional stimulations are pumped at a limited injection rate between 40 and 60 bpm, depending on the diameter of the treating string. Occasionally, higher rates of 70 to 100 bpm can be achieved with 7-in casing or tubing strings. For horizontal wells, as the completion length increases, it becomes difficult to keep a reasonable injection rate for desired stimulation results. A wellbore should be isolated into several intervals where treatments are carried out one interval at a time. The multiple-zone stimulation allows for a much higher injection rate per pay-foot, and therefore better wormhole penetration for matrix acidizing, or longer fracture half-length for acid fracturing resulting in increased stimulation benefits.

The mechanical isolation techniques described here use various methods to divide the wellbore into separated zones, and place isolation between zones. Stimulation is carried out one zone at a time. In this manner, it is possible to stimulate individual zones with the total injection rate and increase the efficiency.

Cemented Liner with Straddle Packers on Workstring or Coiled Tubing

As explained in the previous section, straddle packers are two packers that are joined in the middle by a common pipe (Fig. 5.12). The common pipe has a set of holes that allow the acid to flow out of the straddle packer section and into the wellbore. Straddle packers are run on either a jointed pipe (workstring), or a coiled tubing to the designated locations in a wellbore to selectively isolate and stimulate a small section of the wellbore. After stimulating one section, the packers are released and moved to the next location, and the process is repeated until all of the perfs (or clusters) have been treated. Treatments normally begin at the toe and end at the heel for horizontal wells.

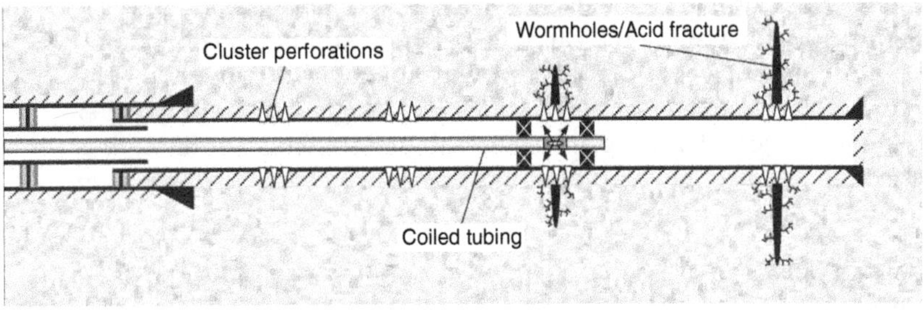

FIGURE 5.12 Straddle packer stimulation with coiled tubing.

Straddle packer stimulation with a jointed pipe has some unique features compared to coiled tubing. Jointed treating strings typically consist of large pipe, such as 5-in drill pipe, which allow for fast pumping rates of 50 to 60 bpm. Because of the larger diameter, however, stimulation with joint pipe is almost always performed prior to running the upper completion (i.e., the completion above the payzone). Workstring and straddle packers are pulled out of the hole and the upper completion is installed after the acid job is finished. As discussed earlier, special well control challenges are anticipated, because fluid losses may be abnormally high and difficult to stop.

Acid stimulation by coiled tubing conveyed straddle packer is performed in almost an identical way as the workstring method. However, the upper completion is typically run prior to perforation and stimulation works, thus eliminating most of the well control risks associated with high fluid losses. Perforating can be done with workstring or coiled tubing allowing the coiled tubing work to be performed off of the rig critical path. The limitation of coiled tubing conveyed operation for acidizing is the injection rate. The treatment rates with coiled tubing are typically limited to 10 to 15 bpm with large-diameter coiled tubing. Most jobs are done with smaller coiled tubing, and the treating rate is 5 to 10 bpm. To overcome this problem, the length of treatment intervals must be reduced, resulting in more stages, and consequently more operating time and higher cost. Because of the increase in treating stages, more repositioning and resettings of the straddle packers are required to treat an equivalent number of perfs. The packers have to be pulled and replaced more frequently due to added wear and tear.

Cemented or Uncemented Liner with Plugs

This method involves running a cemented liner or uncemented liner with openhole packers. Once the liner is positioned and the upper completion has been run, the first zone is perforated with wireline or coiled tubing. The first zone is then acidized at a maximum pump rate. Both matrix acidizing and acid fracturing are possible with this completion technique. When the first stage is completed, a plug is set to isolate the first zone from the rest of the wellbore. The next zone is then perforated and acidized and the process is repeated for all zones. Figure 5.13 illustrates a typical plug, perf, and stimulation operation.

Limited entry perforating can be applied in this method. The number and diameter of perforations are designed for acid to cover the entire interval. Acid placement with limited entry in this case is effective because the interval is short compared with pumping over the full wellbore without zonal isolation.

Nipple Profiles and Plugs When nipple profiles and plugs are used for multistage stimulation, spacing out becomes crucial. A predetermined number of selective-type nipple profiles (Fig. 5.14) must be

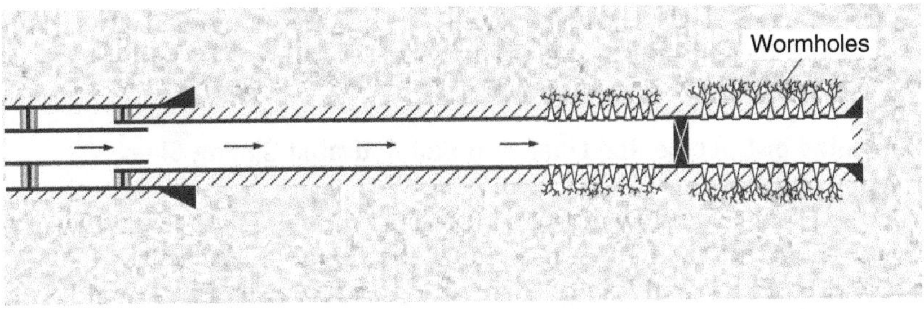

FIGURE 5.13 Multistage stimulation completion using plugs.

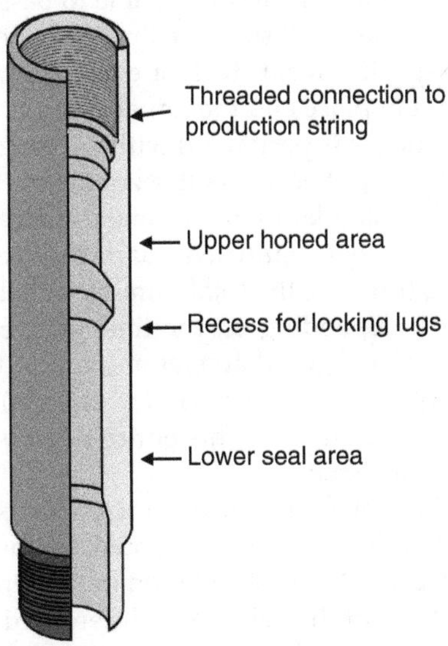

FIGURE 5.14 Selective nipple.

determined prior to running the completion. The nipple provides a seal area and a locking profile for a plug. Cementing the liner in place is required. For cemented liner applications, special wiper plugs are used to ensure that the cement is effectively wiped from the ID of the liner during cement displacement. A dedicated cleanout trip is typically run afterward to remove any remaining bits of cement that might have been left behind. Pressure differentials and crossflow can make plug retrieval a challenging and risky process for this technique.

Drillable Bridge Plugs When drillable bridge plugs are used, after the final zone has been acidized, coiled tubing is used to mill out all of the bridge plugs. Because of high fluid losses, the well must be underbalanced and flowing while circulating to achieve effective cuttings removal. Otherwise, there is a risk of getting stuck. After milling out the drillable bridge plugs, a venturi run may be required to clean the wellbore of cutting debris.

Cemented or Uncemented Liner with Ball-Activated Sliding Sleeves

Integral ball seats are made and built in the liner. For an uncemented liner, openhole packers are also required, and the ball seats should be made along with the packer assembly. With multiple sliding sleeves, the ball seats have different sizes, and are in an increasing order from the first stage to the last stage. In cemented liner applications, since the ball seats have varying IDs, it may be hard to effectively clean out the hole after the cement job. A wiper dart usually is used to clean out the liner. The dart has to be able to pass the smallest ball seat. If the dart is on the small side, the liner may not be cleaned. But if the dart is too large, it can get stuck or even open the seat.

Once the liner is in place and the upper completion has been run, the first zone near the toe is perforated with wireline or coiled tubing. Limited entry can be applied for better zonal coverage. Perforating equipment should be smaller than the smallest ball seat. Because the smallest ball seat must be of sufficient diameter to allow perforating equipment passing through, the total number of ball seats will likely be limited to 4 to 8 depending on the liner size. Several variables, including maximum perf gun diameter, minimum tubing clearance, and ball seat selection (i.e., ¼ in or ⅛ in increments) will dictate how many zones can be established. This entire process should be carefully reviewed as a first step.

After perforating, the first zone is acidized. Once the acid pumping is finished, the second zone is perforated. Following the perforation of the second zone, the first isolation ball is dropped down the tubing (Fig. 5.15). Each of the balls has a different diameter and must be dropped in the correct order.

With the ball in the tubing string, the second zone is acidized. When pumping balls to set on the ball seat, the flow rate should not be too high. Slamming the ball into the seat at a high rate could cause

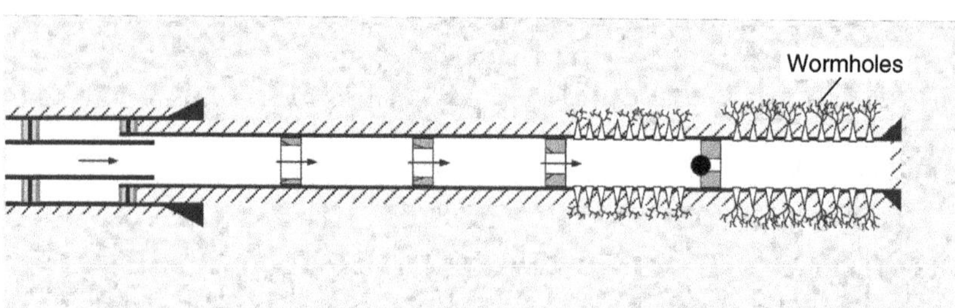

FIGURE 5.15 Multi-zone stimulation completion with ball-activated sliding sleeves.

the ball to shatter and lose zonal isolation. Balls should be seated at sufficiently low rate to keep the balls from breaking. Once seated, the pump rates can be increased to normal levels.

The remaining zones are then perforated and acidized consecutively with additional isolation balls. After the final zone is acidized, the well is put on flow back. The balls are pushed back to the surface and caught at the surface in a special ball-catching device installed in a temporary flowline without any ID restrictions. Ball flowback efficiency depends on ball diameter and specific gravity, well flow potential, and density of the flowing fluid. This must be carefully evaluated during the design. The ball seats can be milled out using coiled tubing or wireline tractor system. However, if the ball seats do not impede production, they can be left in place on a permanent basis.

Cemented or Uncemented Concentric String Completion

This method involves first running a cemented liner or an uncemented liner with openhole packers. This outer liner size typically is 7 in or 7⅝ in. At this point, the tubing-conveyed perforation guns are run in the well on a workstring and all zones are perforated with the desired perforation spacing. The perforation pattern can be uniform for matrix stimulation, or cluster style for acid fracturing stimulation. Once again, fracturing operation may not be feasible with the uncemented liner unless the zone length is relatively short. In any case, depth control is very important—the openhole packers must not be perforated and the blank sections must be at the correct depths for setting the internal packers. After perforating, fluid loss issues are normally controlled by placing a stable, viscosified fluid in the well for running the upper completion.

The next step is to run the internal string on the end of the upper completion. The inner string will typically consist of a 4½-in string with hydraulic set or swell packers and mechanical sliding sleeves or remote-controlled hydraulic valves (intelligent well system) for zone access and isolation.

With the upper and lower completion in place and tested, the well is ready for stimulation. In mechanical sliding sleeve applications, the sliding sleeve for the first zone, nearest toe, is typically a "P sleeve" that is shear pinned and opened by simply pressurizing the well to a predetermined pressure. Remote-controlled hydraulic valves can simply be opened or closed at any time by sending a signal from surface. Once this first sleeve, or valve, is open, the first zone is acidized. Acid flow is through the sliding sleeve or valve and into the inner liner to the outer liner annulus. From here, the limited entry perforation design distributes the flow of acid uniformly to all perforations. To achieve uniform distribution, the friction pressure losses between the inner and outer liners must be much smaller than the perforation friction pressure losses.

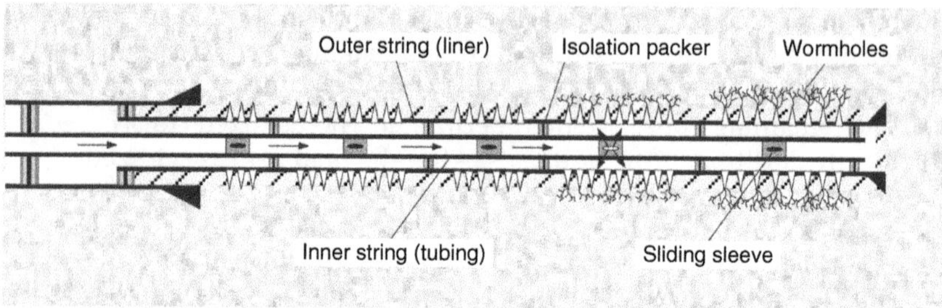

FIGURE 5.16 Concentric string completion for multi-zone stimulation.

For stimulation of the remaining zones, the method of shifting the sliding sleeves depends on the type of sliding sleeve in use. Once again, remote-controlled hydraulic valves can be opened and closed from the surface as desired. Figure 5.16 illustrates a typical stimulation operation for the concentric string completion.

5.4 Injection Design and Pressure Behavior during Acidizing

5.4.1 Sandstone Acidizing

Typical sandstone acidizing treatments consist of preflush, main acid, and overflush stages. HCl is usually used in the preflush and an HCl/HF mixture is used in the main acid stage. Proper acid staging, low-acid concentrations, correct usage of preflushes, and sufficient overflushing are crucial elements to control the precipitates by acidizing in sandstone formations.

Preflush Stage

Before the HF/HCl mixture is injected in a sandstone acidizing treatment, a preflush of hydrochloric acid, most commonly a 15 wt% HCl solution, is usually injected. The primary purpose of the HCl preflush is to prevent the precipitation of species such as calcium fluoride when the HF solution contacts the formation. The HCl preflush dissolves the carbonate minerals, displacing ions such as calcium and magnesium from the near-wellbore vicinity, and lowers the pH around the wellbore.

Main Acid Stage

The HCl–HF mixture used in each treatment should be aligned to the guidelines in Table 5.4. Walsh et al. (1982) suggested that low-HF concentrations should be used to avoid the precipitation of AlF_3 or CaF_2 if the remaining calcite cannot be quantified. They also suggested that 12% HCl–3% HF can be used in low-calcite environments without a

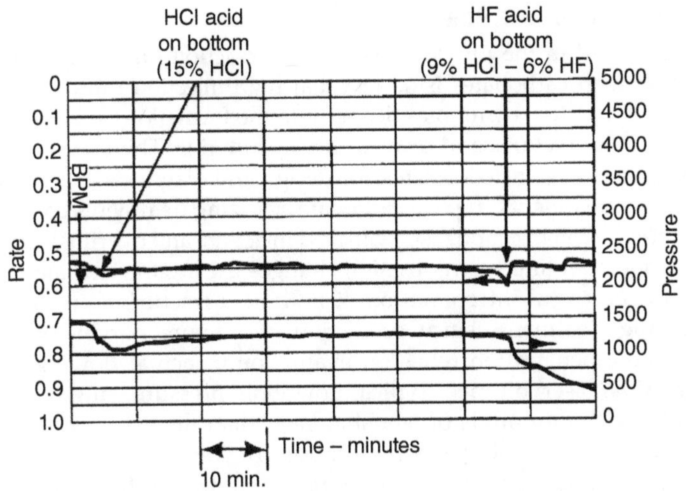

FIGURE 5.17 Pressure behavior during sandstone acidizing (McLeod et al., 1983).

precipitation problem. Some significant problems that may occur in high-clay-content formations include compromised formation integrity and excessive fines migration. When HF concentration is too high, these conditions can happen.

Figure 5.17 shows a pressure behavior during main acid stage (McLeod et al., 1983). When the HCl arrives at the formation, the rate was stabilized at about 0.5 bpm. When the HF acid injection began at the reservoir depth, the pressure slowly declined. This rate was continued as long as the pressure was kept same behavior (decline slowly). This is the type of response that one should observe when a well is treated to remove wellbore damage.

It was recommended that during acid injection of a matrix acidizing job in sandstone formations, the injection rate should be kept as high as it is below the formation fracture pressure. This practice can deliver better stimulation results with lower skin factors achieved (Paccaloni and Tambini, 1993; Kalfayan, 2008).

Postflush Stage

After a sufficient volume of HF/HCl solution has been injected, it is displaced from the tubing and the wellbore with a post-flush. A variety of fluids have been used for the post-flush, including diesel, ammonium chloride (NH_4Cl) solutions, and HCl. The post-flush displaces the spent acid farther from the wellbore so that any precipitates that may form will be less damaging. As a minimum, the post-flush volume should be the volume of the tubing plus twice the volume of the wellbore below the tubing. Because of gravity segregation effects, at least this much post-flush is needed to displace all acid from the wellbore.

5.4.2 Carbonate Acidizing

As discussed in the previous section, the reaction rate of hydrochloric acid with carbonates is so fast that high injection rates are often required to drive wormholes deeper into the formation or move acids through the fractures and toward the fracture tips. This poses unique challenges in the injection schedule design and diagnosis in carbonate acidizing treatments. Typical carbonate acidizing treatments consist of pre-stimulation injection test, main acid injection, and overflush stages.

Understanding tubing friction changes during the treatment is a key to successful interpretation of both acid fracturing and matrix acidizing treatments since prediction of fracture parameters and wormhole penetration relies on evaluation of actual flowing pressure at the reservoir face. Typical downhole pressure gauges are set hundreds or thousands of feet shallower than the injection interval and, as a result, tubing and perforation friction between the downhole gauge and the reservoir-face must be subtracted from the gauge data to arrive at the true injection pressure at the reservoir-face. Friction reducers are widely used in high-rate acid treatments. These additives significantly reduce pipe friction, allowing higher rates of injection than would be attainable for standard, untreated, fluids. However, the effect of the friction reducer can also change during the course of the job due to insufficient control of the friction reducer volume as the injection rate varies during treatment. The best practice to evaluate the effectiveness of the friction reducer during stimulation job is to calibrate tubing friction curves using the actual pressure data measured by surface and downhole gauges.

Prestimulation Injection Test

The prestimulation injection test begins with the step-rate test. Brine or water mixed with a friction reducer is pumped into the well in a series of timed steps with increasing rate. Wellhead and downhole pressures are recorded and a pressure versus rate chart generated. This information is used to determine matrix injection pressures and provide an estimate of fracture initiation and extension pressures prior to acidizing.

At the end of the step rate test, after the final fracture rate step has been completed, injection is stopped suddenly. The pressure falls rapidly to the instantaneous shut-in pressure (ISIP) as fluid movement ceases and frictional pressure losses drop to zero. Comparison of the last pumping pressure with the ISIP shows the amount of friction between the pressure gauge and the reservoir. As shut-in time proceeds, the fluid in the fracture bleeds off into the reservoir while the pressure is recorded. This pressure fall-off is then analyzed to determine fracture closure pressure by the square-root-time plot and/or G-function plot (Barree et al., 2009) and provide an estimate of reservoir permeability within the radius of investigation of the test.

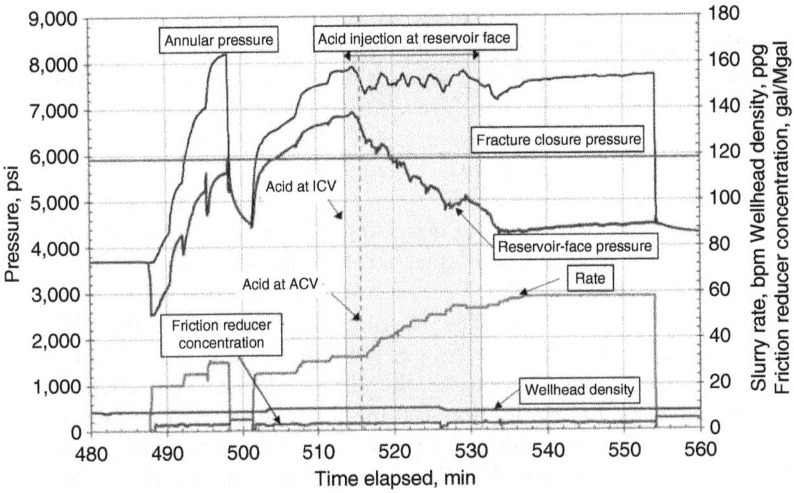

FIGURE 5.18 High-rate carbonate acid stimulation data (Kent et al., 2014).

Main Acid Injection Stage

During the main acid and overflush injection stages, rate and pressure data are recorded to determine wormhole growth rate and associated skin during stimulation. Figure 5.18 shows a field pressure behavior during high-rate carbonate acid stimulation presented by Kent et al. (2014). The main acid treatments consisted of pumping 75 gal of 28% HCl per foot of completion interval at high rate, followed by overflush fluids. The fracture closure pressure is measured from the pre-stimulation injection test. The reservoir-face pressure is extrapolated from the downhole pressure measurements. It should be noted that the reservoir-face pressure shown here is based on the number of perforations open to flow measured from the post-acid ISIP. Therefore, the reservoir-face pressure during the pre-stimulation injection test may not be correct. As shown in the figure, downhole pressures at the reservoir face start out above formation-parting pressure and very quickly drop to less than the fracture pressure as acid enters the carbonate formation. This pressure behavior indicates that the bulk of the acid-stimulation treatment is pumped under matrix-injection conditions conductive to wormholing of the near-wellbore carbonate reservoir. This injectivity improvement can be used to calculate skin evolution during the course of the treatment in a manner similar to that introduced by Paccaloni and Tambini (1993) and later extended by Zhu and Hill (1996) and Furui et al. (2012b).

Overflush Injection Stage and Post-Stimulation Falloff

In matrix acid treatments, the purpose of the overflush is to displace acid from the wellbore and push the acid volume forward, thereby increasing the penetration distance. A high rate is beneficial. Brine or water with friction reducer is commonly used. In acid fracturing, 28% HCl may be pumped at pressures below the fracture gradient after the main acid stage. Acid flows into existing openings between the fracture faces and creates wormholes which generate flow channels improving the connectivity between the acid-etched fracture and the wellbore.

At the end of the overflush period, injection is stopped suddenly similarly to the pre-stimulation injection test. The pressure falls rapidly to ISIP as fluid movement ceases and friction pressure losses drop to zero. Comparison of the last pumping pressure with the ISIP shows the amount of friction between the pressure gauge and the reservoir. The successive pressure falloff data may be used for further skin and permeability evaluation. But in most cases, because of large negative skins, it is very difficult to observe stabilized radial flow in acidized wells unless the well is shut-in for a long time. Therefore the determination of post-stimulation skin factors through the pressure falloff often becomes a challenging task.

Nomenclature

a_1	= constant used for the fracture conductivity correlation
a_2	= constant used for the fracture conductivity correlation
a_3	= constant used for the fracture conductivity correlation
a_4	= constant used for the fracture conductivity correlation
a_5	= constant used for the fracture conductivity correlation
a_6	= constant used for the fracture conductivity correlation
A	= flow area, ft^2, m^2
C	= acid concentration, moles/m^3
C_{fD}	= dimensionless fracture conductivity
C_1	= parameter defined by Eq. (5.48)
C_2	= parameter defined by Eqs. (5.45) and (5.50)
d_{core}	= core diameter, ft, m
$d_{e,wh}$	= effective wormhole diameter, ft, m
d_{wh}	= wormhole diameter, ft, m
E_f	= reaction rate constant, $m^2 \cdot s\ (mol/m^2)^{\alpha}$
h	= formation thickness, ft, m
h_f	= fracture height, ft, m
k	= permeability, mD, m^2
k_f	= fracture permeability, mD, m^2
K_d	= dissociation constant, mol/L
L	= core length, ft, m

m_{wh} = number of dominant wormholes along the angular direction
N_{AC} = acid capacity number
PV = pore volume, ft^3, m^3
PV_{BT} = number of pore volumes to breakthrough, dimensionless
$PV_{opt,BT}$ = optimum pore volumes to breakthrough, dimensionless
q = injection rate, STB/d, m^3/s
Q = production rate for fractured wells, STB/d, m^3/s
Q_o = production rate before fracturing, STB/d, m^3/s
r = reaction rate, mol/s·m^2
r_w = wellbore radius, ft, m
r'_w = equivalent wellbore radius, ft, m
r_{wh} = radius of wormholed zone, ft, m
R = universal gas constant, 10.73 psi ft^3/lb·mol °R
s = skin factor, dimensionless
S_{rock} = rock embedment strength, psi, Pa
T = temperature, °F, °C
v_i = interstitial velocity, ft/s, m/s
$v_{i,opt}$ = optimum interstitial velocity, ft/s, m/s
$v_{i,tip}$ = interstitial velocity at the tip of wormholes, ft/s, m/s
v_{wh} = wormhole propagation velocity, ft/s, m/s
V = total volume of injected acid, ft^3, m^3
w = fracture width, ft, m
w_i = ideal fracture width, ft, m
$\bar{w}$ = average fracture width
x_f = fracture half length, ft, m
α = order of reaction
α = parameter defined by Eq. (5.55)
α_z = wormhole axial spacing, dimensionless
β = gravimetric dissolving power, lb/lb, kg/kg
β = parameter defined by Eq. (5.56)
χ = volumetric dissolving power, ft^3/ft^3, m^3/m^3
ϕ = porosity, fraction
ρ_{acid} = acid density, lb/ft^3, kg/m^3
ρ_F = density of fast-reacting minerals, lb/ft^3, kg/m^3
$\rho_{mineral}$ = mineral density, lb/ft^3, kg/m^3
$\lambda_{D,x}$ = correlation length in the horizontal direction along the fracture
$\lambda_{D,z}$ = correlation length in the vertical direction along the fracture
σ_c = closure stress, psi, Pa
σ_D = dimensionless standard deviation

References

Ali, S. A., Durham, D. K., and Elphingstone, E. A. (1994). "Test Identifies Acidizing-Fluid/Crude Compatibility Problems," *Oil Gas Journal*, Vol. 92 (13), pp. 47–51.

Ali, S. A., and Hinkel, J. (2000). "Additives in Acidizing Fluids," Chap. 15, in *Reservoir Stimulation*, 3rd ed., M. J. Economides and K. G. Nolte (eds.), John Wiley & Sons, Inc., New York.

Ali, S. A., Kalfayan, L., and Montgomery, C. T. (2016). "Acid Stimulation," *Society of Petroleum Engineers*, Richardson.

Barree, R. D., Barree, V. L., and Craig, D. (2009, August 1). "Holistic Fracture Diagnostics: Consistent Interpretation of Prefrac Injection Tests Using Multiple Analysis Methods," *SPE Production & Operation*, Vol. 24 (3), pp. 396–406. doi:10.2118/107877-PA.

Beatty, C. V., Hill, A. D., Zhu, D., and Sullivan, R. B., (2011, January 1). "Characterization of Small Scale Heterogeneity to Predict Acid Fracture Performance," *Society of Petroleum Engineers.* doi:10.2118/140336-MS.

Buijse, M. A., and Glasbergen, G. (2005, January 1). "A Semi-Empirical Model to Calculate Wormhole Growth in Carbonate Acidizing," *Society of Petroleum Engineers.* doi:10.2118/96892-MS.

Chang, F. F., Nasr-El-Din, H. A., Lindvig, T., and Qui, X. W. (2008, January 1). "Matrix Acidizing of Carbonate Reservoirs Using Organic Acids and Mixture of HCl and Organic Acids," *Society of Petroleum Engineers.* doi:10.2118/116601-MS.

Deng, J., Mou, J., Hill, A. D., and Zhu, D. (2012, May 1). "A New Correlation of Acid-Fracture Conductivity Subject to Closure Stress," *SPE Production & Operations*, Vol. 27 (2), pp. 158–169. doi:10.2118/140402-PA.

Economides, M. J., Hill, A. D., Ehlig-Economides, C., and Zhu, D. (2012). *Petroleum Production Systems*, 2nd ed., Prentice Hall, New Jersey.

Economides, M. J., and Nolte, K. (2000). *Reservoir Stimulation*, John Wiley & Sons, New York.

Erbstoesser, S. R. (1980, November 1). "Improved Ball Sealer Diversion," *Journal of Petroleum Technology*, Vol. 32 (11), pp. 1903–1910. doi:10.2118/8401-PA.

Fredd, C. N., and Fogler, H. S. (1999, September 1). "Optimum Conditions for Wormhole Formation in Carbonate Porous Media: Influence of Transport and Reaction," *SPE Journal*, Vol. 4 (3), pp. 196–205. doi:10.2118/56995-PA.

Fredd, C. N., Tjia, R., and Fogler, H. S. (1997, January 1). "The Existence of an Optimum Damkohler Number for Matrix Stimulation of Carbonate Formations," *Society of Petroleum Engineers*, 2–3 Jun. doi:10.2118/38167-MS.

Furui, K., Burton, R. C., Burkhead, D. W., Abdelmalek, N. A., Hill, A. D., Zhu, D., and Nozaki, M. (2012a, March 1). "A Comprehensive Model of High-Rate Matrix-Acid Stimulation for Long Horizontal Wells in Carbonate Reservoirs: Part I—Scaling Up Core-Level Acid Wormholing to Field Treatments," *SPE Journal*, Vol. 17 (1), pp. 271–279. doi:10.2118/134265-PA.

Furui, K., Burton, R. C., Burkhead, D. W., Abdelmalek, N. A., Hill, A. D., Zhu, D., and Nozaki, M. (2012b, March 1). "A Comprehensive Model of High-Rate Matrix-Acid Stimulation for Long Horizontal Wells in Carbonate Reservoirs: Part II—Wellbore/Reservoir Coupled-Flow Modeling and Field Application," *SPE Journal*, Vol. 17 (1), pp. 280–291. doi:10.2118/155497-PA.

Haldar, S., Al-Jandal, A. A., Al-Driweesh, S. M., Said, R., AlSarakbi, S., and Espinosa, M. A. E. (2008, January 1). "Evaluation of Rotary Jetting Tool Application for Matrix Acid Stimulation of Carbonate Reservoir in Southern Area Field of Saudi Arabia," *Society of Petroleum Engineers.* doi:10.2523/IPTC-12023-MS.

Hall, B. E. (1975, December 1). "The Effect of Mutual Solvents on Adsorption In Sandstone Acidizing," *Journal of Petroleum Technology*, Vol. 27 (12), pp. 1439–1442. doi:10.2118/5377-PA.

Hansen, J. H., and Nederveen, N. (2002, January 1). "Controlled Acid Jet (CAJ) Technique for Effective Single Operation Stimulation of 14,000+ ft Long Reservoir Sections," *Society of Petroleum Engineers*, Oct. doi:10.2118/78318-MS.

Houchin, L. R., Foxenburg, W. E., Usie, M. J., and Zhao, J. (1992, January 1). "A New Technique for the Evaluation of Acid Additive Packages," *Society of Petroleum Engineers*. doi:10.2118/23817-MS.

Jinzhou Z., Hai, Y., and Yongming L. (2013, July 1). "China Developing Strategy for Horizontal Fracturing Technology," *Oil and Gas Journal*, Vol. 111 (7), pp. 70–79.

Kalfayan, L. (2008) *Production Enhancement with Acid Stimulation*, 2nd ed., PennWell Books, Tulsa OK. ISBN:159370139X, 9781593701390.

Kent, A. W., Burkhead, D. W., Burton, R. C., Furui, K., Actis, S. C., Bjornen, K., Constantine, J. J., Hodge, R. M., Nozaki, M., Vasshus, A., and Zhang, T. (2014, June 1). Intelligent Completion Inside Uncemented Liner For Selective High-Rate Carbonate Matrix Acidizing. *SPE Drilling and Completion*, Vol. 29 (2), pp. 165–181. doi:10.2118/166209-PA.

King, G. E. (1986, May 1). "Acidizing Concepts—Matrix vs. Fracture Acidizing," *Journal of Petroleum Technology*, Vol. 38 (5), pp. 507–508. doi:10.2118/15279-PA.

King, G. E., and Lee, R. M. (1988, May 1). "Adsorption and Chlorination of Mutual Solvents Used in Acidizing," *SPE Production Engineering*, Vol. 3 (2), pp. 205–209. doi:10.2118/14432-PA.

McDuff, D., Shuchart, C. E., Jackson, S., Postl, D., and Brown, J. S. (2010, January 1). "Understanding Wormholes in Carbonates: Unprecedented Experimental Scale and 3-D Visualization," presented at SPE Annual Technical Conference and Exhibition, Florence, Italy, 19–22 Sep. doi:10.2118/134379-MS.

McLeod, H. O., Ledlow, L. B., and Till, M. V. (1983, January 1). "The Planning, Execution, and Evaluation of Acid Treatments in Sandstone Formations," presented at SPE Annual Technical Conference and Exhibition, San Francisco, CA, 5–8 Oct. doi:10.2118/11931-MS.

McLeod, H. O. and Norman W. D. (2000). Sandstone Acidizing. In *Reservoir Stimulation*, 3rd ed., Economides M. J. and Nolte K. G, Chapter 18, 18-1–18-27. John Wiley & Sons, New York.

Nierode, D. E., and Kruk, K. F. (1973, January 1). "An Evaluation of Acid Fluid Loss Additives Retarded Acids, and Acidized Fracture Conductivity," presented at Meeting of the Society of Petroleum Engineers of AIME, Las Vegas, NV, 30 Sep.–3 Oct. doi:10.2118/4549-MS.

Paccaloni, G., and Tambini, M. (1993). "Advances in Matrix Stimulation Technology," *JPT*, Vol. 45 (3), pp. 256–263. SPE-20623-PA.

Schechter, R. S. (1992). *Oil Well Stimulation*, Prentice Hall, Englewood Cliffs, NJ.

Smith, C. F., Dollarhide, F. E., and Byth, N. J. (1978, May 1). "Acid Corrosion Inhibitors—Are We Getting What We Need?" *Journal of Petroleum Technology*, Vol. 30 (5), pp. 737–746. doi:10.2118/5644-PA.

Taylor, K. C., Nasr-El-Din, H. A., and Al-Alawi, M. J. (1999, March 1). "Systematic Study of Iron Control Chemicals Used during Well Stimulation," *SPE Journal*, Vol. 4 (1), pp. 19–24. doi:10.2118/54602-PA.

Walsh, M. P., Lake, L. W., and Schechter, R. S. (1982, September 1). "A Description of Chemical Precipitation Mechanisms and Their Role in Formation Damage during Stimulation by Hydrofluoric Acid," *Journal of Petroleum Technology*, Vol. 34 (9), pp. 2097–2112. doi:10.2118/10625-PA.

Zhou, Z. J., Gunter, W. O., and Jonasson, R. G. (1995). "Controlling Formation Damage Using Clay Stabilizers: A Review," presented at the Annual Technical Meeting, Calgary, 7–9 June. PETSOC-95-71.

Zhu, D., and Hill, A. D. (1996, January 1). "Field Results Demonstrate Enhanced Matrix Acidizing Through Real-Time Monitoring," presented at Permian Basin Oil and Gas Recovery Conference, Midland, Texas, 27–29 Mar. doi:10.2118/35197-MS.

CHAPTER 6
Intelligent Well Completion: Downhole Monitoring

The terminology of intelligent completion refers to well completions that have abilities to monitor well flow conditions with downhole permanent sensors, and to regulate flow conditions by control equipment installed downhole. Intelligent well technology has broad applications in oil and gas industry. From well performance monitoring, control, and optimization to reservoir surveillance and management, this emerging technology brings a revolution in field development and operations to our industry. This chapter introduces one of the two components for intelligent well completions—the downhole monitoring component. The chapter presents the technology available, and discusses the basic physics of the sensing techniques and their functions in downhole flow monitoring. The field applications in this chapter illustrate how the technology helps to diagnose well operation problems and to improve the efficiency of oil and gas production. Chapter 7 discusses the control component of intelligent completions.

6.1 Closed-Loop Optimization by Intelligent Completion

The purpose of installing an intelligent completion in oil and gas producing or injection wells is to optimize well and reservoir performance. Optimization requires collecting and processing the flowing data, and analyzing the data by models, either analytical or numerical. Based on the outcome of the analysis, operation problems are identified and solutions to improve performance are suggested. The solutions are implemented in the field with the controlling devices of an intelligent completion. New data is collected after implementation of the solutions, and the next cycle of optimization starts. This cycle is often referred to as "closed-loop optimization."

There are two categories of closed-loop optimization, both based on intelligent completions: well performance optimization and reservoir management. Field-scale optimization for reservoir management consists of extensive history matching with numerical reservoir simulation. In this chapter, we focus on well performance optimization. The procedure of closed-loop optimization for single well performance is illustrated in Fig. 6.1.

There are three components in the closed-loop optimization. The optimization loop starts from data collection by downhole sensors. The sensing technology includes fiber optic sensors and crystalline quartz gauges. Fiber optic sensors consist of distributed temperature sensors (DTS), distributed pressure sensors (DPS), distributed acoustic sensors (DAS), and distributed strain sensors (DSS). Pressure and temperature can also be measured by downhole single-point quartz gauges. Downhole pressure and temperature gauges have been existed in the industry for a long time. Downhole pressure gauges are still commonly used. The distributed sensors are fiber optical sensing cables that either are installed permanently downhole, or periodically run downhole during well operations. Downhole temperature and pressure measurements can also be obtained by production logging tools (PLTs), which are "snap shots" at the moment the log is run. The data collected downhole combined with surface flow rate measurements provide the input to the second step in the optimization loop; interpretation and optimization modeling. This step is probably the most challenging step in the optimization loop. The collected data

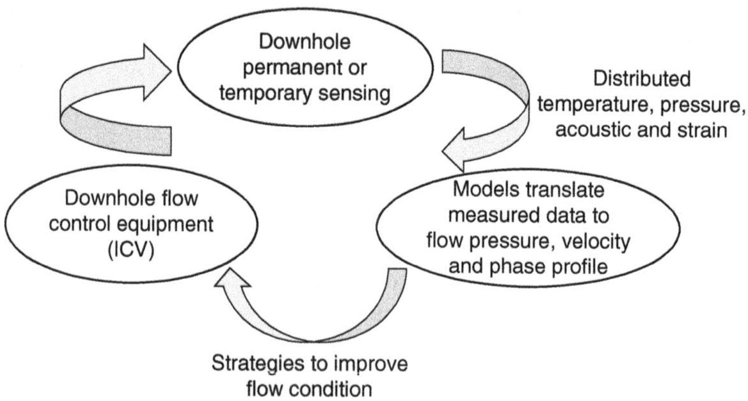

FIGURE 6.1 Closed-loop optimization for well performance by intelligent completion.

are interpreted to downhole flow profiles by forward simulation and inverse error minimization. When the interpretation of downhole flow condition is satisfied, engineers further conduct sensitivity studies to identify production problems and improvement strategies. The output of the second step is the control strategies that enhance the well performance. The third step in the closed-loop optimization is the execution of the optimization strategies by downhole flow control devices (this is discussed in Chapter 7). Finally, to close the loop, new measurements after implementing the strategies are collected and used as the input to the next cycle.

Optimization of well performance is real-time and continuous. When applied correctly, it brings invaluable benefit to well performance diagnosis and improvement. This technology can be used in any phase of well operations, from production or injection, to stimulation or intervention, to enhanced recovery.

6.2 Fundamentals of Fiber Optic Sensing Technology

Fiber optic sensing for distributed temperature measurements has been used in military, health science, lighting, electrical, and electronic industries before it was brought into the petroleum industry. In the early 2000s, the technology was introduced to the oil and gas industry for downhole temperature and pressure monitoring. Since then, the technology found many applications in production monitoring and optimization. The downhole part of fiber optic sensing technology, compared with downhole permanent gauges or PLTs, is simpler and requires less maintenance because it has no moving parts and no downhole power requirement. It provides continuous measurements in time and space with desired intervals. It does not disturb flow of the well while functioning.

The fiber optical sensor consists of a sensing cable made of a glass fiber core. Outside coating materials, also referred to as cladding, of the sensing cables can vary, depending on the environments that the cables are going to be installed in. When a light wave travels through a fiber optic sensing cable, it bounces inside the cladding, and generates backscattered light waves. There are three components in the backscattered waves, Raman Stokes, Raman anti-Stokes, and Raleigh scatterings. Raleigh scatters have the same wavelength as the source light, Raman Stokes have longer wavelength and Raman anti-Stokes have shorter wavelength, comparing with the source light, as shown in Fig. 6.2.

The spectral shift of the backscattering is related to thermal effect. The anti-Stokes wave is dependent to the surrounding temperature, and the Stokes wave is insensitive to the temperature. The ratio of anti-Stokes to Stokes intensities can be used to estimate the temperature at the sensor location (Shiota and Wada, 1991;

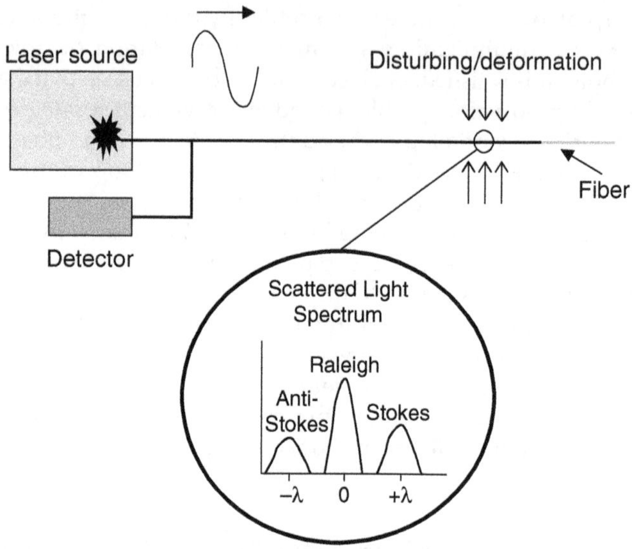

Figure 6.2 The principle of downhole fiber optical sensing technology.

Abhisek et al., 2012). The optical time domain reflectometry method states that (Reyes et al., 2011)

$$\frac{I_{as}}{I_s} \sim \exp\left(\frac{-h\Delta v_R}{kT}\right) \tag{6.1}$$

where I_{as} and I_s are intensity of anti-Stokes and Stokes scattering, h is the Planck number, K is the Boltzmann number, and Δv_R is the frequency separation between anti-Stokes/Stokes and Rayleigh. Figure 6.2 also illustrates how the DTS sensing technology works in the field. At the surface, a laser source is generated and passed through the fiber optic sensing cable, and a receiver collects the back-scatters and translates the scattered wave measurements to temperature. The location of the measured temperature is determined by the travel time of the wave. The temperature can be measured at 1 m in space with a resolution of 0.1°C.

In order for fiber optical core to transport light wave, the refractive index of the cladding must be lower than the one for the core.

6.3 Fiber Optic Sensor for Distributed Temperature

In addition to reservoir surveillance monitoring, more direct and broad applications of downhole sensing technology are single well performance monitoring and diagnosis. This can be for both production and injection wells. When fluids flow through porous media, the pressure drop due to the flow can result in fluid temperature change, and this is called the Joule-Thomson effect. Temperature of liquid phases increases as the fluids flow. When pressure decreases, temperature decreases for gas phases. The Joule-Thomson effect is related to flow velocity. The higher the velocity, the higher the pressure drop, resulting in higher temperature change. Because gas flow usually has higher velocity, temperature change is more pronounced for gas phases when compared with liquid phases. When temperature changes are measured, it possibly provides the information about flow conditions.

For producers, based on the differences of fluid thermal properties of oil, gas, and water, DTS can detect the fluid entries and the types of fluids. The information can be used to make decisions on production control. Temperature signals can easily show cooling effects for gas entries. For water entries, depending on the water source, detecting water entries can be more difficult compared with gas entries. The water from bottom aquifers is more likely with a higher temperature than local reservoir fluids at the wellbore elevation (geothermal temperature increases as depth increases). Injection water may have a lower temperature to start. When cold fluids flow through the reservoir, the temperature increases, and the competition between the cold sources and conductive heating likely makes the temperature difference between water and oil phases subtle.

The applications of DTS include serving as an additional constrain for reservoir permeability characterization, identifying the locations of created fractures and flow capacity of each fractures, diagnosing downhole equipment problems for artificial lift, and evaluating injection efficiency. These examples are presented in the section after reviewing the models for temperature interpretation.

6.4 Interpretation of Downhole Flow Profile with Distributed Temperature

DTS-measured downhole temperatures are direct reflections of downhole flow conditions. With the relationship between temperature and flow velocity, we can generate downhole flow profiles from the measured temperature profiles. The interpretation of temperature to flow requires extensive mathematical models to simulate heat and mass flows, and this becomes even more difficult when considering the effects of well structure, geothermal temperature, and reservoir permeability heterogeneity. The mathematical models, also frequently referred to as forward models, serve as predictive tools to simulate temperature and flow for a defined domain. Then an inversion procedure is used to generate the correct downhole flow profile by minimizing the difference between the simulated temperature and measured temperature. In this section, we focus on the forward modeling.

To model temperature measurements, we first need to define a physical domain. The domain consists of a reservoir and a wellbore. Each component has its own model. Heterogeneity of the reservoir (permeability, porosity, natural, or hydraulically induced fractures) is described in the reservoir model. Well structure, completion configuration, and well function (producing, injecting, or stimulating) are defined in the wellbore model. The two models, reservoir and wellbore models, have two sets of equations each. The first set of equation is based on energy conservation to solve for temperature distribution (often referred to as the thermal model), and the second set is based on mass and momentum conservations to solve for pressure and flow velocity profiles (this is also referred to as the flow model). All equations are coupled at the interface and served as the boundary conditions among each other. Figure 6.3 is a flow chart for the model construction.

To solve the problem, we start from an assumed flow condition within a defined physical domain. We solve the flow model first to generate velocity and pressure profiles, and then use these profiles to solve for the temperature profile in the system. The output of the model is the temperature at the wellbore, where temperature is measured by DTS. We compare the calculated and measured temperatures. If the two sets of temperatures match, we assume that the flow distribution is correctly generated. If the calculated temperature

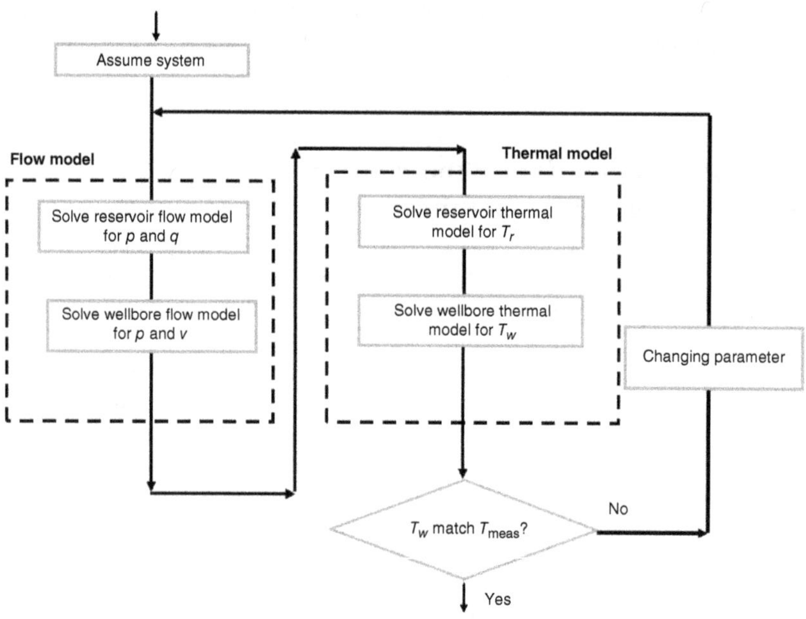

FIGURE 6.3 Flow chart for DTS modeling.

does not match the measured one, we can adjust one set of reservoir parameters and try again. The reservoir parameter can be reservoir permeability, fracture conductivity for fractured wells, steam distribution for thermal recovery wells, and so on. Notice that only one set of parameters can be changed, and the rest of the input is assumed as known. This is because that temperature measurements only bring one set of constrains, and thus only one set of parameters can be estimated from the measurements.

6.4.1 Wellbore Models

Wellbore models are used to obtain pressure, velocity, and temperature distribution along a wellbore. For a wellbore that has communication with the surrounding reservoir (completion locations or openhole wellbores), heat transfer happens in three different manners: heat conduction caused by the temperature difference between the reservoir geothermal temperature and the wellbore fluid temperature, heat convection because of the fluid flow along the wellbore, and heat convection because of the fluid flow from the reservoir into the wellbore. This is illustrated in Fig. 6.4. Through a set of conservation equations, we can estimate the pressure, velocity, and temperature of the wellbore. The derivation of the set of conservation equations and their solutions is presented by Yoshioka et al. (2005) in detail.

Mass Conservation

We start from mass conservation, which simply states that the difference between the mass flows into and out of the element shown in Fig. 6.4 must be equal to the mass accumulation in the element in a given period of time. This, in mathematical form, for phase, i, with the holdup of the phase, y_i, and density, ρ_i, in a multiphase flow system, is

$$\pi R^2 \frac{\partial(\rho_i y_i)}{\partial t} = -\pi R^2 \frac{\partial(\rho_i v_i y_i)}{\partial x} + 2\pi R \gamma \rho_{i,I} v_{i,I} y_{i,I} \tag{6.2}$$

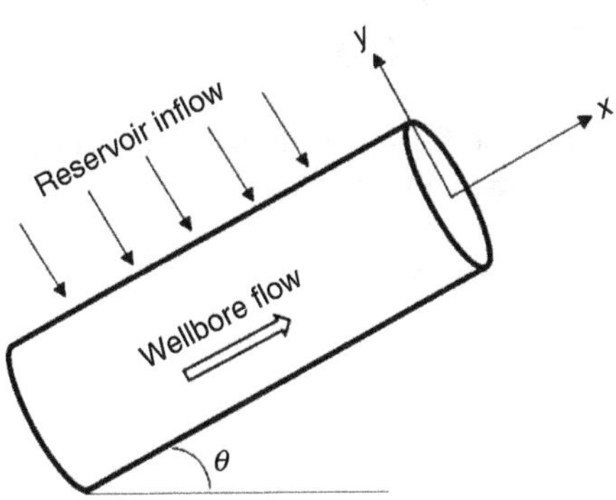

FIGURE 6.4 Schematic of wellbore domain for wellbore temperature modeling.

The three terms in Eq. (6.2) all have physical meanings. The left-hand side term is the differential accumulation. The first term on the right-hand side is the fluid flow along the element, into and out of the element with a velocity of v_i, representing the flow inside the wellbore; and the second term is the flow into the element in the direction perpendicular to the x-axis, representing the reservoir inflow to the wellbore. The inflow velocity is v_I, which is directly related to the inflow rate q_I by

$$q_I = v_I(2\pi R\Delta x) \tag{6.3}$$

The parameter in the second term of the right-hand side in Eq. (6.2), γ, is defined as the pipe opening ratio,

$$\gamma = \frac{\text{Opened pipe surface area}}{\text{Pipe surface area}} \tag{6.4}$$

It is simply set as one (1) if the pipe is open for communication, or zero (0) if the pipe has no communication with the reservoir. γ is a useful variable that can flexibly describe the well completion. For example, for openhole wells, γ is always set to one along the entire wellbore, but for perforated wells, γ is one at the perforation locations, and zero at nonperforated locations. This parameter makes the wellbore model applicable to all completion types.

Momentum Conservation

Momentum conservation applies a mechanical energy balance in the element, and relates pressure to flow. The equation for the defined domain is

$$\frac{\partial(\rho_m v_m)}{\partial t} = -\frac{2\tau_w}{R} - \frac{\partial\left(\rho_m v_m^2 + p\right)}{\partial x} - \rho_m g \sin\theta \tag{6.5}$$

where τ_w in the equation is wall shear stress, commonly expressed in terms of frictional factor, f_m, as

$$\tau_w = \frac{\rho_m f_m v_m^2}{2} \tag{6.6}$$

Notice that for multiphase flow, the mixed properties of fluids, ρ_m and v_m, are used in Eqs. (6.5) and (6.6),

$$\rho_m = \sum_i \rho_i y_i \tag{6.7}$$

and

$$v_m = \frac{\sum_i \rho_i v_i y_i}{\sum_i \rho_i y_i} \tag{6.8}$$

Total Energy Conservation

The last equation is the total energy balance equation, which brings in temperature to the problem. The total energy of phase i, E_i, is balanced over the element during a time period as

$$\sum_i \frac{\partial(E_i y_i)}{\partial t} = -\sum_i \frac{\partial}{\partial x}[(E_i + p_i)v_i y_i] + \sum_i \frac{2}{R}(E_{i,I} + p_{i,I})v_{i,I}\gamma y_{i,I} + \sum_i \rho_i v_i y_i g \sin\theta + \frac{2}{R} q_I (1-\gamma) \tag{6.9}$$

where q_I represents the conductive heat flux. E_i is the sum of kinetic energy and internal energy, defined as

$$E_i = \rho_i \left(\frac{1}{2} v_i^2 + U_i\right) \tag{6.10}$$

In Eq. (6.10), specific internal energy, U_i, is related to specific enthalpy, H_i, and pressure, p, as

$$H = U_i + \frac{p}{\rho} \tag{6.11}$$

Changes of temperature and pressure over a defined domain cause internal energy change, and can be expressed in terms of enthalpy. For example, the special derivative of enthalpy is

$$\frac{dH_i}{dx} = C_{p,i}\frac{dT_i}{dx} + \frac{1}{\rho_i}(1-\beta_i T_i)\frac{dp_i}{dx} = C_{p,i}\frac{dT_i}{dx} + C_{p,i}K_{JT,i}\frac{dp_i}{dx} \tag{6.12}$$

where C_p is heat capacity of fluid, β is thermal expansion coefficient, and K_{JT} is referred as Joule-Thomson coefficient. Solving Eqs. (6.2), (6.5), and (6.9) provides the distribution of pressure, temperature, and velocity over the defined domain.

Working Equations

Solving Eqs. (6.2), (6.5), and (6.9) with the associated equations presented in the previous section generates temperature, pressure, and velocity profiles along a wellbore. For some special cases, the equations can be simplified under certain assumptions. For example, for steady-state condition, the mass balance equation reduces to

$$\frac{d(\rho_i v_i y_i)}{dx} = \frac{2\gamma y_{i,I}}{R}\rho_{i,I}v_{i,I} \tag{6.13}$$

The momentum balance equation in term of pressure gradient is

$$\frac{dp}{dx} = -\frac{\rho_m v_m^2 f_m}{R} - \frac{d(\rho_m v_m^2)}{dx} - \rho_m g \sin\theta \tag{6.14}$$

And the energy balance equation becomes

$$\frac{dT}{dx} = \frac{2}{R}\left[\gamma\frac{(\rho v C_p)_{T,I}}{(\rho v C_p)_T} + (1-\gamma)\frac{U}{(\rho v C_p)_T}\right](T_{\text{well}} - T_I) + \frac{(\rho v C_p K_{JT})_T}{(\rho v C_p)_T}\frac{dp}{dx} + \frac{(\rho v)_T}{(\rho v C_p)_T} g \sin\theta \tag{6.15}$$

where

$$(\rho v)_T = \sum_i \rho_i v_i y_i \tag{6.16}$$

$$(\rho v C_p)_T = \sum_i \rho_i v_i y_i C_{p,i} \tag{6.17}$$

$$(\rho v C_p K_{JT})_T = \sum_i \rho_i v_i y_i C_{p,i} K_{JT,i} \tag{6.18}$$

In Eq. (6.15), U is overall heat transfer coefficient for the heat conduction between the formation rock and the wellbore. This coefficient is defined by the well completion. For example, for a cased and cemented well, this is a combination of formation rock, cement, casing, annulus, and tubing. Obviously, these equations still have to be solved numerically. Because temperature, pressure, and velocity are fully coupled, the equations should be solved simultaneously with iteration. For simplicity, we also can solve flow equations first to obtain pressure and velocity field, and then solve the thermal equations for temperature consecutively.

6.4.2 Reservoir Models

Similar to the wellbore models in Section 6.4.1, reservoir models are based on three conservation equations: mass, momentum, and energy. Reservoir flow problems have been studied thoroughly and the solutions are readily provided by reservoir simulations or analytical/semianalytical equations. To use temperature measurement for flow profiling, a physical model that contains the specific reservoir as the study target needs to be defined first. Formation and fluid properties are critical to the solution of the problems, and need to be described as detailed and accurately as possible. The general equation for multiphase reservoir flow is

$$\frac{\partial}{\partial t}(\rho_i S_i \phi) + \nabla \cdot (\rho_i \vec{u}_i) = 0 \tag{6.19}$$

where S_i is phase i saturation, and u_i is Darcy velocity of phase i. From Darcy's law,

$$\vec{u} = -\frac{\overline{\overline{k}}}{\mu} \cdot (\nabla p + \rho \vec{g}) \tag{6.20}$$

With the boundary conditions that describe the problem, the solution of the reservoir flow equations generates pressure and velocity profile for each phase. We illustrate examples for special cases later in this chapter.

The reservoir thermal model is built from energy conservation. Heat transfer phenomena in a reservoir for flow through porous medium include kinetic energy change; heat conduction because the temperature differences among the moving and stationary fluids, and the formation rock with local geothermal temperature; and convection, conduction, viscous dissipation, and thermal expansion because of fluid flow. Neglecting kinetic energy change, we can write the energy balance equation in a reservoir in three-dimensional (3D) format as (Lake, 1989)

$$\nabla\cdot\left[\sum_{i=1}\rho_i\vec{u}_i(H_i+gD)\right]-\nabla\cdot(K_{Tt}\nabla T)=-\frac{\partial}{\partial t}\left[\phi\sum_{i=1}\rho_i S_i(U_i+gD)+(1-\phi)\rho_s U_s\right] \tag{6.21}$$

Using Eqs. (6.10) and (6.11) for enthalpy and internal energy, and defining for the formation rock,

$$dU_s\approx C_{ps}dT \tag{6.22}$$

the temperature equation for a reservoir with multiphase flow is

$$\begin{aligned}&-\left[\sum_i(\phi S_i\rho_i C_{pi})+(1-\phi)\rho_s C_{ps}\right]\frac{\partial T}{\partial t}+\sum_i\left(\phi S_i\beta_i T\frac{\partial p_i}{\partial t}\right)\\&=\sum_i\rho_i\vec{u}_i\cdot C_{pi}\nabla T+\sum_i\vec{u}_i\cdot\nabla p_i-\sum_i\beta_i T\left(\vec{u}_i\cdot\nabla p_i\right)-\nabla\cdot(K_{Tt}\nabla T)+\sum_i\rho_i\vec{u}_i\cdot\nabla(gD)\end{aligned} \tag{6.23}$$

For single phase flow, the equation reduces to

$$\begin{aligned}&-\{\phi\rho\, C_p+\rho_s C_{ps}(1-\phi)\}\frac{\partial T}{\partial t}+\phi\beta T\frac{\partial p}{\partial t}\\&=\boldsymbol{u}\cdot[\rho C_p\nabla T+(1-\beta T)\nabla p+\rho\nabla(gD)]-K_T\nabla^2 T\end{aligned} \tag{6.24}$$

The significant differences of the above equation compared with conventional reservoir thermal models are the viscous dissipation and thermal expansion terms that are added to the equation. These terms are of small scale and are usually neglected when simulating conventional processes such as thermal recovery. Since these terms are a part of the essential information for flow profiling from temperature measurements on the wells without hot or cold injection, we may not eliminate them. Often, temperature changes are close to or even smaller than 1°F if no significant heat sources are involved.

Since DTS measurements can have a resolution of 0.1°C (0.18°F), subtle temperature changes can be detected, especially when continuous measurements in time and space are available.

Keeping in mind that the fluid velocity in a reservoir is likely lower than the one in the associated wellbore, the temperature changes can be on a different scale for the wellbore compared with for the reservoir. Often consecutively and separately solving the reservoir and wellbore problems is not just yield more accurate results but also necessary. As long as the reservoir and wellbore are connected by the boundary conditions, the solution of the forward model satisfies both subdomains.

6.4.3 Well Temperature from Integrated Wellbore and Reservoir Models

When coupling the reservoir and wellbore thermal models, in the grids that contain the wellbore, the grid temperature from the reservoir model is coupled to the wellbore temperature simulated by the wellbore model through a temperature referred as arriving temperature, T_I. Figure 6.5 illustrates the temperature map in a grid in the near-wellbore region.

If a DTS sensor is installed inside the well, T_{well} is the temperature measured by the sensor. The heat transfer between the reservoir temperature, T_{grid}, and the wellbore temperature, T_{well}, is the combination of heat conduction caused by the temperature gradient and heat convection by flow,

$$\dot{Q}_h\Big|_{\text{res-well}} = -K_{Tt}\frac{\partial T}{\partial r}\bigg|_{r=r_w} + \sum_{i=1}^{N_p}(\rho\vec{u}C_p)_i \cdot (T_{\text{res}} - T_I) \tag{6.25}$$

If we assume fluid properties are constant during the calculation periods, the analytical solution of T_I is (Li and Zhu, 2009),

$$T_I = c_1 r_w^{n_1} + c_2 r_w^{n_2} + b \tag{6.26}$$

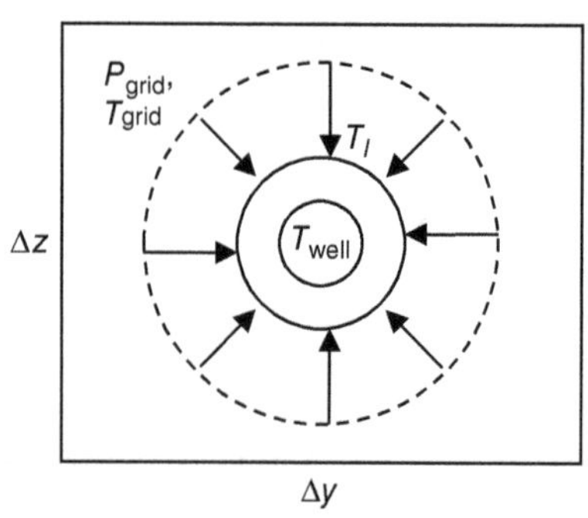

FIGURE 6.5 Schematic of near-wellbore temperature distribution used in the integrated thermal model.

where

$$n_1 = \frac{-\omega}{2} + \frac{1}{2}\sqrt{\omega^2 + 4\frac{k_e a^2}{K_{Tt}} \sum \left(\frac{k_{e,\text{relative},i}}{\mu_i} \beta_i \right)} \tag{6.27}$$

$$n_2 = \frac{-\omega}{2} - \frac{1}{2}\sqrt{\omega^2 + 4\frac{k_e a^2}{K_{Tt}} \sum \left(\frac{k_{e,\text{relative},i}}{\mu_i} \beta_i \right)} \tag{6.28}$$

$$c_1 = \frac{r_{\text{eff}}^{n_2} \frac{\alpha_T}{K_{Tt}}(b - T_w) - \left(\frac{n_2}{r_w} - \frac{\alpha_T}{K_{Tt}} \right) r_w^{n_2} \partial(T_{\text{grid}} - b)}{\left(\frac{n_1}{r_w} - \frac{\alpha_T}{K_{Tt}} \right) r_w^{n_1} r_{\text{eff}}^{n_2} - \left(\frac{n_2}{r_w} - \frac{\alpha_T}{K_{Tt}} \right) r_w^{n_2} r_{\text{eff}}^{n_1}} \tag{6.29}$$

$$c_2 = \frac{\left(\frac{n_1}{r_w} - \frac{\alpha_T}{K_{Tt}} \right) r_w^{n_1} (T_{\text{grid}} - b) - r_{\text{eff}}^{n_1} \frac{\alpha_T}{K_{Tt}}(b - T_w)}{\left(\frac{n_1}{r_w} - \frac{\alpha_T}{K_{Tt}} \right) r_w^{n_1} r_{\text{eff}}^{n_2} - \left(\frac{n_2}{r_w} - \frac{\alpha_T}{K_{Tt}} \right) r_w^{n_2} r_{\text{eff}}^{n_1}} \tag{6.30}$$

$$\omega = \sum \left(\rho_i C_{pi} \left(\frac{k_{e,\text{relative},i}}{\mu_i} \right) \right) \frac{k_e a}{K_{Tt}} \tag{6.31}$$

$$a = \frac{p_{\text{gird}} - p_{wf}}{\ln \frac{r_{e,\text{eff}}}{r_w}} \tag{6.32}$$

and

$$b = \frac{\sum \left(\frac{k_{e,\text{relative},i}}{\mu_i} \right)}{\sum \left(\left(\frac{k_{e,\text{relative},i}}{\mu_i} \right) \beta_i \right)} \tag{6.33}$$

The thermal equation for reservoir temperature, Eq. (6.23), and the thermal equation for wellbore, Eq. (6.15), are solved together with the arriving temperature equation, Eq. (6.26). If DTS cable is outside the casing, the measured temperature is T_I.

6.4.4 Inversion of Temperature Measurement to Flow Field

The forward models presented in Sections 6.4.1 (wellbore model) and 6.4.2 (reservoir model) predict temperature and pressure for a given flow domain. In reality, we measure temperature and pressure with downhole sensors, and flow profiles are inverted from the measured data. To accomplish this, we need an inversion procedure.

Inversion is a procedure that minimizes the difference between a reference parameter and a model predicted parameter. A temperature interpretation model first uses the forward model to calculate temperature and pressure profiles based on an assumed flow field, and then compares the calculated temperature profile with the measured temperature profile measured by DTS. If the calculated temperature does not match the measured temperature, certain parameters can be changed to recalculate temperature and pressure until the model predicted temperature matches the field measurement. Because we only have one set of measurements (i.e., temperature profile), it only provide the solution for one set of parameter.

The inversion procedure starts from defining an objective function that captures the difference between a model-predicted variable, $Y(w)_i$, and the corresponding measured variable, $Y'(w)_i$,

$$E(w) = \sum_{i=1,N} [Y'(w)_i - Y(w)_i]^2 \tag{6.34}$$

where in Eq. (6.34), subscript i represents one temperature measurement in a set of temperature measurements that has total of N measurements. The squared function here is to ensure that the objective function is always positive. w in the objective function (Eq. (6.34)) is the parameter that can be estimated from the inversion. During the process, we first assign a set of value to $w(i)$ and use the forward model to calculate $Y(w)_i$, then we use Eq. (6.34) to evaluate the deviation of model-predicted $Y(w)_i$ to measured Y'_i. A criterion is set for the objective function. If the $E(w)$ value does not satisfy the criterion, then we modify the value of w and repeat the procedure until the criterion of the objective function meets.

The inversion procedure can be for multiple constraints. For example, if we use measured temperature from DTS and measured downhole pressure as two reference variables to interpret the permeability as a parameter by inversion, the objective function can be defined as

$$E(w) = D_{wT} \sum_j (T^{\mathrm{cal}} - T^{\mathrm{obs}})_j^2 + D_{wp} \sum_j (p^{\mathrm{cal}} - p^{\mathrm{obs}})_j^2 \tag{6.35}$$

where D_{wT} and D_{wp} are the weights of temperature and pressure. The weights of each measured data set depend on how influential the data set is in the interpretation. Yoshioka (2007) suggested a method to assign the weights for temperature and pressure measurements depending on the thermal properties of the gas,

$$D_{wT} = (1/K_{JT})^2 D_{wp} = (\rho C_p/(\beta T - 1))^2 D_{wp} \tag{6.36}$$

As defined before, K_{JT} is Joule-Thomson coefficient and β is thermal expansion coefficient.

There are two approaches to find the solution of inversion: gradient methods and global search methods. Gradient methods find the path between the target (measured variable) and prediction (calculated variable), and marches along the slope of the path (derivative of the path) to approach the target. Global search methods randomly sample the population, and at each calculation, the objective function value is compared with the previous sample set, and the minimum objective function of the domain is kept. Gradient methods converge rapidly, but can converge to a local minimum for nonlinear functions and not find the true solution. Global methods can avoid local minimum, but do not guaranty conversion to find the solution, especially when the database is large and computation power is limited. Unfortunately, this is the situation of DTS inversion. The combination of statistically distributed reservoir properties in a large domain and massive data collected by the sensors as functions of space and time makes the interpretation of DTS data extremely challenging with both methods.

Because temperature measurements are along wellbores where the sensors are installed, the measured data is a one-dimension (1D) data set. When interpreting DTS data for the flow profile, we couple a 1D problem along the well with most likely a 3D problem in the reservoir. We have mentioned that the wellbore and reservoir problems do not have the same scale, both in flow and in temperature. Thus, good practice is to prescreen the measured temperature data and also well completion structure to divide the well into segments, preferring that there is only one significant minimum or maximum temperature feature in each segment. Gradient methods can invert temperature to flow profile without being trapped in local minimizations.

The complexity of using DTS to diagnose flow problem makes the interpretation case-dependent. Depending on the problem that needs to be solved, the reservoir model can be different. The wellbore model presented in this section is a general model, but the reservoir model should be built based on the individual cases. The following section shows some applications of using DTS to diagnose production/injection problems.

6.5 Applications of DTS Technology

The examples in this section illustrate how to set up the reservoir models for different problems. Each example uses the same wellbore model presented in the previous section, but builds different reservoir models based on the problems that need to be solved. It also illustrates the procedure of how to apply the technology for well performance diagnosis.

6.5.1 Flow Profiling and Water Entry Identification in Vertical Gas Wells

Multiple-zone, vertical gas-producing wells are good candidates for DTS applications. Gas inflow brings significant cooling effects, and vertical wells bear enough temperature difference along the well because of geothermal gradient. The reservoir temperature equation is modified from Eq. (6.24). For high-rate wells, the volumetric fraction of liquid phase can be small. To allocate production in gas producing wells under stable conditions, we assume single-phase, steady-state flow in the reservoir, and all of the derivative terms respect to time become zero. We also can assume 1D heat transfer in the direction of radial flow to the wellbore. The reservoir thermal equation for this case reduces to

$$\rho C_p \frac{dT}{dr} + (1-\beta T)\frac{dp}{dx} - K_{Tt}\frac{d^2T}{dr^2} = 0 \tag{6.37}$$

Because the cooling effect by gas expansion dominates the heat transfer for low water-cut wells, for simplification, when identifying the water entry locations, we can simulate single-phase gas flow in both the reservoir and wellbore, and compare the generated temperature profile with the measured temperature. When water enters the wellbore, it increases the pressure gradient and also reduces the cooling effect caused by gas inflow.

We use a field example presented by Achinivu et al. (2008) to show how to use temperature measurements to allocate gas flow rate and to identify water entry into a gas producing well.

The well is drilled in a multilayer formation with four perforated zones. The permeability of the reservoir varies along the wellbore ranging from 1 to 680 mD. The well is deviated with a 30° inclination angle. Figure 6.6 shows the well structure and perforation locations. When the well was first put on production, only gas was produced. Several months later, water production started. Downhole temperature, pressure, and surface total flow rate data are used here for two goals: (1) allocate gas inflow along the well, and (2) identify water entry locations after water breakthrough.

A single-phase steady-state reservoir inflow model is used to calculate pressure and gas flow rate for the slanted well, and a negative skin factor is used to count for well deviation. Each perforation zone is treated as individual producing zones, and there is no interference between them. Because perforation zones 3 and 4 are close to each other (see Fig. 6.6), they are grouped as one zone. Average permeability in each perforation zone is the inversion parameter for this problem. Once the permeability distribution is established by the downhole temperature and pressure measurements, the gas inflow distribution is confirmed. The additional reservoir properties used is listed in Table 6.1. The thermal properties used to solve the above equation are also listed.

In addition to the measured temperature, pressure, and surface total flow rate data, we also have the flow rate distribution estimated by production logging to confirm the DTS-inverted production profile.

The interpretation procedure starts from assuming that all zones have the same permeability, and calculating temperature, pressure, and flow rate by the mass, momentum, and energy balance equations

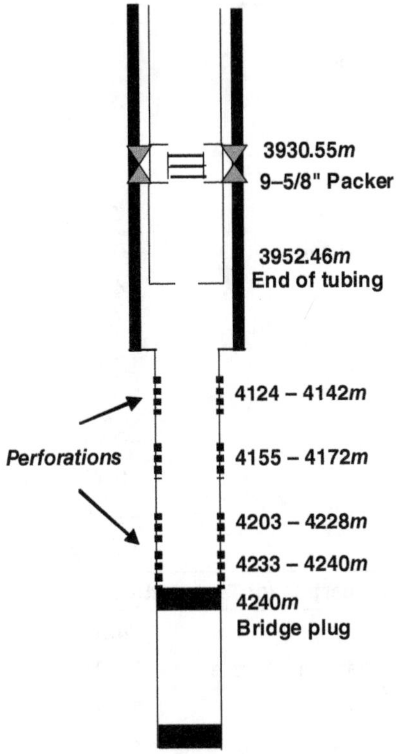

FIGURE 6.6 Well schematic for gas inflow example (Achinivu et al., 2008).

(Eqs. (6.19), (6.20), and (6.37)). Then we use the inversion method (Eq. (6.35)) to match the measured temperature and pressure by changing the permeability distribution among the zones. In each iteration, when the permeability changes, the flow rate distribution changes, resulting in temperature and pressure changes. When the difference between the measured and calculated temperatures is small enough (evaluated by how small the objective function of Eq. (6.35) is), the permeability is close to the real permeability and flow distribution is generated.

Figure 6.7 shows the measured temperature and pressure along the wellbore. The water-gas ratio in this case is 0.04 bbl/MCF; too low to be detected by PLTs. On the temperature plot (Fig. 6.7*a*), the temperature is much higher than the geothermal temperature at the bottom of the well (perforation zones 3 and 4). This is an indication that water production reduced the Joule-Thomson cooling effect of gas production; and also, most likely, the water is from the bottom aquifer with higher temperature. The pressure plot (Fig. 6.7*b*) shows a slope change at the bottom zone. Because heavier phase

Specific gravity of gas	0.88	—
Water salinity (weight percent)	5	%
Pipe diameter	0.15	m
Pipe roughness	0.0006	—
Casing OD	0.168	m
Cement OD	0.216	m
Inclination of pipe from vertical	30	deg
Reservoir pressure	4486	psia
Drainage radius	600	m
Thickness of reservoir	120	m
Slanted borehole skin		—
zone 1	−0.91	
zone 2	−0.78	
zones 3 and 4	−1.01	
Reservoir permeability	1.4–682	mD
Thermal Properties for Temperature Calculation		
Heat capacity of gas	2456	J/kg·K
Coefficient of thermal expansion of water	0.0005	/K
Heat conductivity of gas	0.02	W/m·K
Thermal conductivity of casing material	17	W/m·K
Thermal conductivity of cement	1.7	W/m·K
Reservoir temperature	135	°C
Geothermal gradient	0.031	°C/m
Thermal conductivity of water and matrix	4.33	W/m·K
Thermal conductivity of gas and matrix	2.25	W/m·K

TABLE 6.1 Reservoir Data for Gas Inflow Example

has higher hydrostatic pressure gradient, this supports the idea that water enters the wellbore at the bottom of the well. With this initial analysis, we assign water entry at zones 3 and 4, and interpret temperature (Fig. 6.7*a*) and pressure (Fig. 6.7*b*) with iterations until they match the measured data well. The interpreted flow rate profile shows a satisfactory match compared with the production logging data (Fig. 6.7*c*).

The advantage of using a permanent monitoring system compared to running a PLT is that the permanent monitoring system allows continuous measurements of downhole flow conditions without disturbing well production.

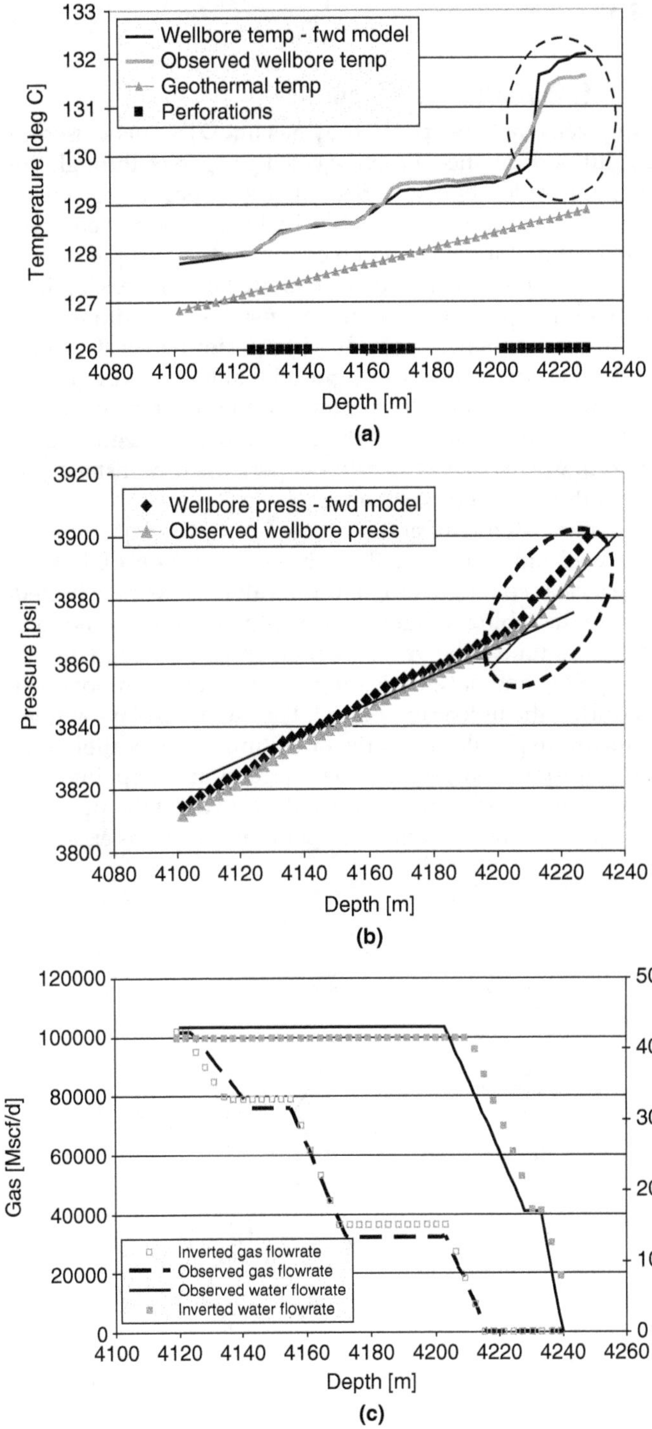

FIGURE 6.7 Water entry at the bottom of the well. (*a*) Temperature predicted, (*b*) pressure predicted, (*c*) flow distribution (Achinivu et al., 2008).

6.5.2 Gas Lift Performance Diagnosis

For producing wells that are operated by gas lift, DTS can be used to diagnose gas lift performance. When injected gas passes through the gas lift valves, the Joule-Thomson effect causes the gas to cool, creating anomalies on the temperature profile. By DTS measurements, we can qualitatively monitor if gas lift valves are working properly.

We use an example of an offshore well located in United Kingdom Continental Shelf in this case. Gas lift was installed for both initiating production and enhancing production performance (Costello et al., 2012). The reservoir produces light oil that has an API gravity of 35. Gas lift design has three lift valves; the top two for unloading the well to initiate production, and the bottom one for regular production (operating valve). A DTS cable is permanently installed in the well to monitor gas lift performance.

Figure 6.8 shows the temperature recorded by DTS at different times during well unloading. At the top valve location of 1200-m depth, we see a clear cooling effect, and the valve works as expected to inject gas into the tubing to lighten the wellbore fluid. On the contrary, at the second unloading valve location of about 2000-m depth, there was no cooling anomaly, indicating that the valve did not open as designed during the unloading period. Figure 6.9 displays a more detailed temperature profile across the operating valve location during the regular production period. Gas injection at this point is to lower the producing fluid density and reduce hydrostatic pressure drop in the wellbore. When passing the operating valve, gas expands,

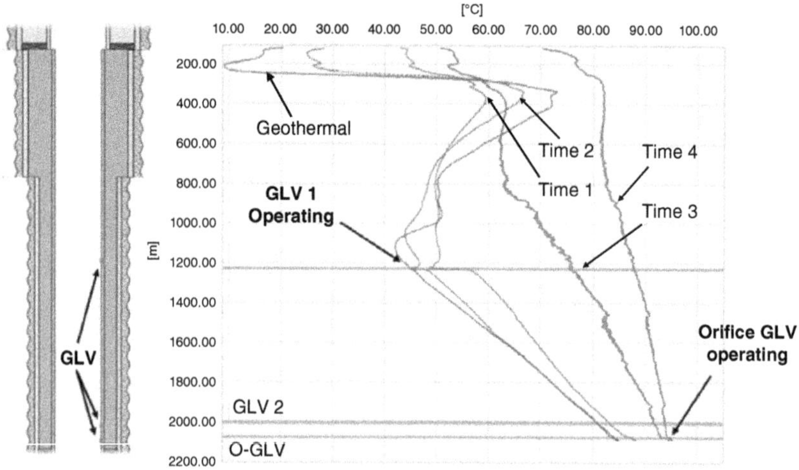

FIGURE 6.8 Temperature record during well unloading for monitoring unloading valve performance (Costello et al., 2012).

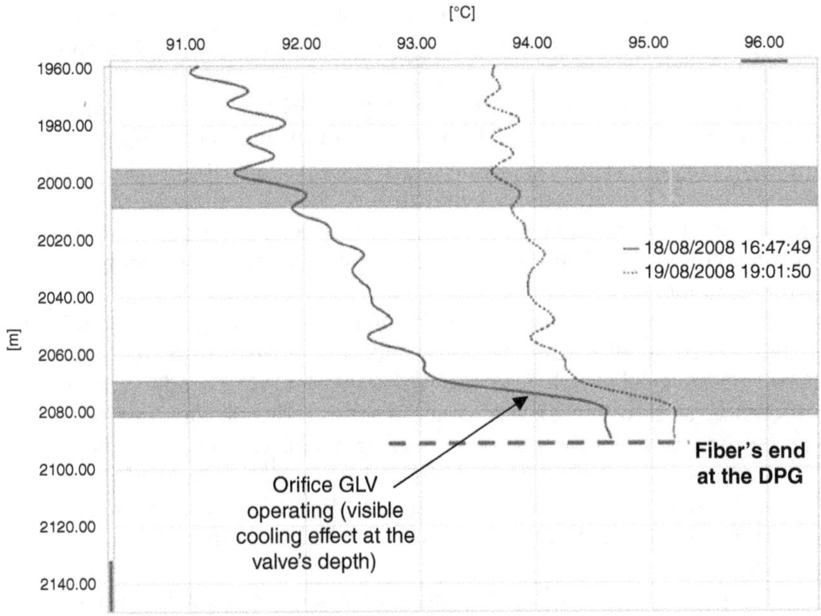

FIGURE 6.9 Temperature record during production to monitor the operating valve performance (Costello et al., 2012).

and the temperature decreases. When reaching the stable condition, temperature decrease caused by the Joule-Thomson effect is less than 1°C. This is enough temperature difference for DTS to detect.

Using DTS to monitor gas lift valves is straightforward. Even though the diagnosis is only qualitative, it brings sufficient information for well operations monitoring.

The same principle can also be used for monitoring electric submersible pumps (ESPs). When ESPs are malfunctioning, it is likely associated with heating, and temperature measurements are a good indicator for early problem identification.

6.5.3 Integrated Reservoir Characterization

History matching of hydrocarbon and water production is a standard method to characterize reservoir properties. When planning field development, we often start from a geological model that is based on seismic investigation, openhole log interpretation and PVT lab measurements from exploration wells. As production data are accumulated, reservoir models are updated and modified by running reservoir simulations with different realizations of reservoir permeability profile against the production history. The permeability realizations that provide satisfactory matching with the production history is considered as close to the permeability distribution in the reality.

Because the large scale of parameters involved, history matching is a statistical procedure, and the solutions are not unique, and likely inexact.

When temperature measurements by DTS are available, the measurements provide an additional constraint that can be used in history matching (Duru and Horne, 2010; Li et al., 2010). DTS measurements are continuous in time but are 1D local measurements along the wellbore. The information serves as a boundary condition for reservoir thermal models requiring the reservoir- and-wellbore coupled model.

As discussed in Section 6.4, the wellbore model (Eqs. (6.14) and (6.15)) can be applied directly in reservoir characterization. The reservoir flow model should be a full-scale, 3D numerical model. Finite difference, streamline, or any other reservoir simulation methods can be used to solve the reservoir flow problem. The reservoir temperature profile is calculated by solving Eq. (6.23) numerically. In the grids that contain a wellbore segment, Eq. (6.26) is used to calculate the arriving temperature required by Eq. (6.25) if the sensor is installed inside the casing.

The procedure of using DTS to assist reservoir characterization is as follows:

1. Grid the wellbore based on well structure, well completion and anomalies on temperature measurements possibility associated with flow.
2. Use the temperature model to generate a coarse permeability distribution by matching the wellbore temperature with measured temperature by DTS. Because the measured temperature is 1D along the wellbore, the estimated permeability is also 1D.
3. Use the coarse permeability distribution as a guideline to build full-scale reservoir geological model. This gives a permeability realization for history matching.
4. With the geological model, use reservoir simulation model to match production history.
5. Use the permeability and flow distribution generated from step 4 to check wellbore temperature. If the temperature converges, the procedure is completed. Otherwise go back to step 2 and iterate again.

Even though the procedure is computationally intensive and highly complex, it is a powerful tool. We use a simple case of a horizontal injector and a horizontal producer to illustrate the application of this procedure. The producing well has a total oil production of 6000 bbl/day. Figure 6.10*a* shows the permeability contour of the example. There is a high-permeability region at 2000 to 2500 ft

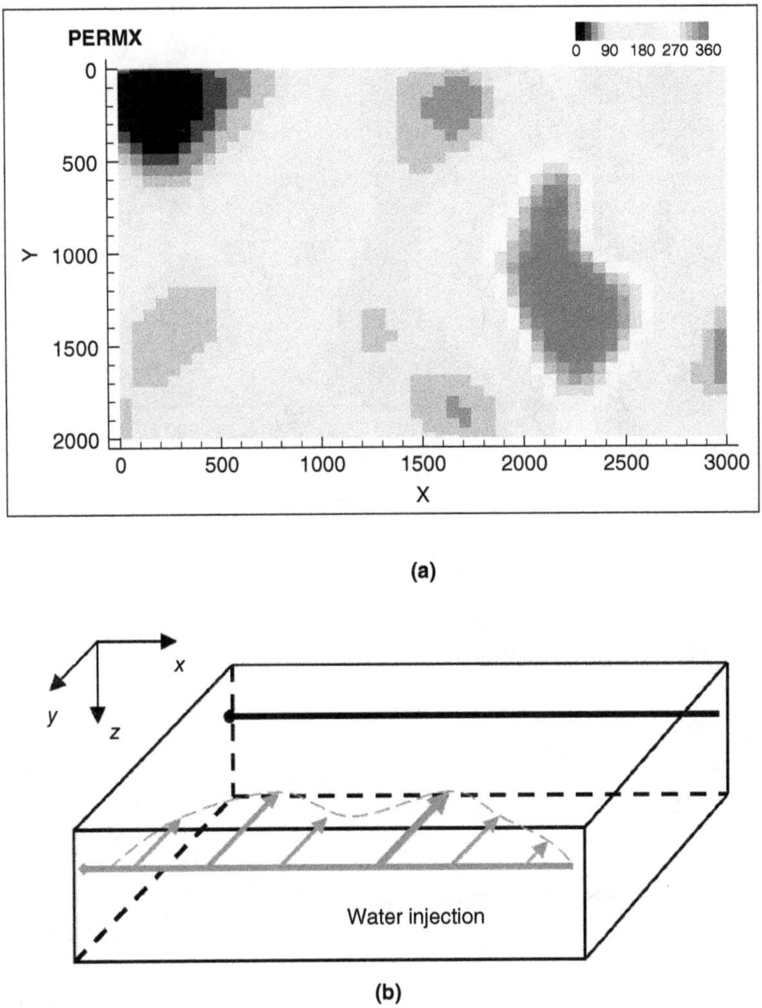

FIGURE 6.10 Example illustrating the procedure of temperature-assisted reservoir characterization (*a*) permeability contour and (*b*) physical model of the example.

location in the *x* direction. Figure 6.10*b* is the physical model used in the problem. In real life, the information that are available would be well rates measured at the surface (Fig. 6.11*a*, the total oil and gas rate and water-cut) and the downhole temperature and pressure measurements (Fig. 6.11*b*). To illustrate the procedure, we start with the known permeability realization (Fig. 6.10*a*) and simulate the temperature and flow distribution along the horizontal well, as shown in Fig. 6.11, using the forward model. These calculation results serve as

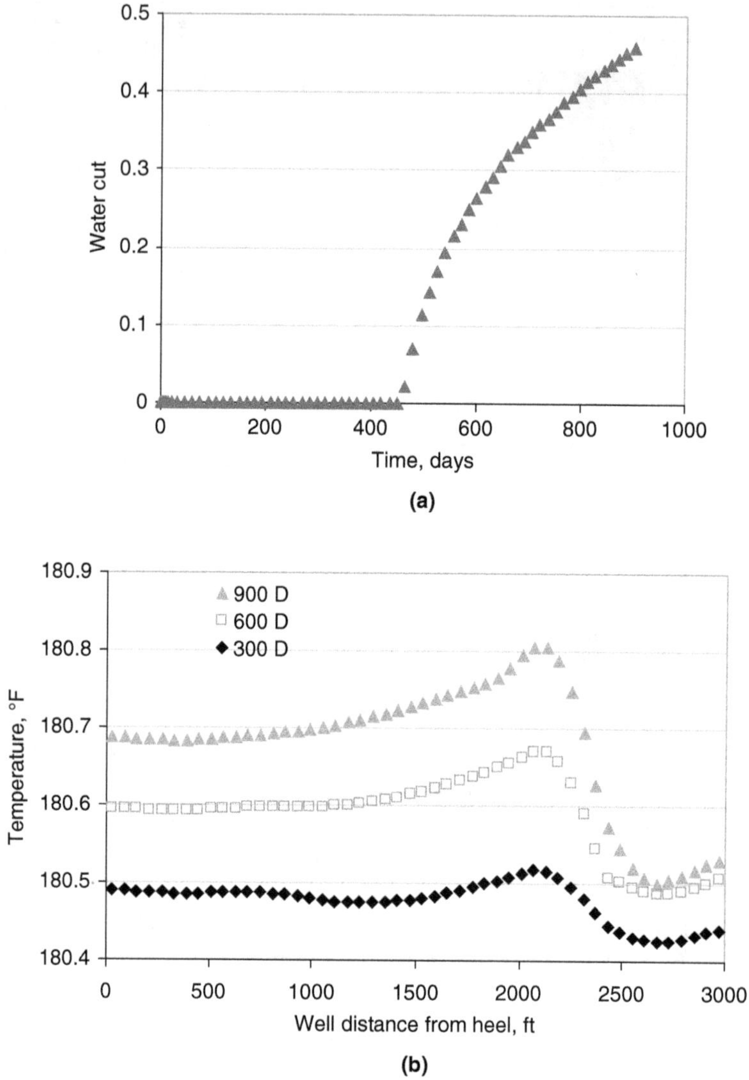

FIGURE 6.11 Simulated horizontal well water-cut history (*a*) and downhole temperature distribution (*b*).

the measured data in the example. The goal here is to use the water-cut history (Fig. 6.11*a*) and the measured temperature profiles at different time (Fig. 6.11*b*) to regenerate the permeability realization (Fig. 6.10*a*) and production distribution along the horizontal wellbore. The input data used in the example is listed in Table 6.2.

By conventional history matching, because of nonuniqueness, many permeability realizations can yield satisfactory match of water-cut.

Reservoir and Wellbore Data			
Reservoir		**Production Wellbore**	
Drainage area, ft × ft	3000 × 2000	Depth of the wellbore, ft	6025
Thickness, ft	50	Length of wellbore, ft	3000
Depth of top, ft	6000	Liquid production rate, bbl/d	6000
Porosity	0.2	Wellbore diameter, ft	0.75
Initial temperature ft, °F	180	Tubing roughness	0.0001
Geothermal gradient, °F/ft	0.01	Tubing diameter, in.	4.5

Fluid Data			
	Oil	**Water**	**Rock**
Density (lb/ft^3)	50	63	165
Viscosity (cp)	1	0.48	
Heat capacity (Btu/lb-°F)	0.516	1.002	0.22
Thermal expansion coefficient × 10^4 (1/°F)	6.79	3.11	
Total thermal conductivity, K_{Tt} (Btu/h·ft·F)		2.0	

Table 6.2 Input Data for Integrated Reservoir Characterization

Figure 6.12*a* and 6.12*b* show two of such permeability fields that provide good matches of water cut (Fig. 6.12*c*), but the predicted temperature deviates from the observed temperature significantly (Fig. 6.12*d*). If we match the production history only, even though the error in history matching is insignificant, the difference in the permeability contour can be critical. Assume that we design inflow control to restrict the reservoir flow at high-permeability intervals according to Fig. 6.11*a* and *b*, these matched permeability contours can lead to incorrect decisions on in which zone(s) the inflow should be shut off or choked back.

Following the procedure discussed above, we use the measured temperature as the first step in the matching. The temperature profile shows a clear anomaly at the high-permeability location (Fig. 6.11*b*), identifying the possible high-permeability region. This information is then used as a guidance to generate the permeability distribution in the reservoir. The final permeability distribution with satisfied matching both in water-cut history and in temperature is shown in Fig. 6.13. Figure 6.13*a* and 6.13*b* are matching results for water-cut and

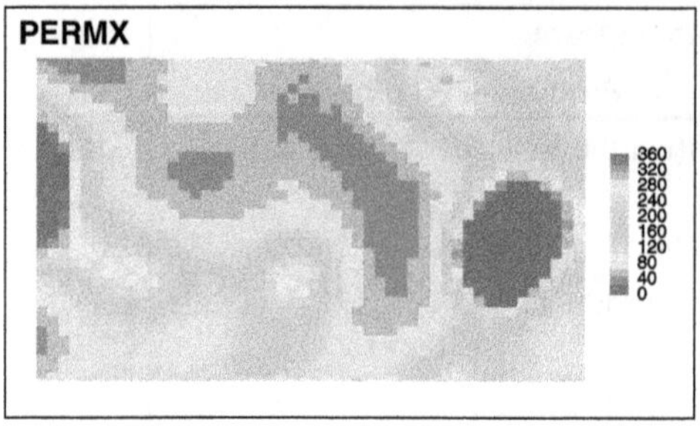

(a)

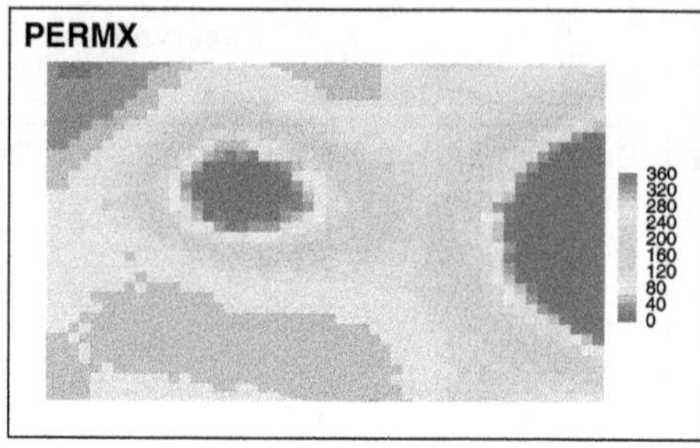

(b)

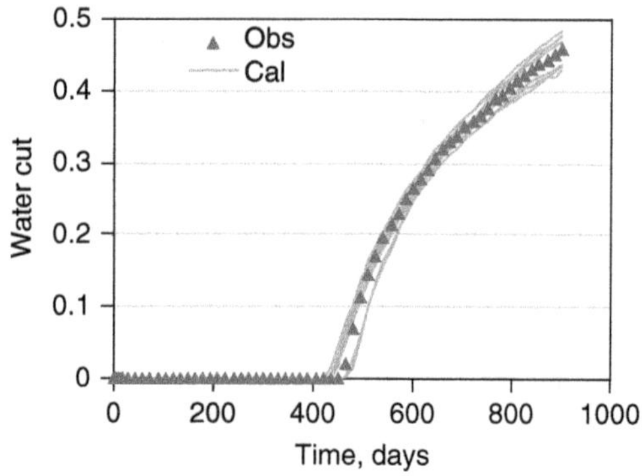

(c)

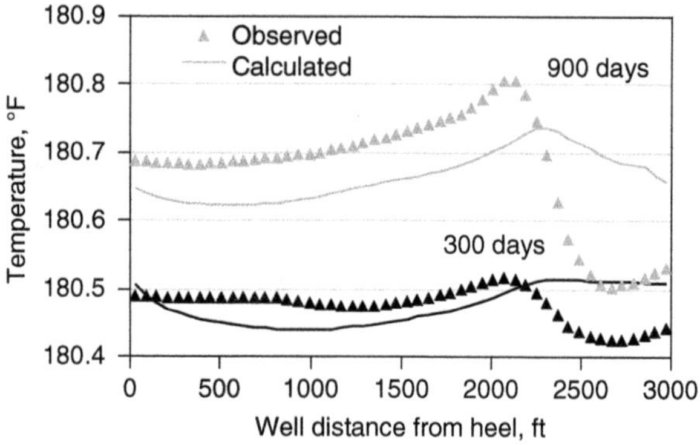

(d)

FIGURE 6.12 Permeability contour generated by history matching (*a*) realization 1, (*b*) realization 2, (*c*) satisfied match of water cut, (*d*) unsatisfied match of temperature.

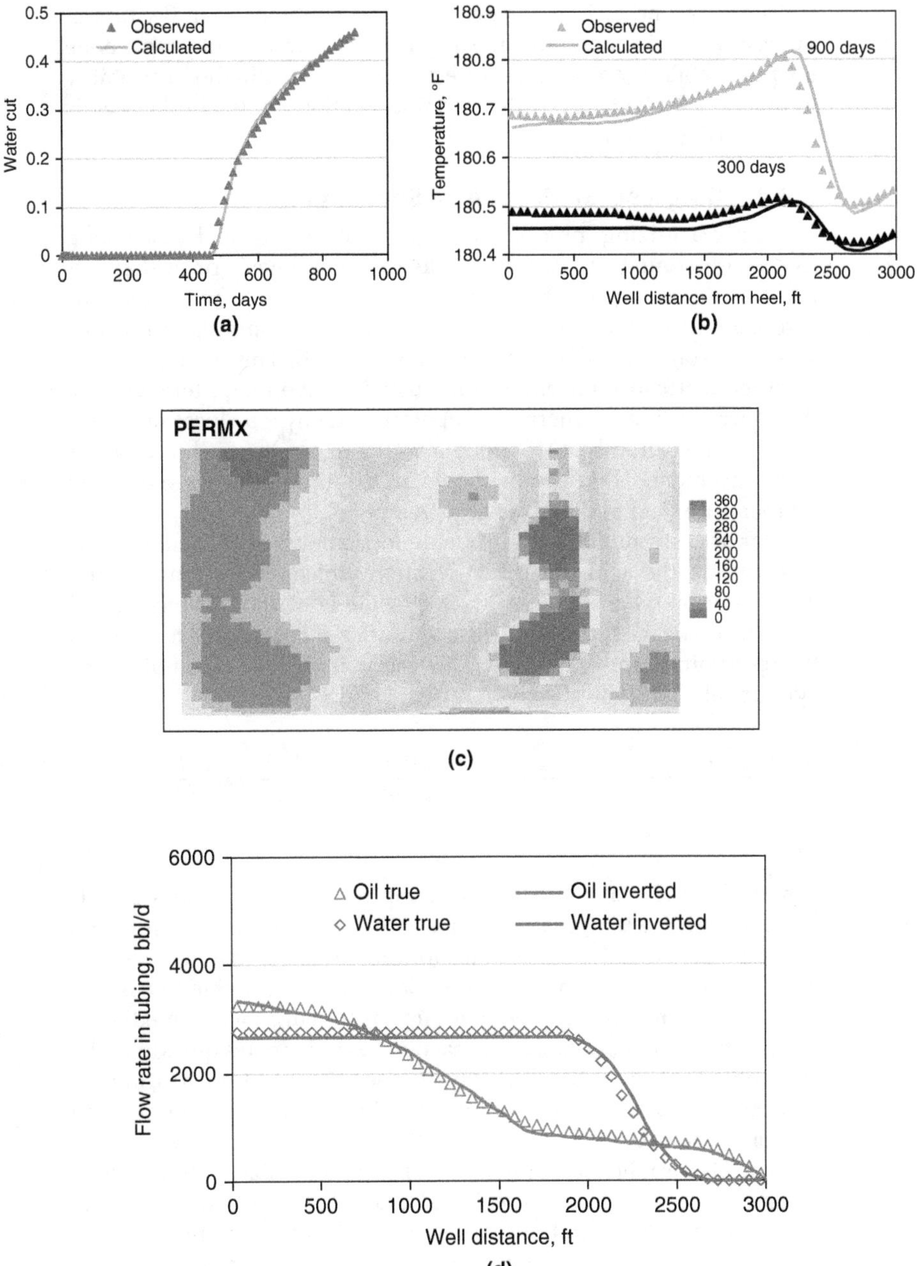

FIGURE 6.13 Integrated match results of water-cut (*a*), temperature (*b*), and inverted permeability field (*c*) and flow distribution (*d*).

temperature, and 6.13*c* is the generated permeability field. Because of the assistance of temperature measurements we successfully regenerate the permeability contour that correctly identifies the high-permeability location. With the additional constraint, the flow profile of the well is inverted (Fig. 6.13*d*).

6.5.4 Diagnosis of Matrix Acid Stimulation

In matrix acidizing, one of the major problems is acid placement. Even with some fluid diversion methods employed for acid placement, acid coverage is often poor, especially for highly heterogeneous carbonate formations. This leads to suboptimum stimulation results where, in some cases, re-stimulation is required. This is discussed in Chapter 5. Because injection fluids usually have lower temperature than surrounding geothermal temperature at payzone locations, temperature measured by DTS along a wellbore can be used to monitor acid placement. This section shows an example of using measured temperature data to diagnose acid treatment.

For acid stimulation in carbonate formations, heat can be generated during the acid (i.e., HCl) reaction which affects temperature interpretation (Glasbergen et al., 2009). Tan et al. (2012) investigated the reaction heat effect by adding an additional heat source term, R_i, to the reservoir thermal equation. The reservoir thermal equation in the cylindrical coordinates is given by

$$\frac{\partial(\rho_s \phi C_{ps} T)}{\partial t} + \frac{\partial[\rho_R (1-\phi) C_{pR} T]}{\partial t} = -\frac{1}{r}\frac{\partial(r\rho_s u C_{ps} T)}{\partial r} + \frac{1}{r}\frac{\partial}{\partial r}\left(rK_T \frac{\partial T}{\partial r}\right) + R_i \tag{6.38}$$

Tan et al. (2012) calculated the reaction heat to be around 4 to 7°F. Even though the temperature increase by reaction heat is significant, the reaction heat only occurs at the location where acid and carbonate minerals react. For carbonate formations, it is reasonable to assume that this is at the tips of wormholes unless flowing back immediately with sufficient rate. Because the reaction heat dissipates quickly in the surrounding formation, we may not see the reaction heating on DTS measurements when monitoring the temperature behavior in carbonate acidizing.

The DTS application of diagnosing matrix acidizing stimulations is illustrated through a vertical well example (Tan et al., 2012). The well was completed and perforated at two locations along the wellbore. The well was constructed with 9⅝-in casing cemented in place to the depth of 5800 ft, and a 7-in casing extended from 5638 ft depth, as shown in Fig. 6.14. There is a 4½-in tubing from the surface to the same depth of the 9⅝-in casing. This construction brings a complication when calculating overall heat transfer coefficient. The input used in the example is listed in Table 6.3.

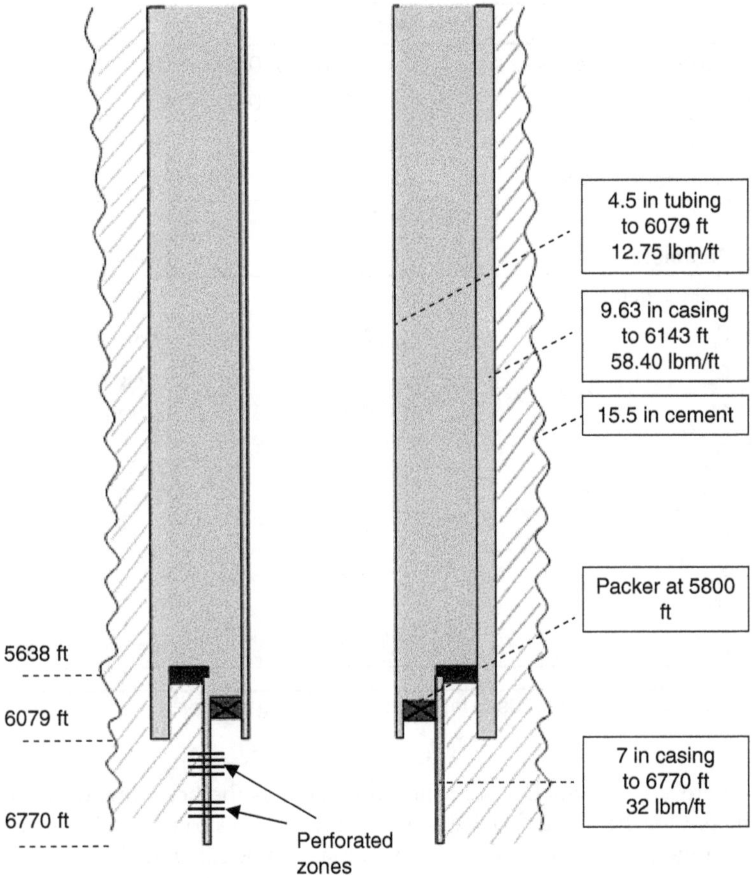

FIGURE 6.14 Well structure for acidizing monitoring example (Tan et al., 2012).

The acid treatment has two stages: an acid wash stage and a main acid stage. Acid was injected through a coiled tubing, which is tailed at the treatment location (6160 to 6370 ft and 6400 to 6500 ft). When calculating the overall heat transfer coefficient, the thermal property of each layer of material is combined. For example, at the interval from 6979 to 6770 ft, we have, from the center of the well outward, the coiled tubing, the casing and cement between the casing and the surrounding formation. There are injection fluids inside both the coiled tubing and the annulus between the coiled tubing and the casing. The overall heat transfer coefficient is (McAdams, 1942),

$$\frac{1}{U}=\frac{A_1}{h_1A_1}+\frac{x_1A_1}{\lambda_s\overline{A}_1}+\frac{A_1}{h_2A_2}+\frac{A_1}{h_3A_3}+\frac{x_3A_1}{\lambda_s\overline{A}_3}+\frac{x_4A_1}{\lambda_s\overline{A}_3}+\frac{x_4A_1}{\lambda_c\overline{A}_4} \tag{6.39}$$

Well Information		
Perforation intervals	6160–6370	ft
	6400–6500	ft
Bottomhole temperature	214	°F
Surface temperature	104	°F
Thermal Properties		
Parameter	**Value**	**Unit**
HCl concentration	20%	Weight fraction
C_{pR}	1040	J/(kg·K)
C_{ps}	4186.8	J/(kg·K)
M_R	0.1	kg/mol
Q_{reac}	4855	J/(mol·HCl)
ρ_R	2710	kg/m³
ρ_s	1080	kg/m³
$V_{i\text{-opt}}$	0.9	cm/min
$PV_{bt\text{-opt}}$	0.95	fraction
ϕ	0.2	fraction

TABLE 6.3 Parameters Used in Acidizing Monitoring Example

where x_i is the thickness of the ith layer, h_i is the heat transfer coefficient for the ith surface, A_i is the area of the ith surface, and $\overline{A}_i$ is the average area of the ith layer. These areas can be calculated as

$$A_i = 2\pi r_i D_{\text{total}} \tag{6.40}$$

and

$$\overline{A_i} = \frac{A_{i+1} - A_i}{\ln(A_{i+1}/A_i)} \tag{6.41}$$

where D_{total} is the total depth of the wellbore. The overall heat transfer coefficient for the entire wellbore is listed in Table 6.4.

	During Injection	**During Shut-In**
0–5638 ft	19.7 Btu/(h·ft²·°F)	14.2 Btu/(h·ft²·°F)
5638–6079 ft	14.6 Btu/(h·ft²·°F)	11.3 Btu/(h·ft²·°F)
6079–6770 ft	18.6 Btu/(h·ft²·°F)	13.7 Btu/(h·ft²·°F)

TABLE 6.4 Overall Heat Transfer Coefficients

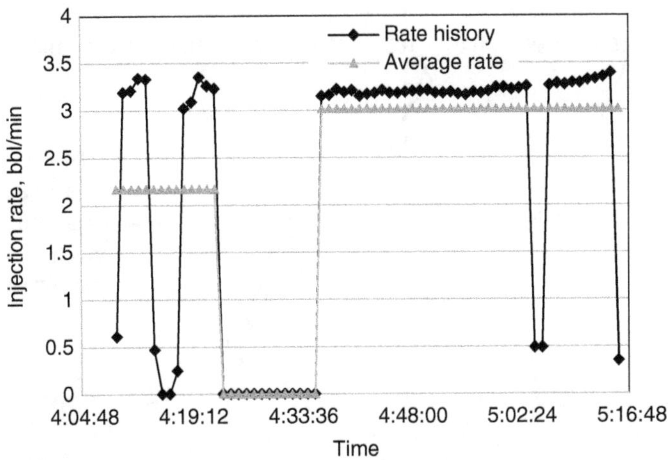

FIGURE 6.15 Injection rate and estimation of acid temperature at the end of the injection for the main acid stage (Tan et al., 2012).

During acid injection, the flow rate was recorded, and is shown in Fig. 6.15. The total acid volume used in this treatment is 156 barrels, and injection lasted 70 min (from 4:05 to 5:15). Notice that the injection rate was fluctuating instead of being a constant. An average rate was used to calculate the wellbore temperature during injection by Ramey's equation (Ramey, 1962),

$$\begin{aligned} T_w(D,t) &= g_G D + T_b - g_G Z + [T_i(t) + g_G Z - T_b]e^{-D/Z} T_w(D,t) \\ &= g_G D + T_b - g_G Z + [T_i] \end{aligned} \tag{6.42}$$

and

$$Z = \frac{wC_{pf}[K + f(t)r_1 U]}{2\pi\lambda r_1 U} \tag{6.43}$$

where T_w is wellbore temperature during injection, g_G is geothermal gradient, D is depth, T_b is geothermal temperature at the depth of zero, $T_i(t)$ is surface injection temperature, w is mass flow rate, C_{pf} is fluid heat capacity, K is thermal conductivity of the formation, r_1 is inner radius of the tubing, U is the overall heat transfer coefficient for the completion, and $f(t)$ is a time function.

During the shut-in period, the wellbore thermal model is simple. Since there is no convection heat transfer by flow nor reaction, the energy balance equation, Eq. (6.38), reduces to

$$\frac{\partial(\rho_s \phi C_{ps} T)}{\partial t} + \frac{\partial[\rho_R(1-\phi)C_{pR}T]}{\partial t} = \frac{1}{r}\frac{\partial}{\partial r}\left(rK\frac{\partial T}{\partial r}\right) \tag{6.44}$$

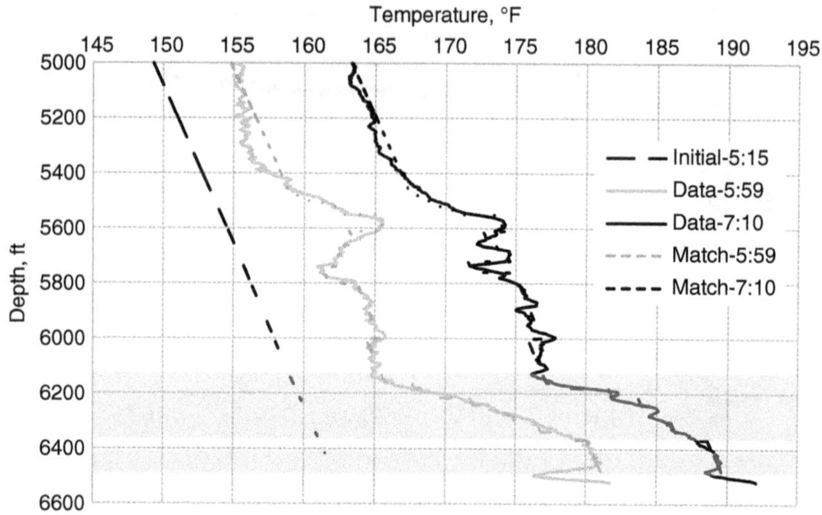

FIGURE 6.16 Temperature matching for multiple zone acid treatment (Tan et al., 2012).

The temperature at 44 min after shut-in (at 5:59) and 115 min (at 7:10) after shut-in is used for treatment diagnosis (Fig. 6.16). The two perforated zones are shaded grey. The temperature behavior is summarized using the temperature at 6200 ft to represent the top-zone temperature, and at 6400 ft for the bottom-zone temperature. In the top zone, the temperature increased 10°F (from 160 to 170°F) during the first 41 min of shut-in and 15°F during the next 71 min of shut-in, from 170 to 185°F. For the bottom zone, the temperature increased about 17°F during the first 41 min of shut-in, and only 8°F during the next 71 min. This inverse temperature behavior in the two zones is caused by the reaction heat effect. The zone that takes more acid has more cold fluid enter the formation, but also has higher reaction heat at the tip of the wormhole, compared with the zone taking less acid. When flowing back, later warm-back is caused by the reaction heat reaching the wellbore. This effect may not be observed on DTS measurements if the wormhole penetrates deeply into the formation because reaction heating is diminishing while the fluid flows back to the wellbore.

Notice that there is an anomalous cooling around the depth of 5600 to 5800 ft. Since there is no perforation, it is assumed that the cement has a channel outside the casing, and acid leaks upward and reacts with the rock along the wellbore. With this initial analysis, the acid distribution is inverted from the measured temperature. Figure 6.16 also shows the model-matched temperature. The acid distribution with satisfactory match is shown in Table 6.5. The result shows that the majority of the injected acid went upward because of

Section	Acid Percentage
5400–5800 ft	26%
5800–6160 ft	42%
6160–6370 ft	12%
6400–6500 ft	20%

Table 6.5 Acid Distribution from the Model for the Main Acid Stage

the channel behind the casing. The two payzones only received total of 32% of the injected acid.

Similar to the application of reservoir characterization, using temperature to diagnose acid stimulation also has high uncertainty and the solution is nonunique. To confirm the interpretation, skin analysis before and after acid stimulation by standard well testing procedures can be used to confirm the interpretation.

6.5.5 Fracture Diagnosis by DTS Measurements

Temperature measurements from PLTs have been used for fracture height estimate before DTS technology was introduced to the petroleum industry. Compared with production logging, DTS can measure temperature during fracturing process, after shut-in, during flow back and during regular production continuously, which provides more information for fracture diagnosis, especially for multiple-stage fracture stimulation along horizontal wells. The technology has been used for diagnosing injection fluid distribution during hydraulic fracturing treatments (Sierra et al., 2008; Huckabee, 2009; Molenaar et al., 2012). When combined with a fracture propagation model, DTS measurements can also help to estimate fracture geometry (Ugueto et al., 2014). For fractured well performance monitoring, DTS data can be used for estimating a flow profile along wellbore identifying the fractures that contribute to production and estimating fracture deliverability decline as a function of production time.

Modeling temperature behavior during fracturing and during production of fractured wells uses the same physical principles, but in different flowing conditions. We will discuss temperature behaviors experienced during different flow periods in multistage fracturing treatments.

During Fracturing and Warm-Back after Shut-In

During fracturing treatments, fracturing fluids that are usually a lot colder that the reservoir temperature flow through the wellbore at high injection rates. Equation (6.42) can be used to calculate injection temperature at the reservoir depth. Once the fracture is initiated, the cold fluid enters the formation as the fracture propagates. To correctly trace the temperature, a fracture geometry model that calculates fracture half-length as a function of injection time is needed. For the formations that have high leakoff, the fracture propagation velocity decreases significantly, and brings the cold fluid into the formation in the vicinity around the fracture; both of these two phenomena have effects on the temperature behavior. Wellbore temperature during injection by itself does not provide sufficient information for fracture diagnosis. It is almost a constant along the entire wellbore (Ugueto et al., 2014), which is the initial condition for the warm-back period or flow-back period.

After pumping stops, the temperature along the wellbore starts to increase. At locations along the wellbore where fractures were created, the temperature will increase more slowly because of the relatively low-temperature fluids in the fractures. Because the rate of temperature recovery depends on how much cold fluids in the fracture and how long the fracture is, from the temperature record, we can estimate the fracture volume.

To model the flow and temperature for one fracture, the physical domain consists of three elements: the wellbore, the fracture, and the reservoir. The three elements are coupled at the boundary. Figure 6.17 shows the schematic of the system. As illustrated, the problem can be modeled

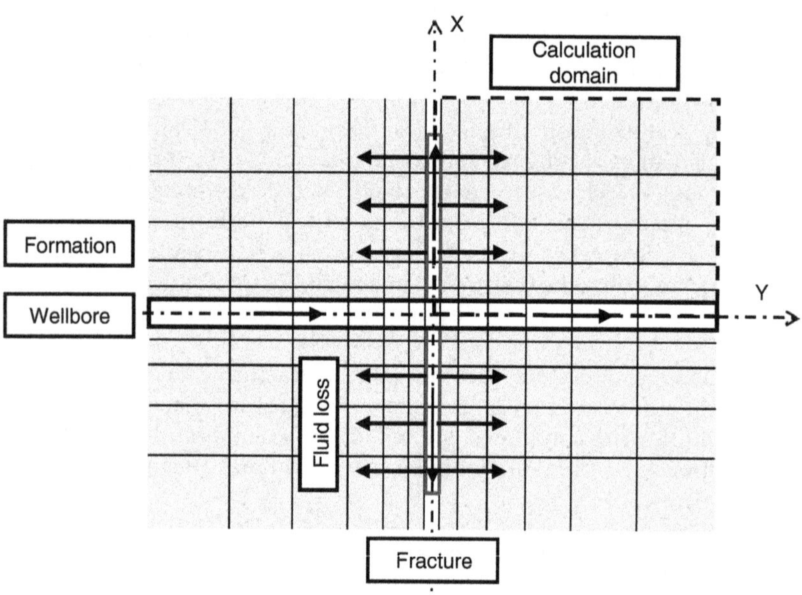

FIGURE 6.17 Flow domain for single fracture problem.

as three linear flow regions. The flow problem during fracturing is a combination of the flow in the wellbore (1D along the wellbore) and the flow in between parallel plates in the fracture with fracture geometry defined by a fracture propagation model. The flow in the reservoir is represented by leakoff from the fracture only. During warm back period, we can assume there is no flow, and only the temperature needs to be modeled.

During injection, the temperature in the wellbore is approximated from the surface injection temperature to the injection depth (Eq. (6.42)). The temperature in the fracture, T_f, is a function of location and time, and can be calculated by the energy balance equation

$$\rho_l \hat{C}_{pl} \frac{\partial T_l}{\partial t} = -\frac{\rho_l \hat{C}_{pl} v_x \partial T_l}{\partial x} - \frac{2\rho_l \hat{C}_{pl} v_{\text{leakoff}} T_l}{w} + \frac{2K_T(T_r - T_l)}{w} \tag{6.45}$$

in the above equation, in addition to the definitions of the parameters mentioned before, K_T is formation thermal conductivity across fracture faces (Zhao and Tso, 1993). Equation (6.45) illustrates that energy accumulation of a differential volume in the fracture over time t is the summation of the heat convection caused by fracturing fluid flow, the convective energy due to fluid loss, and the heat conduction between the fracturing fluid and surrounding formation.

The leakoff velocity in Eq. (6.45), v_{leakoff}, is calculated by

$$v_{\text{leakoff}}(x) = \frac{c_{\text{leakoff}}}{\sqrt{t - \tau(x)}} \tag{6.46}$$

The reservoir thermal model is derived based on the general energy balance equation, and at a local position (x,y), it is

$$\begin{aligned} \overline{\rho \hat{C}_p} \frac{\partial T_r}{\partial t} - \phi \beta T_r \frac{\partial p}{\partial t} &= -\rho_l \hat{C}_{pl} v_{lk} \frac{\partial T_r}{\partial y} + (\beta T_r - 1) v_{lk} \frac{\partial p}{\partial y} \\ &+ \frac{\partial}{\partial x}\left(K_e \frac{\partial T_r}{\partial x}\right) + \frac{\partial}{\partial y}\left(K_e \frac{\partial T_r}{\partial y}\right) \end{aligned} \tag{6.47}$$

where $\overline{\rho \hat{C}_p}$ and K_e are average effective density-heat capacity product and average effective reservoir thermal conductivity, respectively, account for formation fluid and rock. The combination of Eqs. (6.42) (wellbore), (6.45) (fracture), and (6.47) can provide the temperature profile of the interested domain.

Assuming no crossflow occurs during warm back, the wellbore model has the heat conduction only between the reservoir (at non-perforation section), or the fracture (at the perforation locations), and the wellbore,

$$\rho_l \hat{C}_{pl} \frac{\partial T}{\partial t} = \frac{2}{R} U_T (T_{r/f} - T_w) \tag{6.48}$$

The fracture temperature model has the heat conduction between the reservoir and the fracture

$$\rho_l \hat{C}_{pl} \frac{\partial T_f}{\partial t} = \frac{2h_l(T_r - T_f)}{w} \tag{6.49}$$

And the reservoir temperature model is

$$\overline{\rho \hat{C}_p} \frac{\partial T_r}{\partial t} - \phi \beta T_r \frac{\partial p}{\partial t} = \frac{\partial}{\partial x}\left(K_e \frac{\partial T_r}{\partial x}\right) + \frac{\partial}{\partial y}\left(K_e \frac{\partial T_r}{\partial y}\right) \tag{6.50}$$

At the fracture/reservoir and wellbore/reservoir interfaces, we use the following boundary conditions to couple the thermal models,

$$\begin{cases} K_e \dfrac{\partial T_r}{\partial y} = h_l(T_r - T_l), & at\ y = y_{\text{res-frac}}, t \\ K_e \dfrac{\partial T_r}{\partial x} = U_T(T_r - T), & at\ x = x_{\text{res-wb}}, t \end{cases} \tag{6.51}$$

For a multistage fracturing treatment, the problem is modeled consecutively one stage at a time. When pumping of one stage is completed, the stage is isolated and warm back starts. Meanwhile injection moves to the next stage. When modeling temperature behavior, correctly tracking the treatment schedule is important. Figure 6.18 shows the workflow of the modeling procedure. At the moment that the figure is showing, stage 2 is applying the fracture model; stage 1 (the stage upstream toward the toe) is on the warm-back mode; and stage 3 (the stage downstream toward the heel) is on the nonfracture mode, only the wellbore model is needed for heat conduction between the injection fluid and reservoir temperature. The equations used to calculate these three modes are different. For the warm-back mode, there are no convective terms, and heat conduction is between the created fracture and the reservoir, and between the wellbore and the reservoir. For the nonfracture mode, there are no convection terms, and heat conduction is only between the wellbore and the reservoir.

Temperature Behavior during Production

During production, the wellbore models presented before (Eqs (6.14) for flow and (6.15) for thermal) can be directly used. In these equations, γ is one at the fracture locations for open flow, and is zero everywhere else. The flow and thermal equations of the reservoir, Eqs. (6.19) and (6.24), are also applicable. Here we use a semianalytical approach, the tri-linear model, to illustrate how to use temperature data for fracture diagnosis for single-phase wells. Of course, analytical/semianalytical models, finite difference reservoir simulation, streamline reservoir simulation or any other reservoir simulation methods can also be used to obtain the solutions of the flow problem.

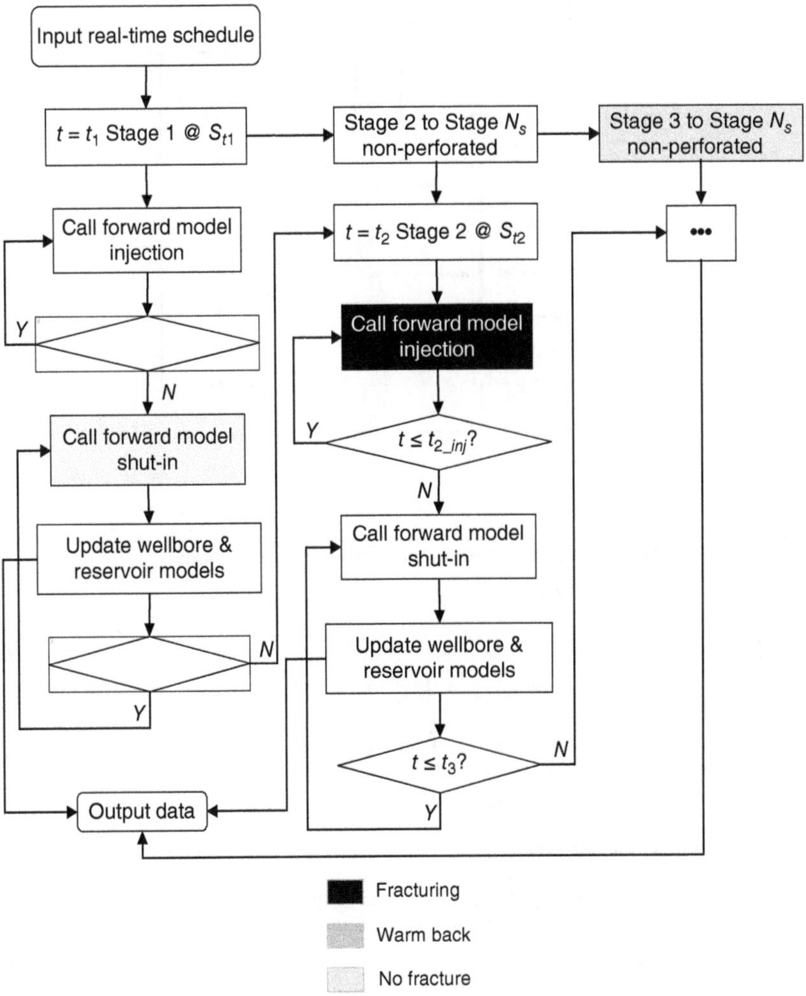

Figure 6.18 Workflow for temperature match during injection.

The tri-linear model assumes that the flow of a fractured well in a reservoir can be simplified as three linear flow regimes combined. This includes linear flow inside the fracture, linear flow from inner reservoir to fracture, and linear flow from outer reservoir to inner reservoir (Meyer and Jacot, 2005), as shown in Fig. 6.19.

The flow equations for the three regions, outer formation, inner formation and fracture, are coupled at the boundaries. The dimensionless format of the flow equation for outer formation is

$$\frac{\partial^2 p_{OD}}{\partial x_D^2} = \frac{1}{\eta_{OD}} \frac{\partial p_{OD}}{\partial t_D}$$

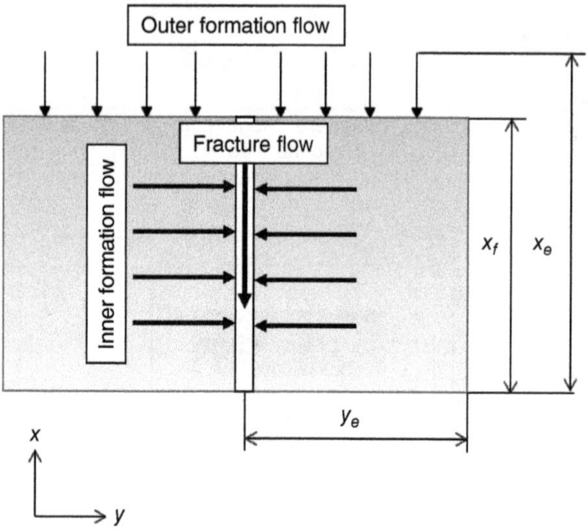

FIGURE 6.19 Schematic of a half of domain for the tri-linear model.

with

$$\begin{cases} (\overline{p}_{OD})_{t_D=0} = 0 \\ (\overline{p}_{OD})_{x_D=1} = (\overline{p}_{ID})_{x_D=1} \\ (\overline{p}_{OD})_{x_D\to\infty} = 0 \end{cases} \tag{6.52}$$

for the inner formation is,

$$\frac{\partial^2 p_{ID}}{\partial y_D^2} + \left(\frac{\partial p_{OD}}{\partial x_D}\frac{\delta y}{\delta x}\right)_{x_D=1} = \frac{\partial p_{ID}}{\partial t_D} \tag{6.53}$$

and for the fracture is,

$$\frac{\partial^2 p_{FD}}{\partial x_D^2} + \left(\frac{2}{F_{CD}}\frac{\partial p_{ID}}{\partial y_D}\right)_{y_D=0} = \frac{1}{\eta_{FD}}\frac{\partial p_{FD}}{\partial t_D} \tag{6.54}$$

Equations (6.52) to (6.54) can be solved by Laplace transforms. In Laplace domain, Eq. (6.52) becomes

$$\frac{\partial^2 \overline{p}_{OD}}{\partial x_D^2} - \frac{l}{\eta_{OD}}\overline{p}_{OD} = 0 \tag{6.55}$$

with the initial and boundary conditions

$$\begin{cases} (\overline{p}_{OD})_{t_D=0} = 0 \\ (\overline{p}_{OD})_{x_D=1} = (\overline{p}_{ID})_{x_D=1} \\ (\overline{p}_{OD})_{x_D\to\infty} = 0 \end{cases} \tag{6.56}$$

The solution of the outer formation in Laplace domain is

$$\bar{p}_{OD} = (\bar{p}_{ID})_{x_D=1} \exp\left(-\sqrt{\frac{l}{\eta_{OD}}}(x_D - 1) \right) \tag{6.57}$$

In the Laplace domain, Eq. (6.51) for the inner formation becomes

$$\frac{\partial^2 \bar{p}_{ID}}{\partial y_D^2} + \left.\frac{\partial \bar{p}_{OD}}{\partial x_D}\right|_{x_D=1} - l\bar{p}_{ID} = 0 \tag{6.58}$$

From Eq. (6.56) the pressure derivative in Eq. (6.57) which connect the outer formation to the inner formation is

$$\left.\frac{\partial \bar{p}_{OD}}{\partial x_D}\right|_{x_D=1} = -\sqrt{\frac{l}{\eta_{OD}}}\, \bar{p}_{ID}\Big|_{x_D=1} \tag{6.59}$$

Assuming a skin factor, s, in the inner formation, the initial and boundary conditions are

$$\begin{cases} (\bar{p}_{ID})_{t_D=0} \\ \bar{p}_{ID} = \bar{p}_{FD} - s\dfrac{\partial p_{ID}}{\partial y_D} \text{ at } y_D = 0 \\ (\bar{p}_{ID})_{y_D \to \infty} = 0 \end{cases} \tag{6.60}$$

If we assume linear flow regime, the pressure gradient in x direction is small compared with the gradient in y direction in inner formation, and can be neglected. The solution for Eq. (6.57) is

$$\bar{p}_{ID} = \frac{\bar{p}_{FD}}{1+\sqrt{\alpha_O}S} \exp(-\sqrt{\alpha_O}\, y_D) \tag{6.61}$$

where $\alpha_O = \sqrt{l/\eta_{OD}} + l$.

In the Laplace domain, the equation for the fracture is

$$\frac{\partial^2 \bar{p}_{FD}}{\partial x_D^2} + \frac{2}{F_{CD}} \left.\frac{\partial \bar{p}_{ID}}{\partial y_D}\right|_{y_D=0} - \frac{l}{\eta_{FD}} \bar{p}_{FD} = 0 \tag{6.62}$$

$$\left.\frac{\partial \bar{p}_{ID}}{\partial y_D}\right|_{y_D=0} = -\frac{\sqrt{\alpha_O}}{1+\sqrt{\alpha_O}S} \bar{p}_{FD} \tag{6.63}$$

The initial and boundary conditions are

$$\begin{aligned} \bar{p}_{FD}\Big|_{t_D=0} &= 0 \\ \left.\frac{\partial \bar{p}_{FD}}{\partial x_D}\right|_{x_D=1} &= 0 \\ \left.\frac{\partial \bar{p}_{FD}}{\partial x_D}\right|_{x_D=0} &= -\frac{\pi}{F_{CD}} \end{aligned} \tag{6.64}$$

The solution for Eq. (6.62) is

$$\bar{p}_{wD}(l) = \frac{\pi / F_{CD}}{l\sqrt{\alpha_F}\tanh\sqrt{\alpha_F}} \tag{6.65}$$

where $\alpha_F = \dfrac{2/F_{CD}\sqrt{\alpha_O}}{1+\sqrt{\alpha_O}S} + \dfrac{l}{\eta_{FD}}$

For constant pressure at the bottomhole

$$\bar{q}_D(l) = \frac{1}{l^2 \bar{p}_{wD}(l)} \tag{6.66}$$

Dimensionless parameters and definitions are shown below:

$$p_D = \frac{141.2kh}{qB\mu}(p_i - p) \qquad \text{(for oil)} \tag{6.67}$$

$$p_D = \frac{kh}{1424qT}(m(p_i) - m(p)) \qquad \text{(for gas)} \tag{6.68}$$

$$\eta = \frac{k}{\varphi\mu c_t} \tag{6.69}$$

$$t_D = \frac{kt}{\varphi\mu c_t x_f^2} = \frac{\eta t}{x_f^2} \tag{6.70}$$

$$x_D = \frac{x}{x_f} \quad x_D = \frac{x}{x_f} \quad y_D = \frac{y}{y_f} \quad F_{CD} = \frac{k_f w}{k x_f} \quad \eta_{FD} = \frac{\eta_F}{\eta} \quad \eta_{OD} = \frac{\eta_O}{\eta} \tag{6.71}$$

Temperature equations are built based on energy conservation. If we ignore the heat transfer in the vertical direction, we can develop a 2D thermal model for fractured wells with three simplified linear flow regions. In the fractures, heat convection dominates over conduction, and the conduction term along x direction can be ignored. The thermal equation for a fracture is

$$\overline{\rho\hat{C}_p}\frac{\partial T}{\partial t} - \varphi\beta T\frac{\partial p}{\partial t} = \frac{k_f\rho\hat{C}_p}{\mu}\frac{\partial p}{\partial x}\frac{\partial T}{\partial x} - \frac{(\beta T - 1)}{\mu}k_f\left(\frac{\partial p}{\partial x}\right)^2 + K_T\frac{\partial^2 T}{\partial y^2} \tag{6.72}$$

For the inner and outer formation, both conduction and convection are considered,

$$\overline{\rho\hat{C}_p}\frac{\partial T}{\partial t} - \varphi\beta T\frac{\partial p}{\partial t} = \frac{k\rho\hat{C}_p}{\mu}\frac{\partial p}{\partial y}\frac{\partial T}{\partial y} - \frac{(\beta T - 1)}{\mu}k\left(\frac{\partial p}{\partial y}\right)^2 + K_T\left(\frac{\partial^2 T}{\partial x^2} + \frac{\partial^2 T}{\partial y^2}\right) \tag{6.73}$$

and for the outer formation

$$\overline{\rho\hat{C}_p}\frac{\partial T}{\partial t} - \varphi\beta T\frac{\partial p}{\partial t} = \frac{k\rho\hat{C}_p}{\mu}\frac{\partial p}{\partial x}\frac{\partial T}{\partial x} - \frac{(\beta T - 1)}{\mu}k\left(\frac{\partial p}{\partial x}\right)^2 + K_T\left(\frac{\partial^2 T}{\partial x^2} + \frac{\partial^2 T}{\partial y^2}\right) \tag{6.74}$$

Even though this is a 2D model, the initial temperature of the entire system is the geothermal temperature calculated corresponding to the wellbore trajectory, which includes the effect of depth on temperature behavior. Because of nonlinearity, the temperature equations need to be solved numerically.

To illustrate the procedure of using measured temperature to interpret flow profile for multi-stage fractured horizontal wells, we use a gas well in the Eagle Ford shale as an example (Cui et al., 2014). The well produces 1600 Mscf/day gas after a multistage fracturing treatment. The well also produces 160 STB/day of water and the surface GOR is about 9000 to 10,000 scf/stb. Since the well is producing under the dew point pressure, there exists retrograde liquid inside the wellbore, which is less than 10% of total volume according to the PVT data. Because the total liquid volume is small, we use the single-phase reservoir model developed in the above section for temperature interpretation. The model neglects the heat transfer caused by phase change. The heat conductivity used in the interpretation is an average value based on the GOR and water cut. With the temperature measurements, we can interpret one set of parameters for the flow problem, and this parameter is the volumetric flow distribution along the well. The fracture stimulation had 15 stages, each stage had four perforation clusters, and the spacing between the clusters is 75 ft. The wellbore trajectory is important in temperature interpretation, especially for two-phase flow wells. The liquid phase tends to accumulate at low spots along the wellbore, and the gas phase accumulates at high spots. A two-phase flow model is used for the wellbore flow model, and the volumetric average heat transfer coefficients are used for the single-phase wellbore thermal model. The wellbore trajectory is shown in Fig. 6.20*a*. The perforation locations are marked along the well trajectory. Figure 6.20*b* shows the geothermal temperature along the horizontal wellbore and the measured temperature as a function of location. All of the other parameters except the volumetric flow rate distribution used for interpretation are shown in Table 6.6.

Formation		Fracture/Wellbore	
Permeability (nD)	583	Fracture spacing (ft)	77
Porosity (%)	4.2	Fracture width (in)	0.24
Pore pressure (psi)	4630	Fracture porosity (%)	25
Reservoir Compressibility (1/psi)	1.68E-04	Fracture permeability (mD)	1250
Gas specific gravity	0.56	Wellbore pressure (psi)	2860

Table 6.6 Input Data in the Field Case for Fracture Diagnosis with Temperature Measurements

The simulated temperature of the well during production is also shown in Fig. 6.20*b*. Because the well is toe-up (Fig. 6.20*a*), the geothermal temperature is higher at the heel. The temperature difference between the toe and the heel is about 1.3°F. Since the temperature change caused by the Joule-Thomson effect is usually smaller than 1°F,

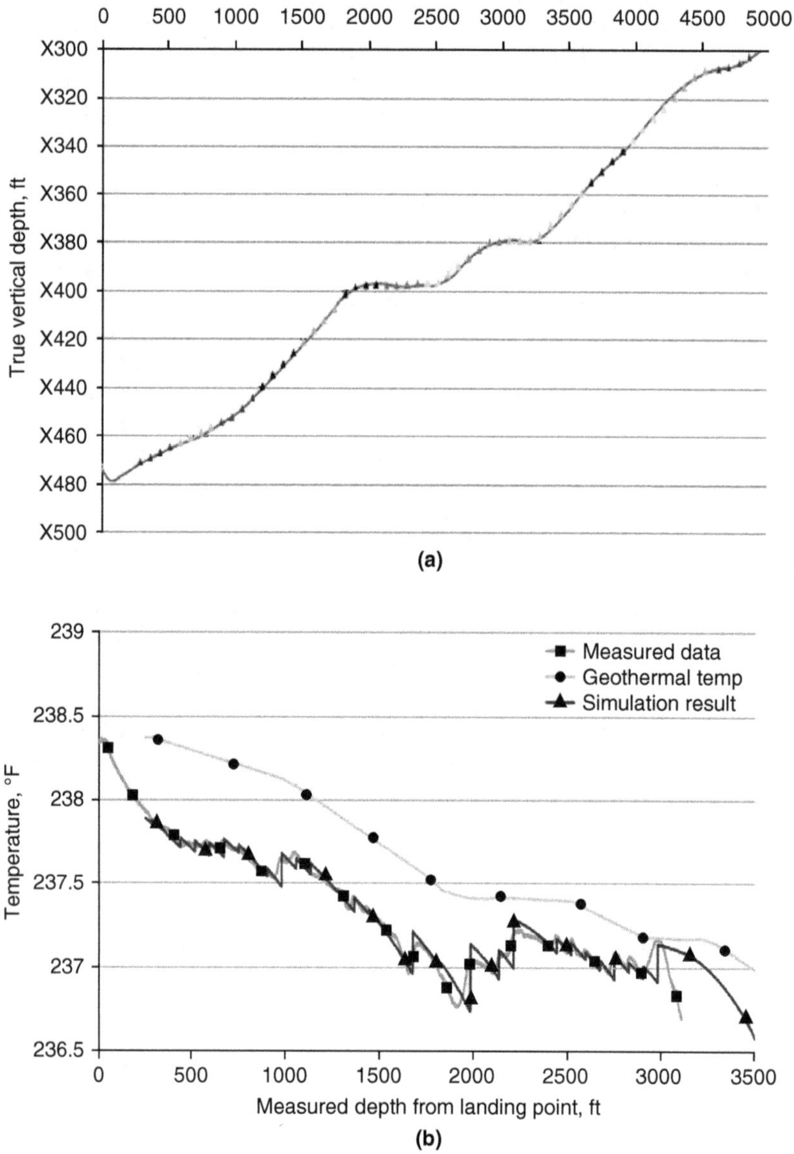

FIGURE 6.20 Simulated and measured temperature data match: (*a*) Wellbore trajectory of the example; (*b*) geothermal, measured, and simulated temperature (Cui et al., 2014).

accurate estimation of geothermal temperature is crucial and should be calibrated carefully before interpreting temperature data. The temperature change at the toe is more obvious compared with the heel. This is because during the fluid mixing procedure, the cooler formation fluid enters into the wellbore and mixes with the warmer fluid inside the wellbore, resulting in a mixed temperature. There is no upstream warmer fluid in the wellbore at the toe, the mixed temperature depends only on the arriving temperature of the fluid from the fracture. As the wellbore fluid flows toward the heel, the total flow rate increases, and the contribution of the fluid from each fracture becomes smaller while the upstream wellbore fluid has stronger influence on the mixture temperature. As the cooling effect is masked by the wellbore flow, the wellbore temperature change becomes smaller toward the heel. It should be pointed out that the mixing procedure takes some distance to stabilize, but the model assumes an instantaneous equilibrium which causes the simulated wellbore temperature to change sharply at the fracture locations (Fig. 6.20*b*).

For a perforated and fractured horizontal well, fractures are generated at the perforation locations. In this example, we assume initially all fractures are equally developed every 77 ft (75-ft spacing plus 2 ft of perforation zone) along the horizontal well. We first use the models to calculate the temperature, pressure, and flow velocity profiles in the system. Then we compare the calculated temperature with the measured temperature, and change the flow distribution along the wellbore to minimize the difference between the two sets of temperatures.

The temperature continuously increases from 1753 to 1984 ft in one fracturing stage. This indicates no gas production in this section, and there were no fractures generated at this stage. The same phenomenon happens between 1444 and 1598 ft. If there is a minor temperature drop in a section instead of a sharp drop, it could be because small fractures with more network pattern are generated instead of one primary fracture.

Using the thermal model developed, we can estimate the flow distribution along the horizontal wellbore. The interpreted flow distribution of this fractured well is shown in Fig. 6.21. This flow distribution can be seen as a product of fracture volume and fracture permeability (or fracture conductivity multiplied by fracture half-length in case a constant height is assumed).

6.5.6 SAGD

For heavy oil production, steam injection is one of the common methods for production enhancement. High-temperature steam injected into the reservoir reduces the reservoir fluid viscosity and increases the mobility of the fluid. The steam-assisted gravity drainage (SAGD)

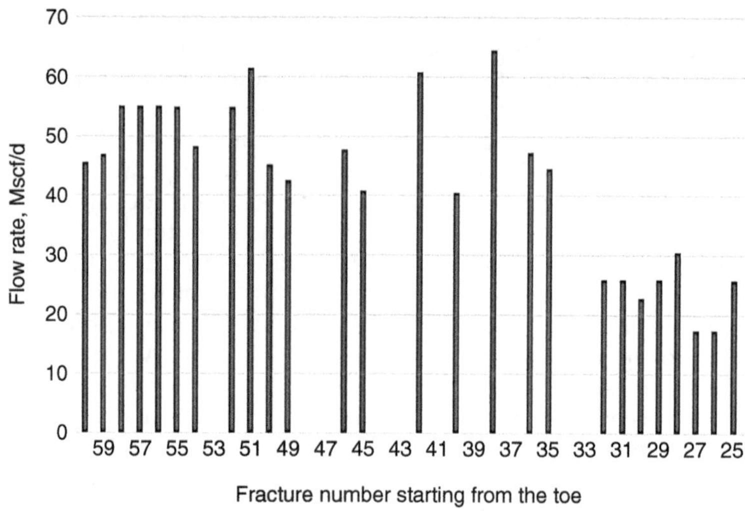

Figure 6.21 Compare a portion of perforation design with temperature data (Cui et al., 2014).

method uses parallel horizontal wellbores, one as steam injector and the other as producer, to produce heavy oil. The distribution of steam along the injector is a measure of efficiency of steam injection. For a heterogeneous reservoir, steam can easily enter the reservoir from high-permeability locations and short cut to the producer. This often results in no steam coverage toward the toe section of injectors. When DTS is deployed, temperature can be used to monitor steam distribution and injection efficiency.

The main challenge of using fiber optic DTS sensors in SAGD application is the harsh environment. The temperature is for injection wellbores averages at 250°C or higher. Both core and cladding materials are made to meet the special requirements of withstanding the high temperature while keeping the quality of light wave transport. There are field test reports showing that the technology can meet the challenge of high-temperature environment of SAGD (Kaura et al., 2008; Hiscock et al., 2015).

For long-term monitoring, temperature data help to diagnose production/injection problems in SAGD wells. Thompson et al. (2015) presented a field application of using temperature data to diagnose the well operation problems. The example showed how to identify the steam leakage along the wellbore. In their example, fiber optic sensor was not permanently installed in the well, but ran as temperature logs every several months. Each time when running a temperature log, well operation had to be disturbed. Using downhole permanent sensors has the advantage of not interrupting well operations.

6.6 Distributed Acoustic Sensing Technology for Production Operation

Acoustic sensing has a long history of being used in the oil and gas industry. One of the examples is noise log, which was first applied in the industry in 1955 (Enright, 1955) to detect leakages behind the casing of a well. Acoustic sensing technology allows us to "listen" to the events and fluid movements that happen thousands of feet downhole along a wellbore with corresponding locations where the events are occurring, and it is a more direct monitoring compared with temperature sensing technology. Examples of distributed acoustic sensing (DAS) applications include completion integrity, production/injection profiling in reservoir management, artificial lift performance evaluation in production, and hydraulic fracturing diagnosis in stimulation.

6.6.1 Fundamentals of DAS Technology

Sound is a form of pressure wave. Similar to DTS, a DAS system uses a fiber optical cable and a laser source to pulse a source light with known wave length along the cable. At the locations where there are acoustic disturbances occurring because of flow or mechanical movements, the sound pressure wave deforms the fiber optic cable, causing the source light wave to backscatter. Unlike the temperature measurement, the backscatter by sound wave is Rayleigh backscatter, which does not have a wavelength shift, as does Raman backscatter in the temperature measurements. Frequency and amplitude, the two parameters that describe an acoustic wave, are related to flow or movement conditions. With this principle, we can tell the location and characteristics of the flow with fiber optical acoustic sensors.

Sound is described by its frequency in hertz (Hz) and sound amplitude in decibels (dB). Frequency describes the number of cycles of a repetitive wave per second of a sound, and amplitude is expressed as the ratio of the pressure generated by the sound, p, to a reference acoustic pressure, p_{ref}. In air this reference pressure is 20×10^{-6} Pa. Sound pressure level, L_{sp}, is a parameter commonly used to describe how strong a sound wave is,

$$L_{SP} = 10\log_{10}\left(\left[\frac{p}{p_{ref}}\right]^2\right) \tag{6.75}$$

Acoustic signals are usually presented in a format of amplitude of the sound in a time domain. Through Fourier transform, this amplitude can be presented in a frequency domain as a complex function,

$$F(\omega) = \frac{1}{2\pi}\int_{-\infty}^{\infty} f(t)e^{-j\omega t}\,dt \tag{6.76}$$

Or a discrete equivalent for a sound signal with N samples is

$$F(k) = \frac{1}{N}\sum_{n=0}^{N-1} F(k)e^{-j\frac{2\pi nk}{N}} \tag{6.77}$$

Figure 6.22 shows a sample acoustic signal recorded in 30 seconds in the original time domain (Fig. 6.22*a*) and transformed into the frequency domain (Fig. 6.22*b*). The frequency peak depicted in Fig. 6.22*b* shows that sound has a mode frequency of 100 cycles per second (Hz). The overall sound amplitude of the signal can be taken by squaring and summing the amplitude of the signal over a frequency range.

$$L_{SP} = 10\log_{10}\left(\frac{\sum p_i^2}{p_{ref}^2}\right),\ p_i = p_1, p_2, p_3 \tag{6.78}$$

Once the signals are presented in the amplitude and frequency format, they are ready to be used for flow diagnosis. As shown in Fig. 6.22*b*, the featured sound has a signature frequency around 100 Hz, with 4.5 Pa peak amplitude.

Before DAS data is processed, a filter is applied to eliminate the data outside the frequency range that is of interest. For fluid flow, a liquid phase creates a lower frequency waves than a gas phase does. Extremely high (above 10,000 Hz) or low (below 100 Hz) frequency data may be more likely related to noise and should be truncated before data processing. Perhaps one of the biggest challenges of DAS technology is the huge amount of data collected during monitoring. The amount of data recorded is determined by the space resolution and time interval demanded. The higher the space resolution, the short the source light length has to be. A single day collection of DAS data can be up to terabytes; storing, processing, and analyzing these data are computational exhaustive (Cannon and Aminzedeh, 2013).

The data collected by DAS are usually presented in 2D contour plots in the field, referred as waterfall plot, with time and distance as two axes, and color (or the intensity of the greyscale) represents the intensity of the signals. Figure 6.23 shows such a plot in greyscale. At any time, the sound pressure level, L_{sp}, at a defined location can be read from the plot.

6.6.2 Correlation Relationship between Flow Condition and Acoustic Measurements

Acoustic behavior can be modeled by a fully computational fluid dynamic approach. Because of the extreme complexity and computational intensity, currently, DAS measurements are limited to qualitative applications in the field. Correlative relationships between characteristics of acoustic signals and flow conditions have been observed in some experimental studies. Early-time noise logging

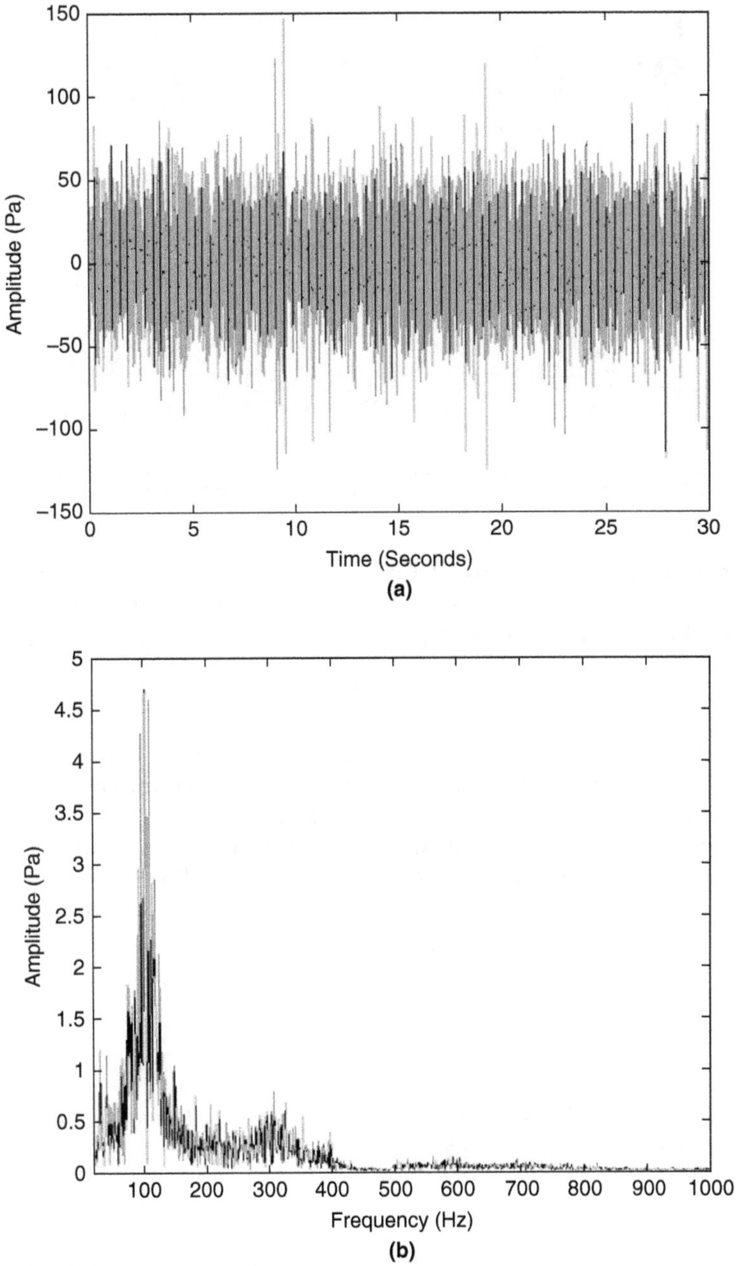

Figure 6.22 Raw data of DAS measurements in (*a*) sampled signal in the time domain and (*b*) time signal transformed into the frequency domain (Martinez et al., 2014).

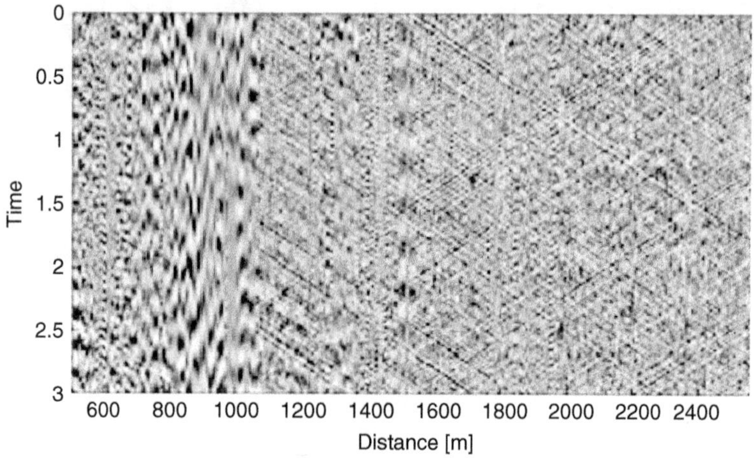

FIGURE 6.23 An example of waterfall plot of DAS (Johannessen et al., 2012).

experiments conducted by McKinley and Bower (1979) showed that for a single-phase gas flowing behind a casing or through a perforation, the following relationship exists

$$N_f^* = C'q^3 \tag{6.79}$$

where N_f^* is the peak to peak noise level above a cutoff frequency in millivolts; a parameter that is similar to sound pressure level. The constant C' in the above equation is determined by the flow path geometry, and it is also a function of fluid properties. Taking 10-based logarithm on both sides of Eq. (6.79), we obtain the following equation

$$\log N_f^* = \log(C'q^3) \tag{6.80}$$

or in terms of flow rate,

$$\log q^3 = \log N_f^* - \log C' \tag{6.81}$$

From Eq. (6.75), since L_{SP} is in the format of 10-based logarithm, we can rewrite Eq. (6.81) as

$$\log(q^3) = A \times L_{SP} + B \tag{6.82}$$

This simple relationship has been observed by other experimental studies (Martinez et al., 2014; Chen et al., 2015). Figure 6.24 shows the results from an experimental study; the bottom picture illustrates the well-defined relationship between sound pressure level (amplitude of acoustic signal) and flow rate, and the right picture shows a good-fit linear function between $\mathbf{log}(q^3)$ and the sound pressure level, L_{SP}.

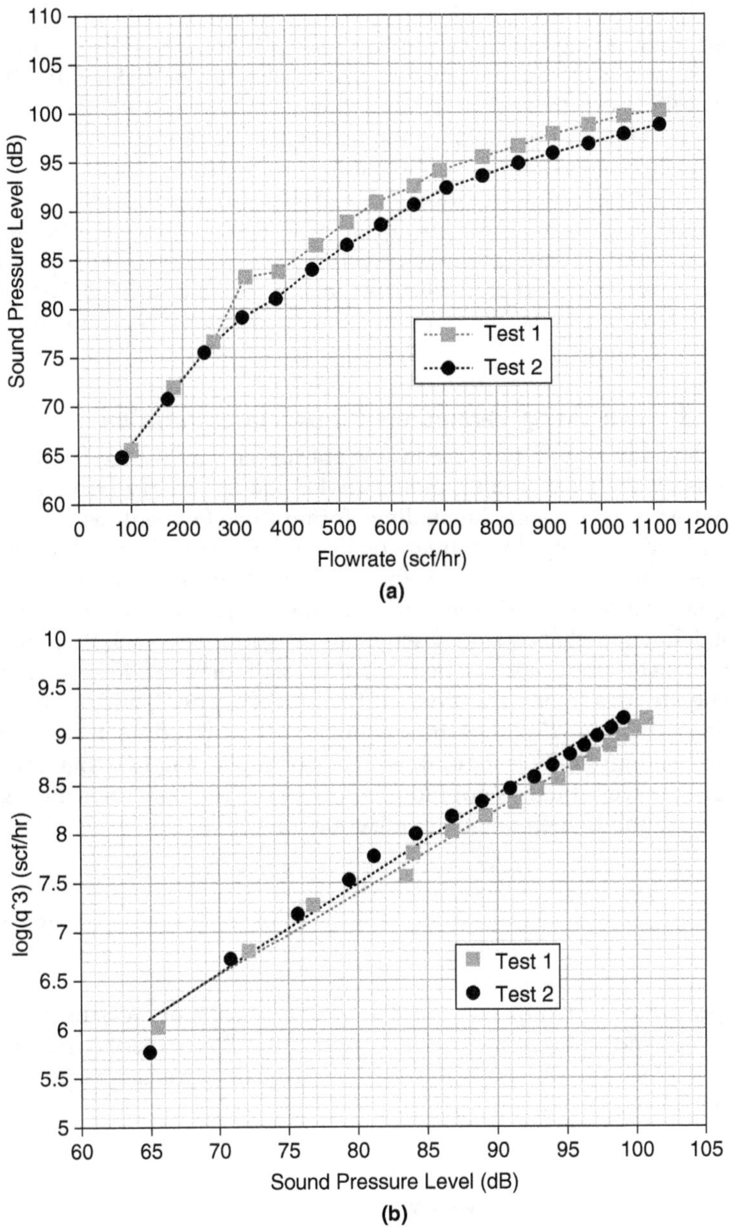

Figure 6.24 (*a*) Sound pressure level versus flow rate curve and (*b*) acoustic signal on dB scale in frequency domain (Chen et al., 2015).

The general format of the correlation is valid for flow through perforations, regardless of the type of well (horizontal or vertical). It also applies to openhole wells, fractured wells, or wells with artificial lift. Even through the correlation equation is straightforward, the constants in Eq. (6.82), A and B, are complex functions of many parameters. For example, these parameters are functions of completion specification (perforation size and density) and fluid properties (density and viscosity) for flow profiling in horizontal wells. They are functions of fracture geometry, fracture fluid, and proppant specifications for fractured wells. Because many variables are involved in these two constants, there are no simple equations to calculate the two constants.

6.6.3 Estimating Parameters for DAS Interpretation

For a wellbore with multiple inflow entries (i.e., multistage fractured wells), if the DAS reading points are far enough apart to isolate the interferences between the readings for multiple flow entries, the last amplitude reading downstream of all fractures could be used to estimate A and B in Eq. (6.82) if there is more than one reading at the same location at a different time. At this point, the flow rate is the total flow rate that can be measured at the surface. At each time step, an acoustic amplitude can be read, along with the total flow rate that generates this signal. This gives us two sets of Eq. (6.82) with known q and L_{SP}. Figure 6.25 shows the schematic of this procedure. If we have two readings at the last point, with its corresponding two total rates from the well (there is not more inflow to the wellbore after the last reading point), then the following equations hold:

$$\begin{aligned} \log(q_1^3) &= A \times (L_{SP})_1 + B \\ \log(q_2^3) &= A \times (L_{SP})_2 + B \end{aligned} \tag{6.83}$$

there are two unknowns, A and B, with two pairs of measurements (q_1, L_{SP1}) and (q_2, L_{SP2}), and A and B can be solved from these two equations. Once the constants A and B are available, we can assign the flow for each entry.

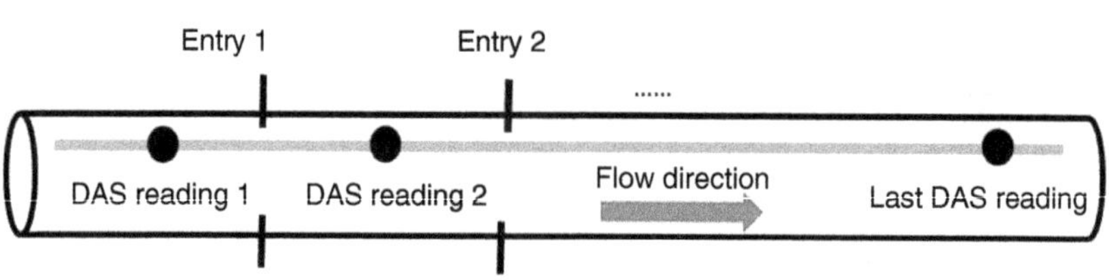

Figure 6.25 Example of flow allocating by DAS measurements.

	LSP (dB)	Rate (MCF/day)
Time 1	110 dB	164
Time 2	120 dB	352

TABLE 6.7 Information for Example 6.1

Example 6.1 We use a two-fracture case to show how the procedure works. We have DAS readings at the two different flow rates downstream of the two fractures and the corresponding total rates, as shown in Table 6.7. If the DAS reading between the two fractures is 100 dB when the DAS reading the total flow rate 110 dB, what is the flow rate distribution in this two-fracture case?

Solution We first calculate the constants A and B, from Eq. (6.83), with the information from Table 6.5,

$$\begin{aligned} \log(164^3) &= A \times (110) + B \\ \log(352^3) &= A \times (120) + B \end{aligned} \tag{6.84}$$

This gives the slope and intercept of the correlation as

$$\begin{aligned} A &= 0.1 \left[\frac{\log(\text{MCF}^3)}{\text{dB}}\right] \\ B &= 0.5\ [\log(\text{MCF}^3)] \end{aligned} \tag{6.85}$$

And this gives the correction for this well at similar conditions as

$$\text{Log}(q^3) = 0.1L_{\text{SP}} + 0.5 \tag{6.86}$$

The flow rate for fracture 1 is

$$q_1 = (10^{0.1\times100+0.5})^{1/3} = 3162 \left[\frac{\text{SCF}}{\text{hr}}\right] = 76 \left[\frac{\text{MCF}}{\text{day}}\right] \tag{6.87}$$

And the flow rate for fracture 2 is

$$164 - 76 = 88\left[\frac{\text{MCF}}{\text{day}}\right] \tag{6.88}$$

6.6.4 Estimating Liquid Holdup by DAS for Two-Phase Flow in Pipe

When a flow stream enters into a wellbore, an acoustic wave is generated with the inflow. The acoustic wave travels in both directions once enters the wellbore, along the flow direction and also against the flow direction. The wave frequency of the upstream and downstream waves is different. The one in the flow direction has higher frequency than the one going against the flow direction, and this is referred to as Doppler Effect. The fluid velocity, v, is related to the

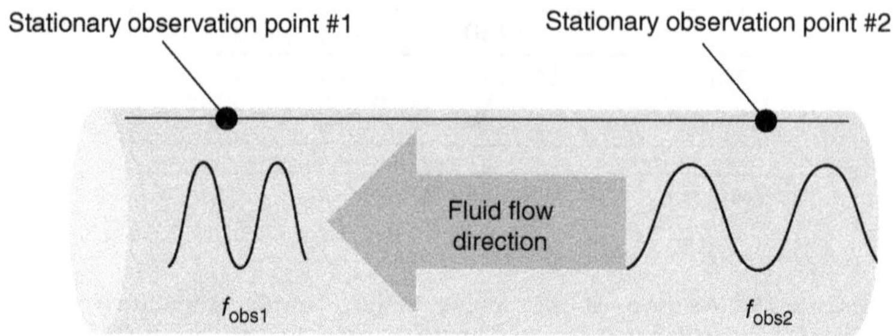

FIGURE 6.26 Illustration of using DAS to profile flow in a pipe (Bukhamsin and Horne, 2014).

upstream sound velocity (with the flow direction), c_u, and downstream sound velocity (against the flow direction), c_d, as

$$v = \frac{c_u - c_d}{2} \tag{6.89}$$

Figure 6.26 illustrates this phenomenon. Knowing the distance between the observation points, DAS can provide the measurement of the speed of sound both upstream and downstream, c_u and c_d. From the measurement, the flow velocity can be estimated.

By definition, the speed of sound for a mixture is (Bukhamsin and Horne, 2014)

$$c_m = \sqrt{\frac{K_m}{\rho_m}} \tag{6.90}$$

where K_m is wave number,

$$K_m = \left[\frac{y_o}{K_o} + \frac{1 - y_o}{K_w}\right]^{-1} \tag{6.91}$$

and

$$\rho_m = y_o + (1 - y_o)\rho_w \tag{6.92}$$

Combining Eqs. (6.91) and (6.92) with Eq. (6.90), we have

$$c_m = \left\{(y_o\rho_o + (1 - y_o)\rho_w)\left[\frac{y_o}{\rho_o c_o^2} + \frac{1 - y_o}{\rho_w c_w^2}\right]\right\} \tag{6.93}$$

Equation (6.93) can be used to estimate oil-phase holdup, y_o. For a two-phase flow system, the single-phase sound speeds, c_o and c_w, are measured in the lab as references. When c_m is measured by DAS, the only variable in Eq. (6.91) that is unknown is the holdup, y_o. This method provides an estimate of phase holdup for flow in a pipe without inflows from the pipe wall.

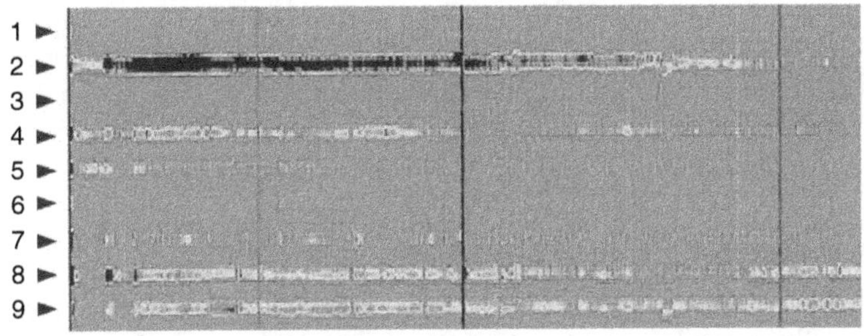

FIGURE 6.27 Using DAS to profile flow (Boone et al., 2014).

6.7 Quantitative Diagnosis with DAS Measurements

DAS measurements along a wellbore are usually presented in a 2D contour plot with length along the well as one axis, and time is on the other axis. The amplitude of acoustic signals in the contour plots is presented in a color scale, with warmer color (darker in the greyscale plots) indicates higher sound energy. This alone can be a qualitative diagnosis tool.

We illustrate some examples of DAS applications. Figure 6.27 is an example of using DAS for flow profiling. There are nine perforation locations along the well (marked on the left side of the figure). In the depth-time plot of DAS measurements, the waterfall plot shows where the inflow comes from. Perforation 2 contributes the most to the production. There is no inflow at perforation 1, 3, and 6. Perforation 4, 8, and 9 also contribute to the flow.

Figure 6.28 shows a casing leakage test results. The fluid movement at the depth of 235 indicates possible problems in wellbore integrity at the location.

These examples are just a few cases to show that the DAS technology provides us rich information about downhole fluid movement.

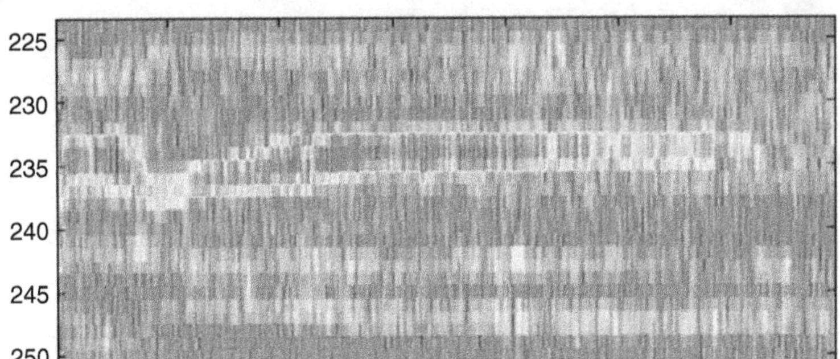

FIGURE 6.28 Detect casing leakage for wellbore integrity (Boone et al., 2014).

The technology has less limitation at downhole condition compared with other measurement means, and it does not interrupt well operation (producing, injecting, fracturing, workover). The technology shows its great potential to assist us in understanding flow conditions in the wells.

6.8 Integrated Application of DTS and DAS

Fiber optical sensing technology for distributed measurements along a wellbore can be used together for diagnosis purpose. Because DTS and DAS are both measurements of flow being correlated to flow velocity, integrated interpretation provides constraints to each other, confirming the interpretations. A common application of combined DTS/DAS monitoring is hydraulic fracture diagnosis for multistage treatments in horizontal wells (Ugueto et al., 2014; Wheaton et al., 2016). An integrated diagnosis consists of DTS/DAS monitoring along with fracture modeling to understand what happened during a fracture treatment. Figure 6.29 shows such an integrated plot (Wheaton et al., 2016). The top picture is DTS monitoring result, the middle is DAS, and the bottom is the injection data (fluid, proppant, and diverter). The data were collected from one stage of a 15-stage fracturing job. There are four perforation clusters, shown on the left side of both DTS and DAS plot. In each stage, proppant concentration ramped twice, and diverters were pumped in between the ramps. The observations

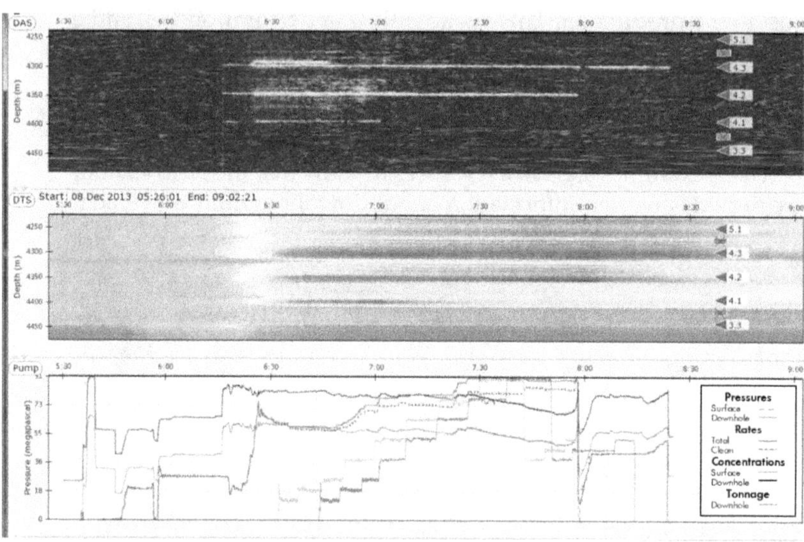

FIGURE 6.29 Integrated fracture diagnosis by DTS, DAS and fracture modeling (Ugueto et al., 2016).

from the DTS/DAS monitoring showed that all perforation clusters were taking fluid/slurry, with the bottom clusters more consistently taking injection fluid compared with the top clusters. The first diversion reduced the flow into the bottom-most cluster, and distributed slurry more evenly toward the end of the stage. Fracture models were used to simulate fracture propagation during the pumping, and the information provided by DTS and DAS were used as guideline/constrains for fluid distribution (Wheaton et al., 2016).

Because DTS and DAS provide the same information of fluid volume distribution, this gives us one more constrain for interpretation. The flow capacity of a fracture is the product of fracture volume and permeability, or in terms of conductivity, a product of conductivity ($k_f w$) and fracture area ($x_f h_f$). With two measurements, we can further interpret one more parameter in addition to fracture volume.

The integrated approach of fracture diagnosis during fracturing follows the procedure as:

1. Analyzing DAS data for fluid and proppant distribution.
2. Analyzing DTS data with the known fluid volume from DAS to match for fracture conductivity.
3. Run fracture model with the total volume to generate fracture geometry.
4. With created fracture, calculate/match measured injection pressure.

This procedure estimates fracture geometry and fracture conductivity. A similar procedure for during-production diagnosis is:

1. Analyze DAS data to allocate inflow distribution.
2. Analyze DTS data to estimate fracture conductivity.
3. Use reservoir flow model (numerical or analytical/semianalytical methods) to calculate pressure response and match pressure and surface total flow rate.

This procedure will generate inflow distribution and conductivity decline as a function of production time.

Nomenclature

A	= area of flow, ft^2
C_D	= valve discharge coefficient, dimensionless
c_d	= downstream sound velocity, ft/s
C_{leakoff}	= leakoff coefficient
C_p	= heat capacity, Btu/(lb · °F)

C_{leakoff} = leakoff coefficient
c_d = downstream sound velocity, ft/s
$c_{u'}$ = upstream sound velocity, ft/s
c_m = mixed fluid sound velocity, ft/s
D = depth, ft
$D_{wT'}D_{wp}$ = weights of temperature and pressure data set in inversion
E_i = sum of kinetic energy and internal energy
f = friction factor
g = gravity constant, ft/h^2
g_G = geothermal gradient, °F/ft
h = Planck number
I_{as} = intensity of anti-Stoke scattering
I_{as} = intensity of stoke scattering
K_{JT} = J-T effect coefficient, °F/psi
K = Boltzmann number
K_T = total heat conductivity, Btu/(hr-ft-°F)
K_m = wave number
k = permeability, mD
L_{sp} = sound pressure level
$\dot{Q}_h$ = heat flux rate, Btu/(hour · ft^2)
N_f^* = peak to peak noise level above a cutoff frequency
q = flow rate, bbl/day
R = pipe radius
r = inner radius of tubing
r_w = wellbore diameter, ft
S = saturation
s = skin factor
T = temperature, °F
t = time, minute
U = overall heat transfer coefficient
V = volume, ft^3
W = mass flow rate
v = velocity, ft/day
v_{leakoff} = leakoff velocity
u = overall heat transfer coefficient
u = Darcy velocity
$Y(w)$ = measured data with w as a parameter
$Y'(w)$ = calculated data set with w as a parameter
y_i = phase i holdup

β = thermal expansion coefficient, 1/°F
γ = open hole rate
ϕ = porosity
μ = viscosity, cp
ρ = density, lb/ft^3
τ_w = wall shear stress, psi

References

Achinivu, O. I., Zhu, D., and Furui, K. (2008, January 1). "Field Application of an Interpretation Method of Downhole Temperature and Pressure Data for Detecting Water Entry in Horizontal/Highly Inclined Gas Wells," *Society of Petroleum Engineers*. doi:10.2118/115753-MS.

Bukhamsin, A., and Horne, R. N. (2014, October 27). "Using Distributed Acoustic Sensors to Optimize Production in Intelligent Wells," *Society of Petroleum Engineers*. doi:10.2118/170679-MS.

Boone, K., Ridge, A., Crickmore, R., and Onen, D. (2014, January 19). "Detecting Leaks in Abandoned Gas Wells with Fibre-Optic Distributed Acoustic Sensing," *International Petroleum Technology Conference*. doi:10.2523/IPTC-17530-MS.

Chen, K., Zhu, D., and Hill, A. D. (2015, September 28). "Acoustic Signature of Flow From a Fractured Wellbore," *Society of Petroleum Engineers*. doi:10.2118/174877-MS.

Clanton, R. W., Haney, J. A., Pruett, R., Wahl, C. L., Goiffon, J. J., and D. Gualtieri, D. (2006). "Real-Time Monitoring of Acid Stimulation Using a Fiber-Optic DTS System," Paper presented at the SPE Western Regional/AAPG Pacific Section/GSA Cordilleran Section Joint Meeting, Anchorage, Alaska, 8–10 May. SPE-100617-MS.

Cannon, R. T., and Aminzadeh, F. (2013, March 5). "Distributed Acoustic Sensing: State of the Art," *Society of Petroleum Engineers*. doi:10.2118/163688-MS.

Costello, C., Sordyl, P., Hughes, C. T., Figueroa, M. R., Balster, E. P., and Brown, G. (2012, January 1). "Permanent Distributed Temperature Sensing (DTS) Technology Applied in Mature Fields—A Forties Field Case Study," *Society of Petroleum Engineers*. doi:10.2118/150197-MS.

Cui, J., Zhu, D., & Jin, M. (2014, October 27). "Diagnosis of Multi-Stage Fracture Stimulation in Horizontal Wells by Downhole Temperature Measurements," *Society of Petroleum Engineers*. doi:10.2118/170874-MS.

Datta-Gupta, A., and King, M. J. (2007). *Streamline Simulation: Theory and Practice*, Textbook Series, SPE, Richardson, TX, Vol. 11, pp. 110–129.

Duru, O. O., and Horne, R. N. (2010, January 1). "Joint Inversion of Temperature and Pressure Measurements for Estimation of Permeability and Porosity Fields," *Society of Petroleum Engineers*. doi:10.2118/134290-MS.

Enright, R. J. (1955). Sleuth for Downhole Leaks. *Oil & Gas J.* (Feb. 28, 1955), pp. 78–70.

Glasbergen, G., Gualtieri, D., Van Domelen, M., and Sierra, J. (2009). "Real-Time Fluid Distribution Determination in Matrix Treatments Using DTS," *SPE Production and Operations*, Vol. 24 (10), pp. 135–146. SPE-107775-PA.

Hiscock, B., Hinrichs, L., Banack, B. M., & Rapati, B. (2015, June 9). "Methodology for In-Well DTS Verifications in SAGD Wells," *Society of Petroleum Engineers*. doi:10.2118/174487-MS.

Huckabee, P. (2009). "Optic Fiber Distributed Temperature for Fracture Stimulation Diagnostics and Well Performance Evaluation," Paper SPE 118831 presented at the SPE Hydraulic Fracturing Technology Conference, The Woodlands, TX, USA, 19–21 January.

Johannessen, K., Drakeley, B. K., and Farhadiroushan, M. (2012, January 1). "Distributed Acoustic Sensing—A New Way of Listening to Your Well/Reservoir," *Society of Petroleum Engineers*. doi:10.2118/149602-MS.

Kaura, J. D., Sierra, J., and Parks, B. (2008, January 1). "New High-Temperature Optical Fiber Technology Successfully Provides Continuous DTS Data under Harsh SAGD Environment," *Society of Petroleum Engineers*. doi:10.2118/114458-MS.

Yoshioka, K., Zhu, D., Hill, A. D., Dawkrajai, P., and Lake, L. W. (2005). "A Comprehensive Model of Temperature Behavior in a Horizontal Well," Paper SPE 95656 presented at SPE Annual Technical Conference and Exhibition, Dallas, TX, 9–12 October. doi:10.2118/95656-MS.

Lake, L. W. (1989). *Enhanced Oil Recovery*, 33. Prentice-Hall Inc., New Jersey.

Li, Z., and Zhu, D. (2009, January 1). "Predicting Flow Profile of Horizontal Wells by Downhole Pressure and DTS Data for Infinite Waterdrive Reservoir," *Society of Petroleum Engineers*. doi:10.2118/124873-MS.

Li, Z., Yin, J., Zhu, D., and Datta-Gupta, A. (2010, January 1). "Using Downhole Temperature Measurement to Assist Reservoir Optimization," *Society of Petroleum Engineers*. doi:10.2118/131370-MS.

Martinez, R., Hill, A. D., and Zhu, D. (2014, February 4). "Diagnosis of Fracture Flow Conditions With Acoustic Sensing," *Society of Petroleum Engineers*. doi:10.2118/168601-MS.

Martinez, R., Chen, K., Santos, R., Hill, A. D., and Zhu, D. (2014, October 27). "Laboratory Investigation of Acoustic Behavior for Flow from Fracture to a Wellbore," *Society of Petroleum Engineers*. doi:10.2118/170788-MS.

Meyer, B. R., and Jacot, R. H. (2005, January 1). "Pseudosteady-State Analysis of Finite Conductivity Vertical Fractures," *Society of Petroleum Engineers*. doi:10.2118/95941-MS.

McAdams, W. H. (1942). *Heat Transmission*, McGraw-Hill Book Co., New York, pp. 135–137.

McKinley, R. M., and Bower, F. M. (1979, November 1). "Specialized Applications of Noise Logging," *Society of Petroleum Engineers*. doi:10.2118/6784-PA.

Molenaar, M. M., Fidan, E., and Hill, D. (2012, January 1). "Real-Time Downhole Monitoring of Hydraulic Fracturing Treatments Using Fibre Optic Distributed Temperature and Acoustic Sensing," *Society of Petroleum Engineers*. doi:10.2118/152981-MS.

Ramey, H. J. Jr. (1962). "Wellbore Heat Transmission," *J. Pet Tech*, Vol. 14 (4), pp. 427–435; *Trans.*, AIME, 225. SPE-96-PA.

Reyes, R. P., Yeager, V. J., Glasbergen, G., Parrish, J. L., and Tucker, R. (2011, January 1). "DTS Sensing: An Introduction To Permian Basin With A West-Texas Operator," *Society of Petroleum Engineers*. doi:10.2118/145055-MS.

Salah, M., Bereak, A., Gabry, M., Shaheen, T., and Kewan, M. (2015, November 9). "A Holistic Stimulation Approach to Unlock Potential of Horizontal Open Hole Completion in Tight Carbonate: Optimization of Acidizing Treatment and Successful Diversion in Real-Time Using Distributed Temperature Sensing," *Society of Petroleum Engineers*. doi:10.2118/177800-MS.

Schechter, R. S. (1992). *Oil Well Stimulation*, Prentice Hall, Englewood Cliff, N.J.

Shiota, T., and Wada, T. (1991). "Distributed Temperature Sensors for Single Mode Fibers," *Proc. SPIE*, 1586, pp. 13–18.

Sierra, J. R., Kaura, J. D., Gualtieri, D., Glasbergen, G., Sarker, D., and Johnson, D. (2008, January 1). "DTS Monitoring of Hydraulic Fracturing: Experiences and Lessons Learned," *Society of Petroleum Engineers*. doi:10.2118/116182-MS.

Tan, X., Almulla, J. M., Zhu, D., and HIll, D. (2012, January 1). "Field Application of Inversion Method to Determine Acid Placement with Temperature Profiles," *Society of Petroleum Engineers*. doi:10.2118/159296-MS.

Thompson, S., Bello, A., Medina, M., and Greig, T. (2015, June 9). "Well Integrity Evaluation Using Distributed Temperature Sensing (DTS) on an Operating SAGD Injector Well Influenced by Neighboring Steam Chambers," *Society of Petroleum Engineers*. doi:10.2118/174504-MS.

Thiruvenkatanathan, P., Langnes, T., Beaumont, P., White, D., and Webster, M. (2016, November 7). "Downhole Sand Ingress Detection Using Fibre-Optic Distributed Acoustic Sensors," *Society of Petroleum Engineers*. doi:10.2118/183329-MS.

Ugueto, G. A., Ehiwario, M., Grae, A., Molenaar, M., Mccoy, K., Huckabee, P., and Barree, B. (2014, February 4). "Application of Integrated Advanced Diagnostics and Modeling to Improve Hydraulic Fracture Stimulation Analysis and Optimization," *Society of Petroleum Engineers*. doi:10.2118/168603-MS.

Ugueto C., G. A., Huckabee, P. T., Molenaar, M. M., Wyker, B., & Somanchi, K. (2016, February 1). Perforation Cluster Efficiency of Cemented Plug and Perf Limited Entry Completions; Insights from Fiber Optics Diagnostics. *Society of Petroleum Engineers*. doi:10.2118/179124-MS.

Ukil, Abhisek, Braendle, H., and Krippner, P. (2012). "Distributed Temperature Sensing: Review of Technology and Applications," *IEEE Sensors Journal*, Vol. 12 (5), pp. 885–892.

Wheaton, B., Haustveit, K., Deeg, W., Miskimins, J., and Barree, R. (2016, February 1). "A Case Study of Completion Effectiveness in the Eagle Ford Shale Using DAS/DTS Observations and Hydraulic Fracture Modeling," *Society of Petroleum Engineers*. doi:10.2118/179149-MS.

Yoshioka, Keita (2007). Detection of Water or Gas Entry into Horizontal Wells by Using Permanent Downhole Monitoring Systems. PhD Dissertation. Texas A&M University.

Zhao, J., and Tso, C. P. (1993). "Heat Transfer by Water Flow in Rock Fractures and the Application to Hot Dry Rock Geothermal Systems," *International Journal of Rock Mechanics and Mining Science and Geomechanics Abstracts*, Vol. 30 (6), pp. 633–641.

CHAPTER 7

Intelligent Well Completion: Downhole Flow Control

Downhole flow control devices complete the intelligent systems as a companion to downhole monitoring discussed in Chapter 6. The ultimate goal of downhole flow control is to regulate and optimize flow performance of individual wells and recovery from a target reservoir. This chapter presents functions, design, and applications of downhole flow control valves to achieve these goals under various well architectures and production scenarios.

7.1 Introduction

An intelligent well completion is an advanced well completion that employs permanent downhole sensors and flow control valves (sliding sleeve valves) that are remotely operated from the surface. The technology was first introduced to the industry in the late 1990s, and the installation has been steadily increasing since then. According to the information by service providers, as of 2012, the number of intelligent completion wells installed around the world reached beyond a thousand wells. Intelligent well system technology has been used in both on-land and offshore applications. The well type split for intelligent completion wells is 24% for on-land, 37% for subsea, and 39% for platform installations, respectively. About 90% of intelligent wells are producers. The technology has been applied around the world, with Middle East and Europe (North Sea) as the leading regions, but also in Asia-Pacific, North America (include Gulf of Mexico), South America, and Africa. Incremental initial-well capital costs for intelligent completion systems vary from U.S. $200,000 for a permanent downhole-gauge system to U.S. $2,500,000 for a fully equipped multi-zone remote controlled completion (Robinson, 2003).

There are certain environments where commingled production (or injection) may cause production operation problems because of the reservoir properties such as dissimilar pressure gradients, unfavorable fluid compatibility, and high permeability theft zones. Long horizontal wells and multilateral wells provide much larger reservoir contacts, resulting in higher well productivity, but also possible higher heterogeneity along the wells. These wells may also have increased risks of early gas and/or water breakthrough, which pose significant production challenges over time. Remote flow control and downhole monitoring capabilities of intelligent completion can detect these production problems, shut off unwanted fluids without well intervention, and thus can accelerate production even with relatively large reservoir uncertainties in place.

During the late 1990s and early 2000s, simple binary on/off valves were employed for reducing intervention, especially in offshore platform and subsea environments. This is relatively easy to justify economics of employing intelligent completion. Now the technology has advanced, and multi-position choking valves have become popular and allow us to control injection and production allocation in complex reservoirs. The technology can bring significant benefits for many enhanced oil recovery (EOR) projects. Also, downhole flow control valves in intelligent completions have been implemented with many types of conventional wells such as horizontal/multilateral wells, sand control wells, and the wells with electrical submersible pumps (ESPs), resolving interface and integration issues along with these systems. Many operators stress the criticality of adequate preparation time and effective project management in order to maximize the probability of successful implementation. Operational issues associated with running packers, control lines, and related well control concerns must also be addressed when deploying intelligent completion technology.

The value of the intelligent well technologies comes from the ability to actively modify the well zonal completions and performance through flow control and to monitor the response and performance of the zones through real-time downhole data acquisition, thereby maximizing the value of the asset. The oil and gas industry has only begun to realize the potential of intelligent well technology in their contribution to efficiency and productivity. In addition to the attraction of interventionless completions in the high-cost arena of subsea and deepwater wells, intelligent well technology can deliver improved hydrocarbon production and reserves recovery with fewer wells. Intelligent well technology is also effective in waterfloods and gasfloods in heterogeneous or multilayered reservoirs when applied to injection wells, production wells, or both. The production and reservoir data acquired with downhole sensors can improve the understanding of reservoir behavior and assist in the appropriate selection of infill drilling locations and well designs. Intelligent well technology enables a single well to conduct the job of several wells, whether through controlled commingling of zones, monitoring and control of multiple laterals, or even allowing a well to take on multiple simultaneous functions—injection well, observation well, and production well.

Intelligent well technology allows the operator to monitor wellbore mechanical integrity or the environmental conditions under which the completion is operating and to modify the operating conditions to maintain them within an acceptable integrity operating envelope.

7.2 Intelligent Well Equipment

The flow control valves of intelligent wells initially were based upon technology used by conventional wireline-operated sliding-sleeve valves. These valves were reconfigured to be operated by hydraulic, electrical, and/or electrohydraulic control systems providing on/off and variable position choking. Further development has resulted in the choke devices being configured for high-pressure differential service and to some degree being resistant to erosional effects for high rate services. Intelligent well requires some special downhole equipment including feed-through packers, splices, splice subs, flat packs, and downhole flow control valves. In this section, we discuss how the components of the intelligent completion operate. Figure 7.1 illustrates basic components of the intelligent well system.

7.1.1 Production and Isolation Packers

Packers employed for intelligent well completions have certain differences from conventional production packers. The production packer is a primary well barrier. Isolation packers divide the contact reservoir interval into a number of individual sections. First of all, the packers require penetrations from multiple control lines. Also, the packers are exposed to reservoir and intervention fluids above and below the packer. Many downhole flow control completions require the well to be killed prior to running the completion. A possible mitigation of the formation damage that can incur is an acid wash. The packer should be resistant to any such intervention fluids. The packer can be exposed to loads from below the packer while most production packers only experience significant loads from the tubing above the packer. They are required to set when the tubing below the packer is not necessarily free to move.

There are a number of methods of setting packers. The valves themselves can be used to seal the completion and allow packers to

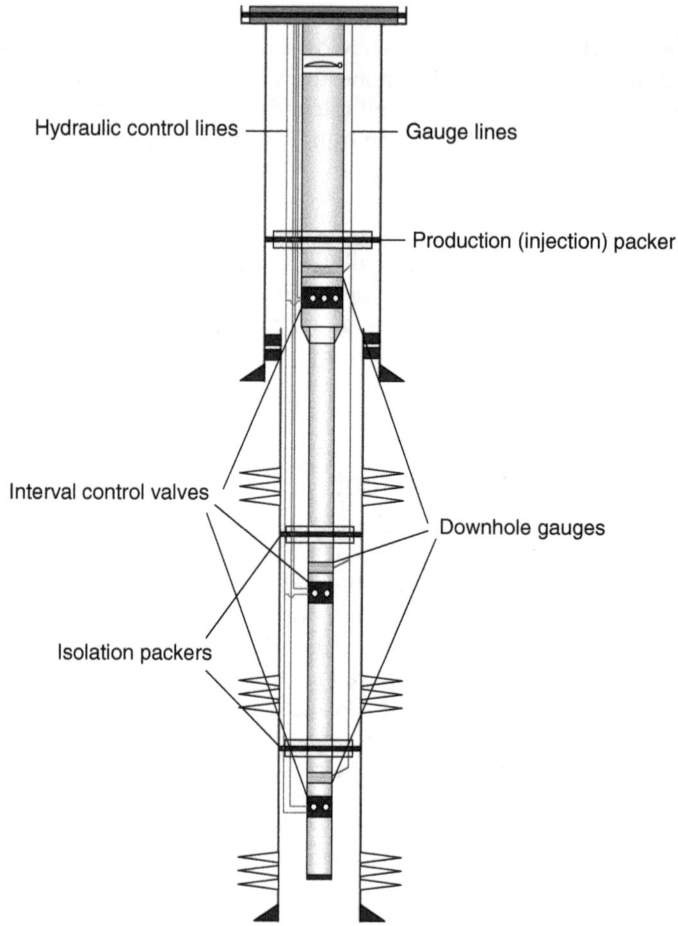

Figure 7.1 Basic components of an intelligent well system.

be set hydraulically by pressurizing the entire completion. In order to reduce surge/swab effects when running in hole, most completions are installed with the interval control valves open, although circulating fluid to the toe of the well may be required for removal of debris. Once the completion has been run, the valves are closed and the well pressure tested. Increasing the pressure sets the packers. It is possible to sequentially set packers by varying the setting pressure.

It is important to note that packer is not a splice point. This means that control lines are continuous and fed through the packer. The control line bypass located under elements and slips maintains tension and compression properties, which can be verified by externally testable connections.

Key points for intelligent completion packers are no mandrel movement allowed during setting, and dynamic parts must not damage control lines while setting. Packer initiation and full pack-off pressures are important setting pressures to know for well design.

7.1.2 Splices and Splice Subs

One of the biggest challenges in intelligent well systems is installation and management of control lines. Feeding control lines through packers and connecting them to inflow control valves and downhole gauges are time consuming but critical for appropriate operation of the valves in the future. Where possible, control line connections to these devices should be made up as subassemblies and tested prior to transport to the wellsite. This is aided by the use of splice subs. The splice sub is typically positioned above each packer. The sub provides a protected groove for the location of the control line connections (Fig. 7.2). The lines penetrating the packer terminate at the splice sub and are preinstalled and tested in the completion module prior to shipment to the wellsite.

7.1.3 Hydraulic Control Line and Flat Pack

Hydraulic control lines are used to provide actuation of hydraulic sliding sleeves. The standard size (outer diameter) of control lines is ¼ in with a wall thickness of 0.049 in. Control lines are commonly manufactured from Grade 316L stainless steel. Until the development of inflow control valves in intelligent wells, the use of control lines exposed to reservoir fluids was rare. However, with intelligent wells, exposure of control lines to reservoir fluids must be considered. This leads to pushing most completions to use 825 Nickel Alloy for intelligent completion applications.

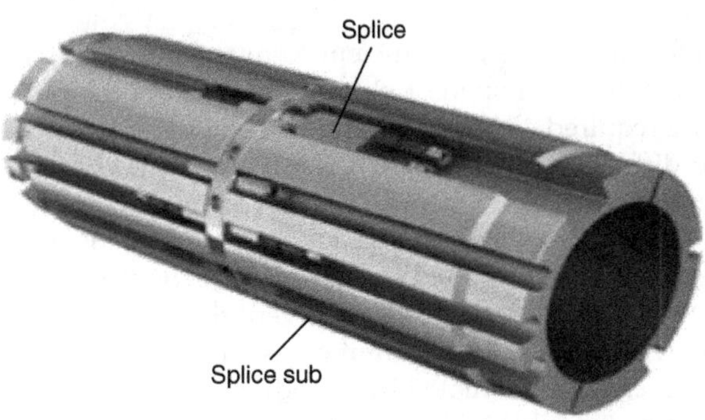

FIGURE 7.2 Splice and splice sub (*Courtesy of Halliburton*).

Figure 7.3 Flat pack for control lines in intelligent completion (*Courtesy of Halliburton*).

A flat pack is an encapsulation to bundle and protect hydraulic control lines and data acquisition cables. Flat packs can incorporate bumper bars and are most robust (but harder to manipulate) than a single hydraulic control line (Fig. 7.3). The flat pack is available in a variety of different configurations to suit the application and well environment. The specification and materials are selected based on the well temperature and pressure and wellbore fluids.

7.1.4 Interval (Inflow) Control Valve (ICV)

Inflow control valves can be hydraulic (Hodges et al., 2000), all electric (Moreira, 2004), or electrohydraulic (the electrics diverting hydraulic power to actuate the (valves) (McLauchlan and Nielsen, 2004). For controlling flow from annulus to the tubing, a sleeve type valve is required. For controlling tubing flow, a ball valve or shrouded sleeve can be used. All types of valves should be selected, positioned, and operated to avoid problems such as wax, scale, asphaltene, and erosion.

A differential pressure sufficient to overcome piston and sleeve friction is required to open or close the sleeve. As the piston is balanced, the required surface pressure to open or close a valve is this pressure differential. Figure 7.4 illustrates a basic configuration of the balanced piston design of the inflow control valve. With scale, asphaltene, or other downhole deposits, the required pressure may increase, but the force applied to open or close the sleeve can be large depending on the piston area. In the event that the valves cannot be operated remotely by the hydraulic system, many downhole flow control valves can be actuated by a landing nipple profile and either slickline or coiled tubing. This could be effective if the hydraulic system leaks.

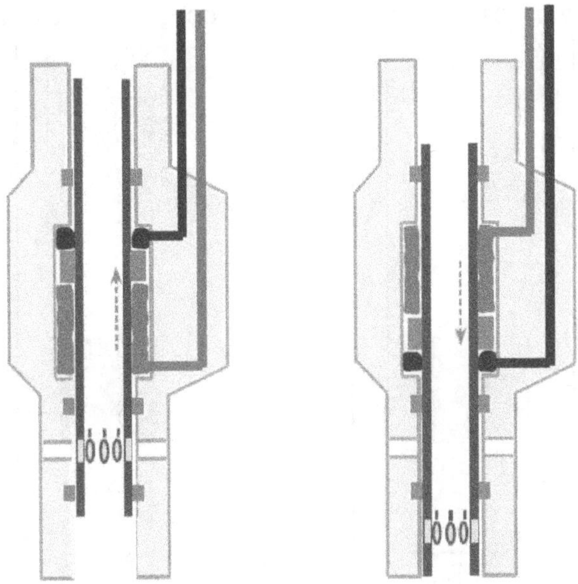

FIGURE 7.4 Balanced piston design of inflow control valve (Hodges et al., 2000).

Figure 7.5 shows typical control line connections of hydraulic inflow control valves. In the basic configuration, two control lines are required to control one valve. For multiple valves, it is possible to use the common close configuration as shown in the figure. In the figure, if all the valves need to be opened, then pressure is applied to the O1, O2, and O3 lines but not to the CC line. If the top valve needs to be opened, but the other two left closed, then pressure is applied to the O1 line, but not to the other lines. If all the valves need to be closed (for example, to pressure test the completion or to set packers), then only CC line is pressurized.

The limit for the number of control lines is usually determined by the tubing hanger. Multiple hydraulic control lines for inflow control valves, plus sensor cables, chemical injection (often multiple lines), and the safety valve control line(s) can be an issue to accommodate through the hanger and wellhead or tree.

In order to reduce the number of control lines, a variety of digital decoders are available (Skilbrei et al., 2003). The decoders convert pressure signals (pressure on or off) applied down multiple control lines into applied pressure for a single valve—for example, three control lines can control up to six inflow control valves making a large number of tubing hanger penetrations unnecessary.

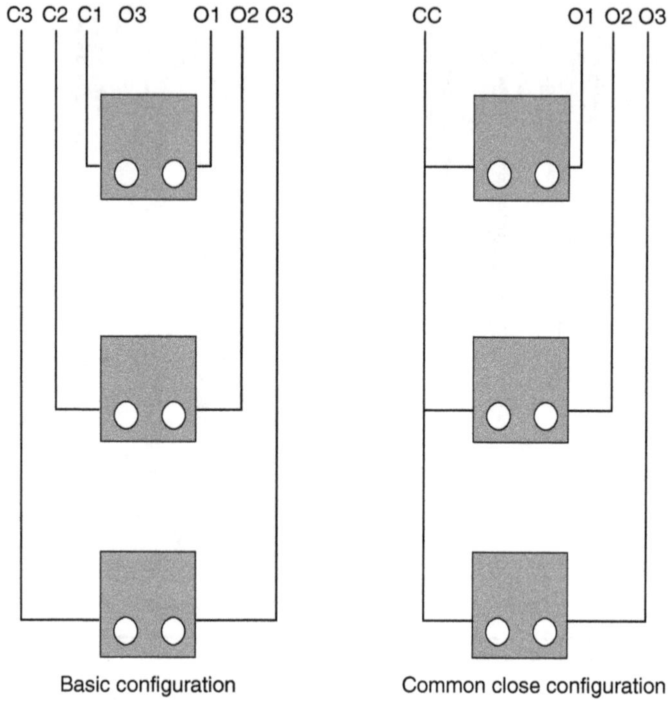

Figure 7.5 System architectures for basic and common close designs.

7.3 Basic Architectures of Intelligent Well Completions

In this section, we describe basic well completion designs of intelligent wells and how intelligent well systems can be integrated with conventional completions, help manage uncertainties, and improve project economics under various production scenarios.

7.3.1 Selective Completion in a Layered Reservoir

At present, with conventional completions and surveillance techniques, accurate or even meaningful measurement of individual zonal properties is often impossible. When producing reservoir fluids from multiple zones, a main concern is the interference between layers (i.e., crossflow caused by unbalanced pressures from different layers). In some countries, governmental regulations prevent operators from commingling production from multiple zones in the same well without properly allocating production from each zone (Puckett et al., 2004). An intelligent well system allows temporary isolation of layers while surveys are performed (Fig. 7.6). Zonal flow data may be obtained continuously from individual sensors or by controlling the position of the inflow control valves to alter or isolate the flow from certain layers.

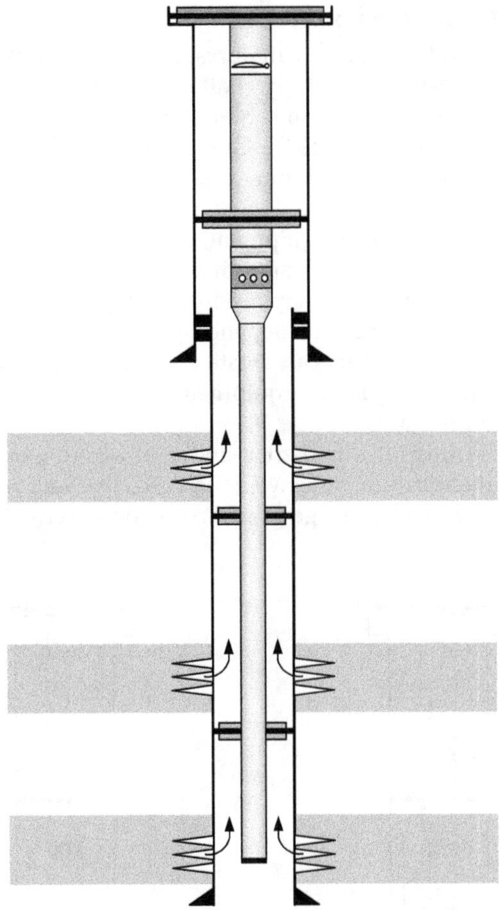

FIGURE 7.6 Multi-zone intelligent well system in a layered reservoir.

The routine conventional field management cycle of proposal, obtaining data, interpretation, then planning, and implementing remedial action can lead to delays of months or even years before beneficial results are achieved. In some cases, the cycle may not even occur because of the costs and risks involved and the lack of a firm case on which to undertake the first step in the process. Remote completion control allows zonal data to be obtained at no incremental operational cost, and remedial action can be taken immediately in the form of zonal isolation, choking, or treatment. Changes to the well inflow profile can thus be temporary and adjusted for water or gas breakthrough trends or surface handling constraints.

7.3.2 Sand Control Completion

As intelligent well technology matures, the field of application continues to expand to increasingly challenging environments such as poorly consolidated, high-permeability, and high-productivity clastic reservoirs that are common to the Gulf of Mexico, offshore Southeast Asia, offshore west Africa, offshore Brazil, and the North Sea (Moore et al., 2002).

Intelligent flow control equipment, transducers, gauge mandrels, and flat packs all take significantly more space than a conventional completion equipment and may need to be deployed directly inside the sand control equipment, depending on the number of zones required to complete. This can create conflicts when attempting to keep casing and completion equipment sizes within conventional casing design programs.

Figure 7.7 illustrates two-zone and multi-zone intelligent sand control completions. For the two-zone case, the well is completed with a conventional two-stage gravel pack (or screens) isolating the

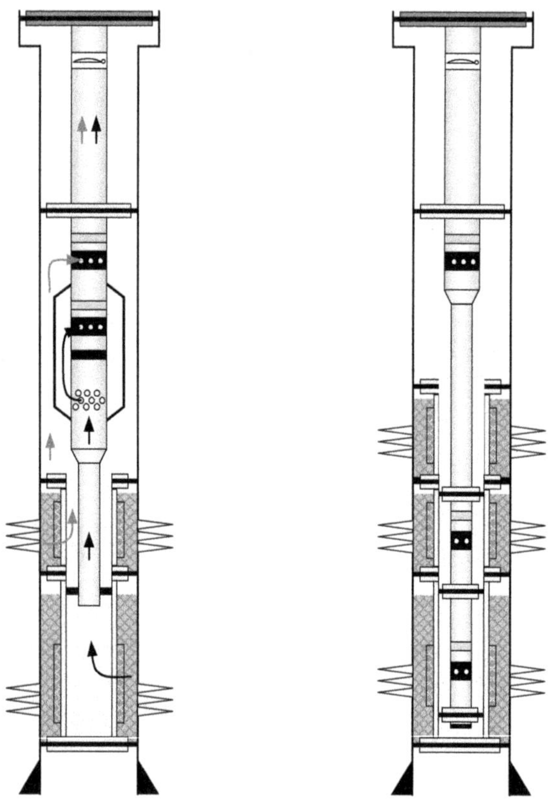

FIGURE 7.7 Two-zone and multi-zone intelligent sand control completions.

zones from each other with a section of blank pipe and a packer. The completion is composed, top down, of the production tubing, feed-through production packer, gauge mandrel, an inflow control valve, a shrouded inflow control valve, and clip tube with seal assembly which stings into a sealbore in the packer isolating the two zones. Production from the lower zone flows through the dip tube and through the shroud on the lowermost valve, entering the production tubing. On the other hand, production from the upper zone flows in the annular area between the upper gravel pack screens, in the annular area between the shroud and the production casing, and enters the production tubing through the uppermost valve.

For three-zone or more gravel packs, inflow control valves are placed within the zones themselves. Placement of the sleeves within the zones means that the sleeves must pass through, and into, the sand control completion system. This limits the size and flow area of the sleeves. Seal assemblies designed to allow hydraulic control lines to pass through are placed between zones. Despite the above challenges, multi-zone intelligent sand control completion system can greatly increase the number of accessible zones and maximize well productivity.

7.3.3 Multilateral Completion

A multilateral well is a single well with one or more lateral wellbores extended from the main bore. It may be as simple as a vertical wellbore with one sidetrack or as complex as a horizontal, extended-reach well with multiple laterals. A multilateral well that replaces several wells can reduce overall drilling and completion costs, increase production, and provide more efficient drainage of a reservoir. Despite these advantages, there are concerns about higher initial costs and the risk of possible interference of laterals with each other, crossflow and difficulties with production allocations in multiple branches. Installation of an intelligent well system into a multilateral well gives the ability to monitor and control each lateral separately, which can resolve these concerns while maximizing well productivity.

When intelligent completion is applied to dual lateral wells, the shrouded system can be used to control flow from each lateral. For multilateral wells with more than two laterals, inflow control valves must be placed at the mouth of each lateral as shown in Fig. 7.8. Injection treatments such as scale inhibition squeezes and stimulations are typically more problematic in a multilateral well. However, remote zonal flow control enables treatments to be diverted as required into each branch, if not into each zone.

Haugen et al. (2006) presented a field application of intelligent multilateral wells at Gullfaks South Statfjord field. They concluded that intelligent multilateral wells accelerated the production in the field while gaining important reservoir information for better understanding the reservoir.

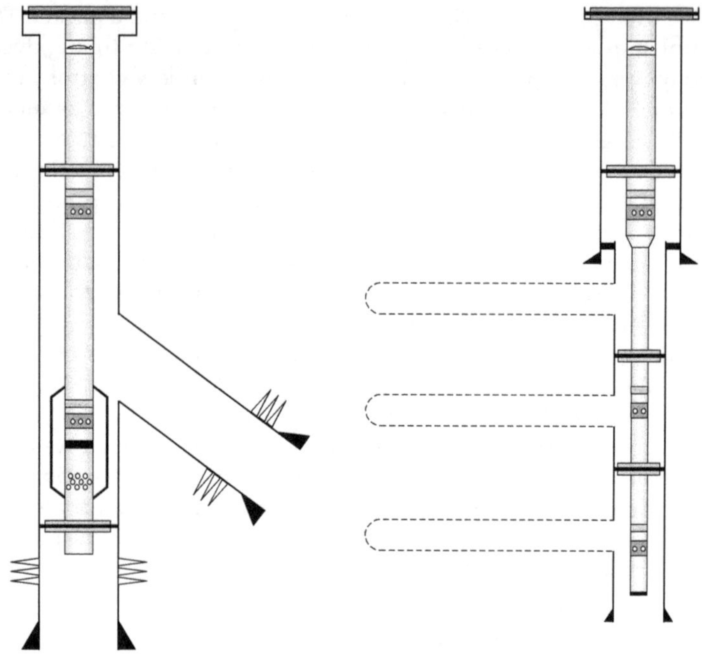

FIGURE 7.8 Dual-lateral and multilateral wells with the intelligent well system.

7.3.4 Multi-Zone ESP Completion

Less than 10% of the world's producing oil wells flow naturally, and the remaining 90% are produced by some form of artificial lift. Many of these wells are equipped with ESPs (Ali and Shafiq, 2008). Intelligent well monitoring and downhole flow control technology complements the capabilities of the ESP systems by providing the ability to balance the production from multiple zones, restrict or shut off zones with high water or gas production, and selectively test zones to monitor reservoir performance.

Application of ESP artificial lift to intelligent well completions requires special considerations. In many cases, the well needs to have the capability to disengage from the intelligent well completion downhole to change pumps while maintaining well control and minimizing formation damage through shut-in at the reservoir face.

Figure 7.9 illustrates multi-zone ESP wells with the intelligent well system. The challenge is to handle the control lines for the flow control valves and downhole sensors. The simplest solution is to run the ESP inside of tubing and run the control line outside the tubing (Vachon and Bussear, 2005). The problem with this approach is that the size of the ESP is constrained. The alternative solution is to deploy a hydraulic wet-connect allowing the downhole completion

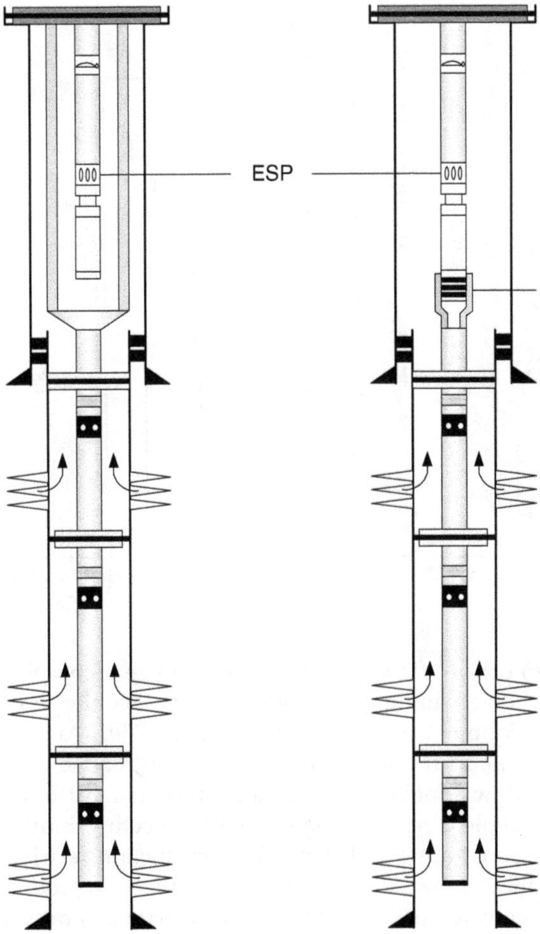

Figure 7.9 Multi-zone ESP completions with the intelligent well system.

equipment to be reintegrated with the intelligent well control system at surface once the ESP is back in place. The limitation of this approach is the required reliability of the wet-connect. If it fails, the valve cannot be reopened.

7.3.5 Intelligent Well System with Adjustable Choking Valves

Downhole flow control valves may be binary (on–off) or multi-position chokes. For multi-position chokes, the control trim profile can be custom designed for different reservoir applications. The selection of the right flow control design is critical, as it may have an impact on the number of zones or intervals that can be effectively controlled in one well.

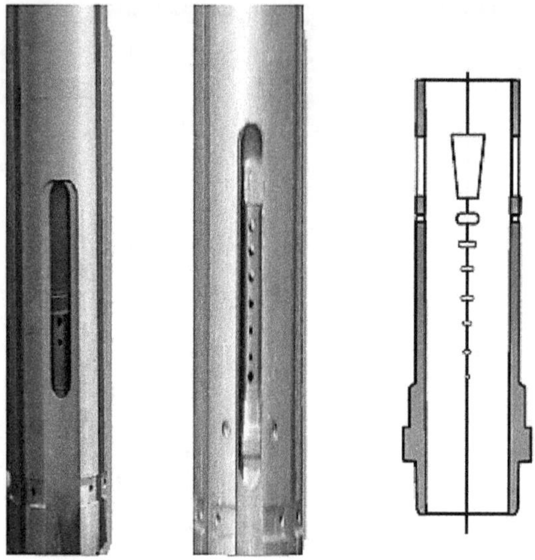

FIGURE 7.10 Inflow ports of a multi-position choking valve (Al-Mubarak et al., 2008).

The ability to control flow in multi-zone intelligent completions has been shown to accelerate production and enhance the recovery efficiency of development projects (Brouwer et al., 2001; Arashi and Konopczynski, 2003; Zhu and Furui, 2006). In multi-zone applications, downhole flow control valves balance production from individual zones to deplete reservoir layers while meeting individual zone production constraints and regulatory production restrictions. Variable flow control allows the operator to optimize the production from the well when water or gas begins to encroach on oil production. Multi-zone injection wells depend on variable flow control for the appropriate distribution of the injected fluid to ensure uniform pressure maintenance or balance voidage placement.

Figure 7.10 shows inflow ports of a multi-position choking valve with 10 discrete positions (Al-Mubarak et al., 2008). The flow area at each position is the cumulative area of the exposed ports. The first picture to the left shows the valve in position 2 where only two ports are open to flow entering the tubing. The middle picture shows the valve in the fully open position. Note that actual choke designs and geometry of flow ports can be different for different valve manufacturers.

7.4 Using Pressure Drop through Inflow Control Valves for Flow Control

The design for inflow control valves is to be able to balance the inflow and control the rate of the unwanted fluids (for example, water). The sliding sleeve valves used for flow control are specified by the manufactures based on the valve design. The flow restriction can be expressed in the term of pressure drop through the valve. The inflow control valves in the intelligent well control are designed differently compared with surface chokes. Inflow control valves are sliding sleeve valves that regulate the flow by the opening area of the slots to the fluid flow, and the opening of surface production chokes is circular orifices. Regardless the opening shape difference of inflow control valves and surface chokes, the flow rate and pressure drop relationship follows the same principle. The well-developed choke equations can be used to design inflow control valves. For multiple position valves, if the fluid is slightly compressible or incompressible, we can assume that the flow rate is linearly proportional to the opening area. For example, if the maximum rate is defined by 100% fully open, the 40% of maximum rate is allowed to be through the valve if the opening is set at 40%. The pressure and rate relationship for single phase at subcritical condition is adopted from the choke equation as (Economides et al., 2012)

$$q_l = CA_{\text{choke}}\sqrt{\frac{2g_c\Delta p}{\rho_l}} \tag{7.1}$$

where C is choke flow coefficient, ρ_l is fluid density, and A_{choke} is cross-sectional area of the choke. Equation (7.1) can be written in field units as

$$q_l = 22{,}800C(D_2)^2\sqrt{\frac{\Delta p}{\rho_l}} \tag{7.2}$$

where D_2 is the restriction diameter. The choke flow coefficient, C, can be calculated by

$$C = \frac{C_d}{\sqrt{1-\beta^4}} \tag{7.3}$$

In Eq. (7.3), C_d is the discharge coefficient and β is the ratio of small to large diameters in the restriction, and this is the ratio of the choke orifice diameter to the diameter of upstream pipe. To use this for sliding sleeve valves, the opening area can be converted to an equivalent diameter.

Single-phase liquid flow through restrictions is also described in terms of the valve flow coefficient, C_v, which is determined in a test bench using water flow. C_v is described in units of volume/time/psi$^{1/2}$ and is defined as

$$C_v = q_l\sqrt{\frac{\gamma_l}{\Delta p}} \tag{7.4}$$

The pressure drop and flow rate relationship through the flow ports (i.e., flow restrictions) can be significantly different for different valve designs. The discharge coefficient C_d or the flow coefficient, C_v, must be carefully selected for flow characterization. The coefficients are provided by the manufactures based on the design of each type of valves. These coefficients depend on Reynolds number and flow port opening areas where a special calibration may be required at certain valve positions (Williamson et al., 2000; Sun et al., 2008, 2011).

Example 7.1 **Use ICV to balance multilateral well production.**
We use a trilateral well as an example here to illustrate how inflow control valves can help to balance the production from three isolated reservoir compartments that have different properties (Zarea and Zhu, 2011). Figure 7.11 shows the schematic of the well, and Table 7.1 lists the reservoir properties in each compartment. The parameters used in this example are the same for all three laterals, except the permeabilities, in both the horizontal and vertical directions.

We assume an average reservoir pressure of 2500 psi for all three reservoir compartments and no formation damage (s = 0) for all three laterals. We use the same fluid properties for all the reservoirs also. The formation volume factor is 1.4 and the viscosity of the fluid is 1 cp.

Solution For Lateral-1, if we use a p_{wf} at the toe of the lateral of 2300 psi, we can use Babu and Odeh's model with segmented wellbore to calculate a flow rate from each lateral with the effect of frictional pressure drop on the flow rate. The flow rate by the segment method with the wellbore divided into 15 segments is 18,451 bpd, and the pressure drop inside the well because of the friction is 64 psi. The readers can refer *Multilateral Wells* (Hill et al., 2008) for the details of the segment approach for the flow rate from a horizontal lateral. To balance the flow at the junction, the wellbore pressure at the junction (or at the heel of each lateral) needs to be close for each lateral. The calculated results of wellbore

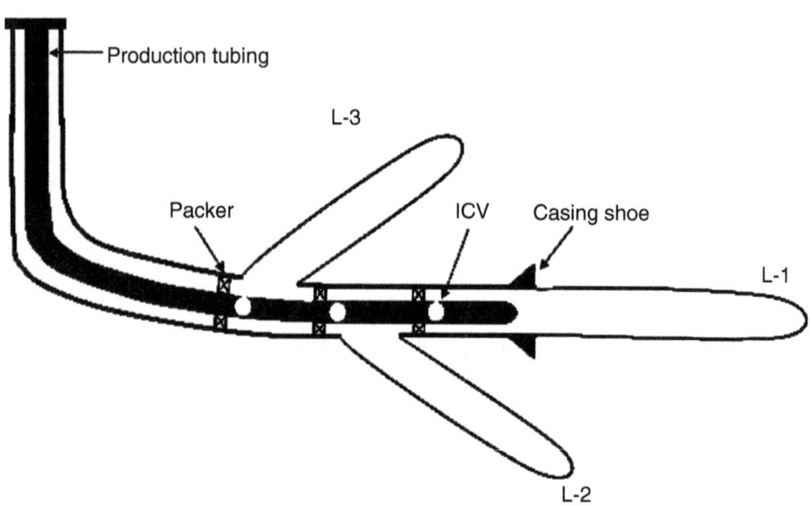

FIGURE 7.11 Schematic of Example 7.1.

	Lateral-1	Lateral-2	Lateral-3	
Porosity, ϕ	0.24	0.24	0.24	
Permeability, k_H	60	20	30	mD
Permeability, k_V	6	2	3	mD
Reservoir thickness, h	50	50	50	ft
Wellbore radius, r_w	0.255	0.255	0.255	ft
Wellbore length, L	5500	5500	5500	ft
Viscosity, μ	1	1	1	cp
Volume factor, B	1.4	1.4	1.4	bbl/stb
Fluid density, ρ	55.0	55.0	55.0	lbm/ft^3
Drainage length	6100	6100	6100	ft
Drainage width	600	600	600	ft
Wellbore relative roughness, ε	0.01	0.01	0.01	
Openhole diameter, D	6.125	6.125	6.125	in
Press at the toe, p_{wf}	2300	2300	2300	psi

Table 7.1 Reservoir and Wellbore Properties for the Multilateral Well Example

pressure drop, flow rate, and junction pressure are summarized in Table 7.2. Noted the large difference of the flow rates from each lateral. This rate difference is caused by permeability difference and also by wellbore pressure drop.

Obviously, this is not an efficient way to produce the multilateral well. We now add the ICVs to each lateral to balance the flow rate. The pressure and flow rate values in Table 7.2 can be the conditions upstream of each corresponding ICV prior to commingling the mainstream production. Based on the flow performance of each lateral, the pressure drop across each ICV is calculated for each valve position using Eq. (7.4) with the given C_v values. Table 7.3 lists the characteristics of the three ICVs at each position. The corresponding pressure drop at each valve opening position is also shown in the tables. For example, the flow rate of Lateral-1 at the valve open position 9 is 90% of the maximum

	Lateral-1	Lateral-2	Lateral-3	
Wellbore press drop, Δp	63.8	5.8	12.9	psi
Total oil flow rate, q	18451.1	5268.1	7963.3	bbl/day
Flowing press at the heel, p_{wf}	2236.2	2294.2	2287.1	psi

Table 7.2 Summary of Results for the Coupled Reservoir and Wellbore Model

			Lateral-1		Lateral-2		Lateral-3	
ICV Position	Open area (in^2)	Cv	rate (bbl/d)	Dp (psi)	rate (bbl/d)	Dp (psi)	rate (bbl/d)	Dp (psi)
1	0.04	0.35	1845	21086.5	580	2085.2	878	4779.5
2	0.08	1.07	3690	8859.2	1160	867.1	1756	2008
3	0.15	2.41	5535	3953.7	1740	391	2634	896.2
4	0.23	4.30	7380	2207.1	2320	218.3	3512	500.3
5	0.31	6.65	9225	1443.8	2900	142.8	4390	327.3
6	0.47	11.08	11070	748.6	3480	74	5268	169.7
7	0.83	21.25	12915	276.8	4060	27.4	6146	62.7
8	1.22	37.93	14760	113.5	4640	11.2	7024	25.7
9	2.52	99.68	16605	20.8	5220	2.1	7902	4.7
10	4.06	175.33	18450	8.3	5800	0.8	8780	1.9

Table 7.3 ICV Performance for Lateral-1

rate, which is 16,606 bbl/day (18,451 × 90% = 16,606 bbl/day). This gives a pressure drop across the ICV for Lateral-1 when the ICV is at position 9 as

$$\Delta p = \gamma_l \left(\frac{q_l}{C_v}\right)^2 = \frac{55.0}{62.4}\left(\frac{16606}{99.7}\frac{42}{24\times 60}\right)^2 = 20.8 \text{ psi} \tag{7.5}$$

Using Table 7.3, the ICV position for each valve can be adjusted to satisfy the balance pressure in the main wellbore with best flow performance out of all three laterals. The design also needs to consider the pressure drop in the main wellbore. This example only serves the purpose of illustrating how ICV can optimize a multilateral well performance. Other applications are qualitatively discussed in the following sections.

7.5 Field Applications for Downhole Flow Control by Intelligent Well Completions

7.5.1 Commingled Production

A first example of the use of an intelligent well system is the deployment of downhole flow control valves to allow commingled production from multiple zones with different formation pressures and reservoir properties (Fig. 7.12). In a conventional approach, an individual well dedicated to each layer may be drilled and completed to prevent zonal interference although the cost of drilling and completion can be dramatically increased. Using downhole flow control valves, the operator may be able to complete all three hydrocarbon-bearing formations with a single wellbore where adjustable choking

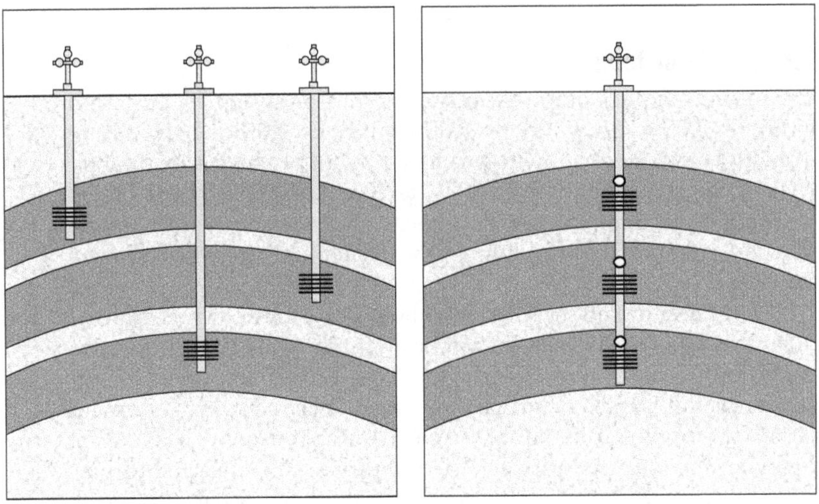

FIGURE 7.12 Production from multiple layers with conventional wells versus single intelligent well.

valves can restrict inflow from the highest pressured zone avoiding crossflow to the lower pressure zones.

Technically, a multi-zone completion with mechanical sliding sleeves could also be used to separately produce the zones through shifting of a sleeve on slickline or coiled tubing. However, the workover-based operation can be costly and risky, especially for subsea and remote environments. The remote actuation of downhole flow control valves can be applied immediately without having to mobilize intervention equipment and, in worst cases, a rig. Additional benefit is water shutoff with remote actuation when the water cut becomes too high for operational limits such as lifting, separation, filtering, and disposal.

Downhole flow control valves can also improve the understanding of flow in layered formations. Zonal well testing can be conducted by sequencing intervals open and closed. Dedicated downhole pressures, flow meters, pressure drops through sleeves (internal and external pressure gauges), and temperature differentials can be used to estimate zonal flow and fluid content leading to better reservoir management. Reservoir communication between zones can also be assessed by shutting in one interval and flowing adjacent zones, which provides valuable information in reservoir characterization. In assessment of the well integrity, zonal isolation can be effectively evaluated by multiple, zone-specific, pressure and temperature gauges (Nozaki et al., 2015).

Lastly, during the life of production, wells must be treated by acids (i.e., restimulation) or scale inhibitors. With remotely operated downhole flow control valves, acid stimulations can be performed or scale inhibitors deployed solely into intervals at risk of scaling.

7.5.2 Waterflood

The second example of the use of intelligent well system is illustrated in Fig. 7.13. A reservoir composed of multiple hydrocarbon bearing formations where each layer is isolated by a flow barrier is drained with a single production well. Water breakthrough in the layers does not occur simultaneously because of permeability differences. This is a common concern in many waterflooded reservoirs around the world.

Using a completion with downhole flow control valves, injection rates into reservoir layers can be adjusted while allocating more water into lower permeability layers. This leads to improved sweep efficiency and increased ultimate recovery. For producers, a segment of the wellbore can be shut off when water breaks through, thus reducing the amount of water to be processed at surface and preventing early lift-die out of the well. Therefore, water handling costs can be reduced.

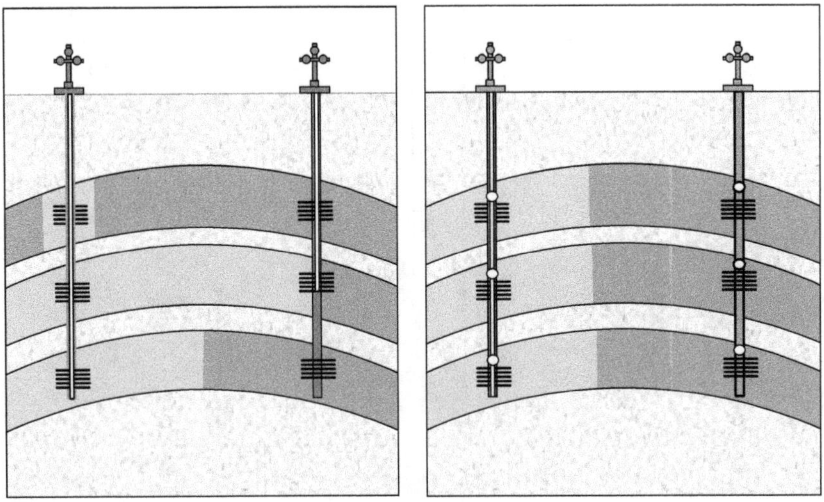

FIGURE 7.13 Waterflooding with an intelligent well system.

Detection of the water could be done, in theory, by using real-time data measured from zonal pressure and temperature sensors. A similar solution could of course be used to shut off early gas influx from a gas injector. In terms of measurement and control, this example relates to daily production optimization, as well as asset management.

7.5.3 Horizontal Producers

The advantages of horizontal wells include increased production rate because of the greater wellbore length exposed to the payzone, larger and more efficient drainage pattern leading to increased overall reserves recovery, and reduced water and gas coning because of reduced drawdown in the reservoir for a given production rate. However, because of reservoir heterogeneity and pressure gradients across a reservoir, caused by production and injection operations, breakthrough of gas or water often occurs in a small section of a horizontal well. The resulting production of large amounts of gas or water with the oil increases processing cost and depletes reservoir energy.

In relatively homogeneous, high-permeability reservoirs, the flux will be higher at the heel of horizontal wells because of frictional pressure losses resulting from cumulative flow from the toe to the heel (Fig. 7.14). This will cause higher flux at the heel region and faster movement of the water or gas fronts toward the wellbore, which will cause earlier water or gas breakthrough at the heel. Choking of the heel region should correct the imbalance and reduce the flux to what it would have been without the frictional pressure loss.

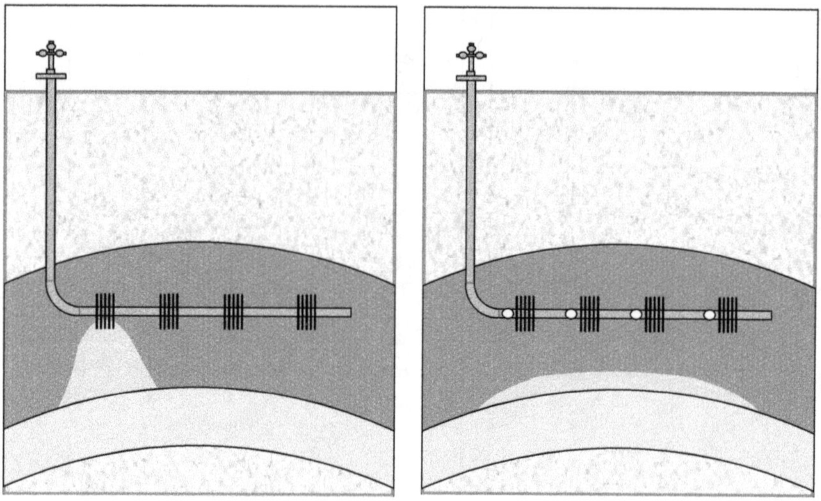

Figure 7.14 Horizontal producers with an intelligent well system.

Another concern in employing long horizontal wells is well clean-up. For example, the toe of a long, high-angle well can be selectively produced using the intelligent well system and thus provide a greater drawdown and better clean-up characteristics than commingled flowback.

7.5.4 Auto-Gas Lift

Auto-gas lift systems use gas from a gas bearing formation or gas cap to lift an oil-producing zone artificially (Fig. 7.15). Unlike conventional gas lift in which gas is pumped down the annulus from surface, an auto-gas lift well uses in situ formation gas for gas lifting. The flow of gas into the production tubing is controlled by a downhole flow control valve with a capability to adjust the flow area from the surface.

Certain well conditions must exist to justify auto-gas lift (Konopczynski, 2017):

- The pressure of the gas reservoir must exceed the kick off pressure of the well.
- The gas reservoir must produce sufficient gas for effective lift at moderate drawdown pressures.
- The volume of gas reserves from the gas bearing formation must be large enough to maintain sufficient pressure and productivity under a variety of conditions as the oil zone is depleted and water cut increases.

In many cases, the productivity of the gas source is uncertain and pressure and productivity can change over the life of the well.

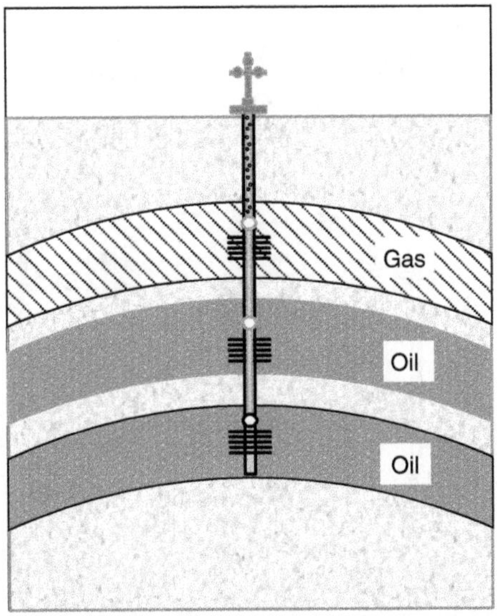

FIGURE 7.15 Intelligent well system utilizing in situ gas zone for auto-gas lift.

The design of the downhole flow control valve for the gas zone must be properly sized for the range of gas and oil reservoir performance, fluid compositions, and reservoir pressure over time.

Auto-gas lift systems can generate significant values by lower gas lift investment in required infrastructure, accelerated oil production, lower operating expenditure. It has been reported that an estimated 60 auto-gas lift systems have been implemented most of which are installed in the Scandinavian sector of the North Sea (Vasper, 2008).

7.5.5 Dumpflood

Waterflooding is one of the most widely applied EOR techniques in oil fields. However, it will require a costly and long-lead-time surface infrastructure system. When a conventional waterflooding project cannot economically be justified, a dump-flood concept can be implemented. In cases in which a high pressure water-bearing formation near the producing zone is available, the water can be produced from the aquifer and injected into the oil-bearing interval as shown in Fig. 7.16. This is often called a dumpflood well providing pressure support and maintenance in the depleting formation.

An intelligent well system also can be used for dumpflooding of in situ water zones. The adjustable control valve regulates the water flow rate from the lower zone remotely, and therefore controls the water injection rate into the upper zone. The injected water helps

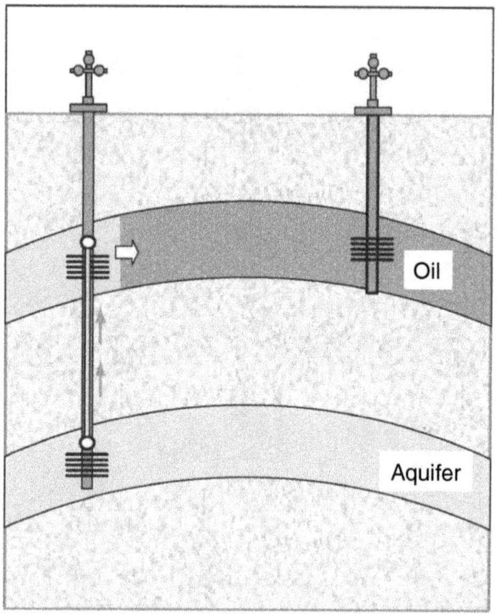

Figure 7.16 Intelligent well system utilizing active aquifer for controlled dumpflooding.

increase oil recover from the upper zone. This innovative approach is successfully applied in Nigeria (JPT , 2005) and West Kuwait (Rawding et al., 2008) and allows for saving the costs of drilling an injection well as well as expenses associated with building additional surface facilities related to water treatment.

Nomenclature

A_{choke} = cross-sectional area of the choke
C = choke flow coefficient
C_v = valve flow coefficient
D_2 = restriction diameter
C_d = discharge coefficient
β = ratio of small to large diameters in a restriction
γ = specific gravity
ρ_l = fluid density

References

Al-Mubarak, S. M., Sunbul, A. H., Hembling, D. E., Sukkestad, T., and Jacob, S. (2008, January 1). "Improved Performance of Downhole Active Inflow Control Valves through Enhanced Design: Case Study," *Society of Petroleum Engineers*. doi:10.2118/117634-MS.

Ali, M. A., and Shafiq, M. (2008, January 1). "Integrating ESPs with Intelligent Completions: Options, Benefits and Risks," *Society of Petroleum Engineers*. doi:10.2118/120799-MS.

Arashi, A., and Konopczynski, M. (2003, January 1). "A Dynamic Optimisation Technique for Simulation of Multi-Zone Intelligent Increase through Water Flooding with Smart Well Technology," *Society of Petroleum Engineers*. doi:10.2118/68979-MS.

Economides, M. J., Hill, A. D., Ehlig-Economides, C., and Zhu D. (2012). *Petroleum Production Systems*, 2nd ed., ISBN: 0-13-703158-0, Prentice Hall. Englewood Cliffs, N.J.

Haugen, V. E. J., Fagerbakke, A. -K., Samsonsen, B., and Krogh, P. K. (2006, January 1). "Subsea Smart Multilateral Wells In Well Systems in a Reservoir Development," *Society of Petroleum Engineers*. doi:10.2118/83963-MS.

Brouwer, D. R., Jansen, J. D., van der Starre, S., van Kruijsdijk, C. P. J. W., and Berentsen, C. W. J. (2001, January 1). "Recovery creases Reserves at Gullfaks South Statfjord," *Society of Petroleum Engineers*. doi:10.2118/95721-MS.

Hill, A. D., Zhu, D., and Economides, M. J. (January, 2008). "Multilateral Wells," *Society of Petroleum Engineers*. Richardson. ISBN-13 DIGIT: 9781555631383.

Hodges, S., Olin, G., and Sides, W. (2000, January 1). "Hydraulically-Actuated Intelligent Completions: Development and Applications. Offshore Technology Conference," 1–4 May, Houston, TX. doi:10.4043/11933-MS.

JPT Staff (2005, September 1). "Adjustable Downhole Choke Technology Facilitates Intelligent-Completion Push," *Society of Petroleum Engineers*. doi:10.2118/0905-0020-JPT.

Konopczynski, M. R. (2017, January). "Intelligent well technology can control oil reservoir inflow, auto-gas lift system," *Offshore Magazine*.

McLauchlan, A., and Nielsen, V. J. (2004, January 1). "Intelligent Completions: Lessons Learned From 7 Years of Installation and Operational Experience," *Society of Petroleum Engineers*. doi:10.2118/90566-MS.

Moore, W. R., Konopczynski, M. R., and Nielsen, V. J. (2002, January 1). "Implementation of Intelligent Well Completions Within a Sand Control Environment," *Society of Petroleum Engineers*. doi:10.2118/77202-MS.

Moreira, O. M. (2004, January 1). "Installation of the World's First All-Electric Intelligent Completion System in a Deepwater Well," *Society of Petroleum Engineers*. doi:10.2118/90472-MS.

Nozaki, M., Burton, R. C., Furui, K., and Zwarich, N. R. (2015, September 28). "Review and Analysis of Zonal Isolation Effectiveness in Carbonate Reservoirs Using Multi-Stage Stimulation Systems," *Society of Petroleum Engineers*. doi:10.2118/174755-MS.

Puckett, R., Solano, M., and Krejci, M. (2004, January 1). "Intelligent Well System with Hydraulic Adjustable Chokes and Permanent Monitoring Improves Conventional ESP Completion for an Operator in Ecuador," *Society of Petroleum Engineers*. doi:10.2118/88506-MS.

Rawding, J., Al-Matar, B. S., and Konopczynski, M. R. (2008, January 1). "Application of Intelligent Well Completion for Controlled Dumpflood in West Kuwait," *Society of Petroleum Engineers*. doi:10.2118/112243-MS.

Robinson, M. (2003, August 1). "Intelligent Well Completions," *Society of Petroleum Engineers*. doi:10.2118/80993-JPT.

Skilbrei, O., Chia, R., Schrader, K., and Purkis, D. (2003, January 1). "Case History Of A 5 Zone Multi-Drop Hydraulic Control Intelligent Offshore Completion in Brunei," *Offshore Technology Conference*. doi:10.4043/15191-MS.

Sun, K., Coull, C., and Constantine, J. J. (2008, January 1). "A Practice of Applying Downhole Real Time Gauge Data and Control-Valve Settings to Estimate Split Flow Rate for an Intelligent Injection Well System," *Society of Petroleum Engineers*. doi:10.2118/115135-MS.

Sun, K., Omole, O. A., Saputelli, L. A., and Gonzalez, F. A. (2011, January 1). "An Application Case of Transferring Intelligent Well System Triple-Gauge Data into Real-Time Flow Allocation Results," *Society of Petroleum Engineers*. doi:10.2118/144259-MS.

Vachon, G., and Bussear, T. R. (2005, January 1). "Production Optimization in ESP Completions with Intelligent Well Technology," *Society of Petroleum Engineers*. doi:10.2118/93617-MS.

Vasper, A. C. (2008, February 1). "Auto, Natural, or In-Situ Gas-Lift Systems Explained," *Society of Petroleum Engineers*. doi:10.2118/104202-PA.

Williamson, J. R., Bouldin, B., and Purkis, D. (2000, January 1). "An Infinitely Variable Choke for Multi-Zone Intelligent Well Completions," *Society of Petroleum Engineers*. doi:10.2118/64280-MS.

Zarea, M. A., and Zhu, D. (2011, January 1). "An Integrated Performance Model for Multilateral Wells Equipped with Inflow Control Valves," *Society of Petroleum Engineers*. doi:10.2118/142373-MS.

Zhu, D., and Furui, K. (2006, January 1). "Optimizing Oil and Gas Production by Intelligent Technology," *Society of Petroleum Engineers*. doi:10.2118/102104-MS.

CHAPTER 8

Completion Designs for Production and Injection Optimization

With the advances in completion and stimulation technologies, many reservoirs that were uneconomical to develop before have become the main players in today's energy supply. To develop these reservoirs, completion design, installation, and operation to fully execute the completion functions are crucial. Six field case examples are used to demonstrate production and injection optimization to combat unique challenges encountering in different types of reservoirs using various well completion and stimulation techniques.

The first field case example is an application of intelligent well system (IWS) to a horizontal openhole producer in a mature waterflooded chalk reservoir, presented by Kent et al. (2014). In this example, multizone openhole completion was selected allowing high-rate matrix acidizing treatment in a high-porosity chalk formation. Also IWS is employed for production conformance, selective stimulation, scale squeeze treatment, and zonal production testing.

The second example is an integration of multilateral well and inflow control device (ICD) applied to a reservoir that consists of interbedded sandstone and shale, presented by Chowdhuri et al. (2015). In this example the multilateral drilling maximizes reservoir exposure and increases well productivity, while ICD regulates inflow profile along each producing lateral, lowering the risk of early water breakthrough.

The third example is an integration of the maximum reservoir contact, multilateral, and IWS applied to Ghawar oilfield, presented by Al-Mubarak et al. (2008). Compared to the second example, this case highlights how water production risks can be mitigated using real-time production data obtained from the IWS permanent monitoring systems.

The fourth example covers a successful application of acid fracturing technique in a deep, hot carbonate presented by Galarraga et al. (2005). In this example, the well production was dramatically improved by the application of HCl-formic acid system in acid fracturing compared to those stimulated by organic acids systems.

The fifth example presents a horizontal openhole gravel pack completion in an unconsolidated formation with high fines content, presented by Hadfield et al. (2007). In this example, the well was successfully completed at minimum completion skin (less than 5) while producing at minimal fines production (less than 0.05 g/s) at the target rate of 60 MMscf/d.

The last example presents production optimization results in multistage hydraulic fracturing stimulation in a shale formation presented by West et al. (2014). In this example, adjusting the horizontal lateral position relative to the geological formations and closely spacing hydraulic fractures improved the amount of stimulated reservoir volume (SRV) and the well productivity dramatically.

8.1 Multistage Matrix Acid-Stimulation in a High-Porosity Chalk Reservoir with Openhole Intelligent Completion

The first field case example discusses the design, testing, installation, and performance of a horizontal openhole well using an intelligent inner completion inside an uncemented liner with openhole packers for zonal isolation. The openhole completion was selected to achieve a full reservoir access along the reservoir pay length for production. Selective stimulation is achieved using the intelligent well systems where high-rate matrix acid stimulation using limited entry is employed for fluid diversion within well segments. The study is presented by Kent et al. (2014).

8.1.1 Ekofisk Field Background

The Ekofisk field is in the central part of the North Sea, off the southwestern coast of Norway. The Ekofisk field produces from the naturally fractured Ekofisk and Tor chalk formations of the early Paleocene and late Cretaceous. The reservoir chalk has high porosity but low matrix permeability, with an oil column of approximately 1000 ft at depths ranging from 9000 to 11,000 ft below sea level. Typical oil gravity is between 30 and 40° API. Ekofisk was originally developed by pressure depletion. Large-scale water injection started in 1987. Scale-inhibitor treatments have been performed regularly since 1996 to combat the high barium sulfate scaling potential related to floodwater and formation-water incompatibility.

Completion practices at Ekofisk have evolved over time. Initial vertical wells were cased, perforated, and stimulated by acid fracturing. As horizontal wells were introduced to the field, the long horizontal sections were also cased and cemented. The wells were stimulated with acid by bullheading alternating stages of pad and acid with ball sealers for diversion. Production logging results indicated that many of the perforation clusters were not effectively stimulated, with some clusters, typically at the heel of the wells, receiving far more acid than required and others receiving little or no acid. The large volumes of acid placed into a limited number of perforation clusters weaken the rock and coupled with reservoir compaction and associated chalk movement, lead to liner buckling. Liner buckling then led to the loss of zonal access and, in extreme cases, liner failure and loss of the well. Restricted access to the reservoir liner impeded well intervention and remedial opportunities.

8.1.2 Well Design Requirements

A number of functional requirements were established to improve well performance, reservoir management, and reservoir-liner integrity in a mature waterflooded reservoir. These functional requirements include:

- Uncemented reservoir liner with openhole packers, providing effective hydraulic isolation of water-swept intervals while allowing high-pressure/high-rate acid stimulation, and oil production from target oil zones.
- Full completion of entire length of productive zones, by use of high-rate matrix acid stimulation to improve post-acid completion skin, improve well productivity, and minimize liner deformation for extended well life.
- Surface controlled access to productive zones in the horizontal lateral to allow opening and closing zones without wireline or coiled-tubing intervention, flow control into and out of each zone during stimulation, production, and scale-inhibitor squeezes, well-test flexibility, measurement of zonal pressure/temperature data.

8.1.3 New Well Design

Figure 8.1 is a schematic of the well structure and completion. The 7⅝ in, 51.2 lbm/ft Q-125 openhole liner system runs to the total depth. Two toe isolation valves are run in series at the shoe of the liner. The main function of both toe valves was to provide permanent hydraulic isolation in both directions once closed. The openhole packers were

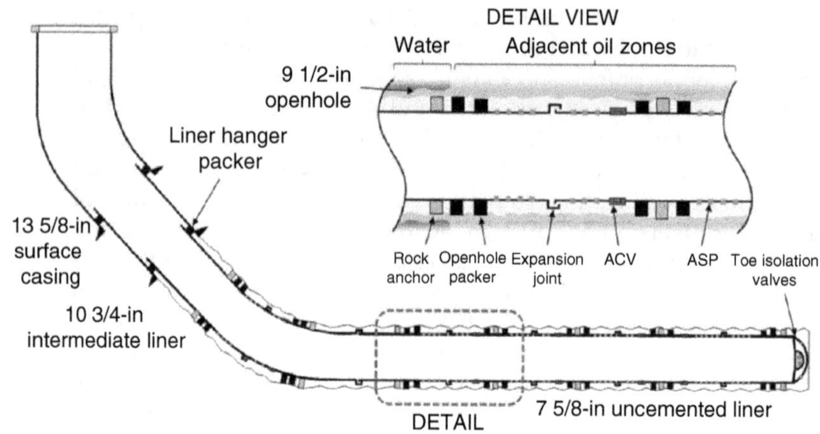

FIGURE 8.1 Openhole uncemented liner employed in the Ekofisk field (Kent et al., 2014).

designed to provide hydraulic isolation between all zones during the acid-stimulation and production phases. The well is divided into five zones (A, B, C, D, and E) according to the reservoir characteristics, illustrated in Fig. 8.2. Because of the tremendous forces exerted by the large thermal and hydraulic loads during acid stimulation, special consideration was given to allowing for liner contraction and stabilization of the openhole packers. Rock anchors were designed to minimize liner and packer movement during stimulation and production while expansion joint was run near the middle of each zone to relieve tensile loads that were expected to develop during stimulation because of thermal contraction and ballooning.

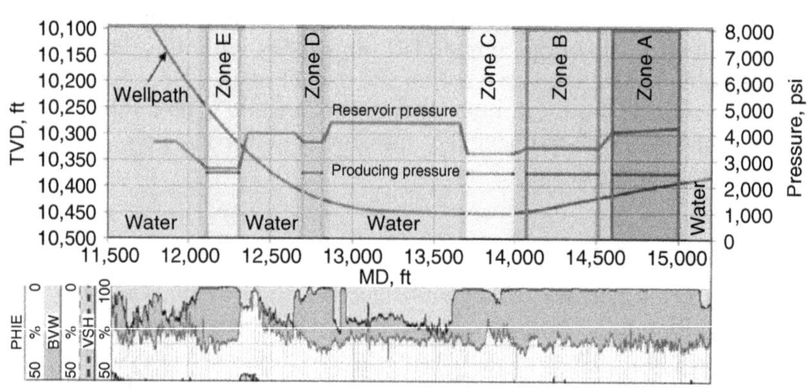

FIGURE 8.2 Well trajectory and saturation profiles from log data (Kent et al., 2014).

The openhole liners were pre-perforated and plugged by acid soluble plugs. These plugs allow for running of a pressure-tight liner, thus enabling the pressure activation of all liner components and preventing fluid loss while running the intelligent inner-completion string. The liner also has acid circulating valves included at one end of each treatment zone to allow for acid to be circulated into the wellbore after running and setting the intelligent completion system.

The ability to selectively shut off water-swept sections was considered critical for well performance and reserves recovery, and has proved to be invaluable. This entailed use of a dedicated intelligent well string with multiple control valves and packers to be run inside the 7⅝-in production liner (Fig. 8.3). A 10¾ × 5½-in feedthrough production packer was installed above the uncemented liner while 7⅝ × 3½-in feedthrough zonal isolation packers were set in blank sections between the openhole packers separating reservoir net pay intervals.

Quartz oscillator transducer gauges were installed to measure both tubing and annulus pressure and temperature in each zone. This allowed reservoir characterization and stimulation diagnostics to be performed after acid injection. In addition, the gauges provide early identification of water breakthrough, allowing valve positions to be configured for optimal producing conditions. Variable-position sliding sleeve valves were installed in each zone. Full tungsten carbide trim was used for tolerance of extreme-stimulation internal-flow velocities of more than 130 ft/s and port-flow velocities of more than 150 ft/s in fully open position. These inflow control valves operate by indexing sleeve actuation, providing 14 valve positions from fully opened to fully closed.

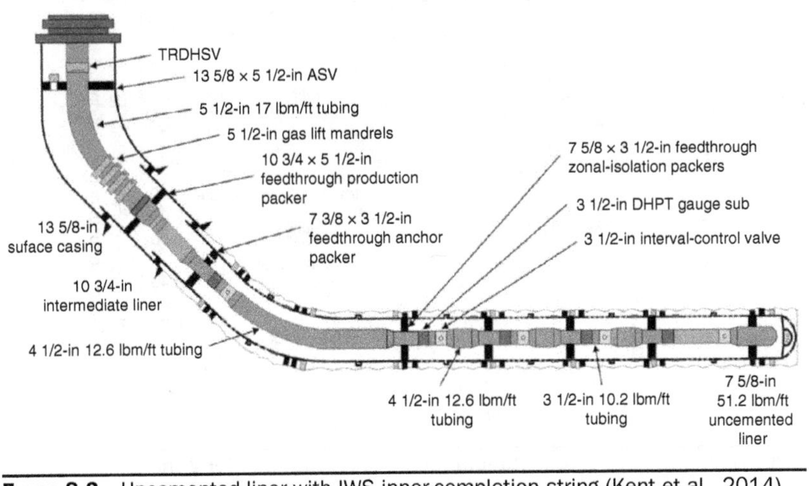

FIGURE 8.3 Uncemented liner with IWS inner-completion string (Kent et al., 2014).

8.1.4 Stimulation and Production Results

Each completion interval (A, B, C, D, and E) was selectively stimulated starting from Zone E (heel zone) to Zone A (toe zone) by use of the intelligent well flow-controlled valves. Twenty-eight percent HCl was used for the main acid treatment. The acid exists the 7⅝-in casing/intelligent well liner string annulus through the 0.20-in diameter holes. As in a conventional cased-hole limited entry treatment, frictional pressure losses resulting from high velocity flow through the small plug holes limit the amount of fluid exiting through any one point in the liner, thereby distributing flow more evenly along the openhole treatment interval. At designed acid rates of 0.7 to 1.3 bbl/min per hole, average fluid velocities discharging through the 0.20-in holes in the liner and into the openhole range from 300 to 550 ft/s and provide perforation pressure drops in the 1000 to 3000 psi range. These high perforation pressure losses caused by the flow along the annulus of 7⅝-in liner/4½-in intelligent well string provide uniform injection along the treatment interval.

A pressure plot of a typical zone treatment is provided in Fig. 8.4. The main acid treatments consisted of pumping 75 gal of 28% HCl per foot of interval, followed by approximately 150 gal of freshwater overflush per foot of interval. At designed pump rates, downhole pressures at the reservoir face typically start out at or above formation-parting pressure and very quickly drop to less than the fracture pressure as acid enters the chalk formation. In the example case presented in the figure, it took about 6 min for the injection pressure to drop to less than the fracture closure pressure. This pressure behavior

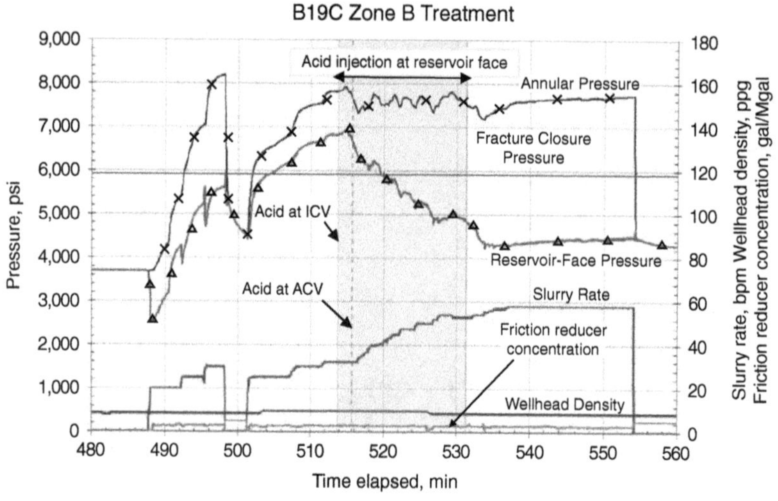

Figure 8.4 High-rate matrix acid stimulation data versus time for Zone B (Kent et al., 2014).

indicates that the bulk of the acid-stimulation treatment is pumped under matrix injection conditions.

As the acid wormholes grow in the formation, reservoir-face pressure continues to decline as pump rate increases. This injectivity improvement can be used to calculate skin evolution during the course of the treatment. In this case, the effective wellbore radius is estimated to be 33 ft corresponding to a skin of –4.4, at the end of the treatment. Production stabilized at approximately 3500 STB/D after 1 year, and water cut has steadily increased since then, from 5% to 25%.

8.1.5 Value Realized

An openhole uncemented completion with a five-zone IWS was successfully installed in the Ekofisk field. In terms of the completion objectives, the well has delivered better-than-projected production results despite lower-than-expected reservoir pressures and net-pay lengths; provided water shutoff in Zone D to reduce scaling potential; allowed interventionless stimulation of each fully completed interval with zonal isolation; and produced with no indication of deformation or reduced productivity because of liner-stability issues. Downhole flow control valves allowed selective stimulation of each fully completed interval. Zone tubing and annulus gauges provided real-time differential pressure, which was valuable when attempting to maximize stimulation rate without exceeding burst pressure ratings of the tubulars in isolated zones.

8.2 Improved Production Using Multilateral Well with ICD

Multilateral technology has been employed to increase the wellbore contact with reservoir improving well productivity. In coning situations, such as in the production of oil reservoirs with a bottom aquifer, implementing ICD can be used to delay water or gas breakthrough by reducing annular velocity across a selected interval. Chowdhuri et al. (2015) presented successful integration of the first multilateral well with a Level-4 junction with an ICD employed in the Upper Burgan reservoir of Raudhatain field, north Kuwait.

8.2.1 Field Background

The Raudhatain oilfield structure is one of several developed along a prominent anticlinal ridge that plunges gently north of Kuwait. The Raudhatain structure is a faulted anticlinal dome with 65 to 90 ft of topographical relief. The 3D seismic has defined the major faulting in the northern part of Raudhatain as northwest trending, whereas in the southwestern part of the field the faults trend northwest. Fault throws are highly variable in the structure and range from less than 30 ft to as much as 150 ft.

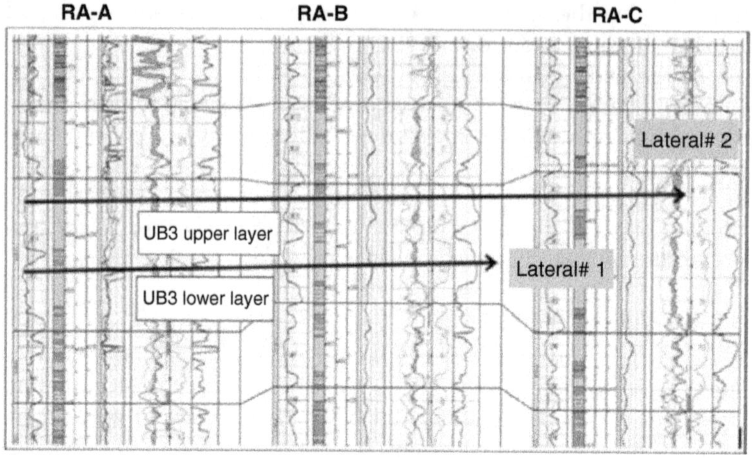

FIGURE 8.5 Location of upper and lower laterals in UB3 sands (Chowdhuri et al., 2015).

The location for the multilateral well is determined by considering the reservoir pressure, fluid properties, oil saturation, and oil water contact. The new well is planned in the eastern flank of the field with good oil saturation, viscosity of 4 to 5 cp, and 2300 psi reservoir pressure, which is far above the bubble point pressure of 1300 to 1400 psi. The Upper Burgan formation is a layered sandstone-shale sequence deposited in deltaic settings having very fine to fine grained marine influenced channel sand as reservoir rock. It was decided to exploit UB3 upper and UB3 lower sands with two vertically staggered laterals with commingled production (Fig. 8.5). In the eastern part of the Raudhatain field, the channel sandstones are particularly well-developed, with thicknesses of 25 to 40 ft. The range of permeability varies widely, from a few hundred mD to 3.5 Darcies. An impermeable shale layer separates the two sand layers.

8.2.2 Drilling a Multilateral and Geosteering Challenges

Geosteering and evaluating the well is challenging without using a proper logging while drilling (LWD) technology. Indeed, the resistivity anisotropy is a major issue, especially if it occurs with influence of other bed boundary effects like resistivity of adjacent beds or polarization horn effects. Water coning issues in the field make it even worse to interpret the resistivity data as they become spiky.

A pilot hole of 8½ in was drilled with 64.6° inclination for evaluating the sublayers of UB3 by use of an LWD tool and a formation pressure tester. At the beginning of drilling the upper lateral, it was sidetracked for geological reasons helping to geosteer the well into the good quality sand reservoir and avoid penetrating the shale barrier between the upper and lower UB3 layers. The second lateral was placed in UB3 upper zone. The length of the upper lateral in UB3 upper was 2145 ft in good quality reservoir sand whereas the length of the lower lateral was 1757 ft in the UB3 lower channel sand of equally good quality reservoir rock. The well was drilled and completed successfully in 72 days, that was 8 days less than the planned rig days.

8.2.3 New Well Design

The objective of the completion was to have a pressure integrity at the junction with a Level-4 type of multilateral completion assembly, to achieve the optimum production target from both laterals simultaneously. Hence, the well was completed with a 5½-in openhole ICD for the first lower lateral and a 4½-in openhole ICD for the second upper lateral. Figure 8.6 illustrates the well schematic of the dual lateral well. The production is commingled using an electrical submersible pump as the pressure difference between the two laterals was less than 100 psi.

8.2.4 Production Results

The initial production results indicated that the multilateral well is expected to produce approximately 5322 BOPD with 0% water cut

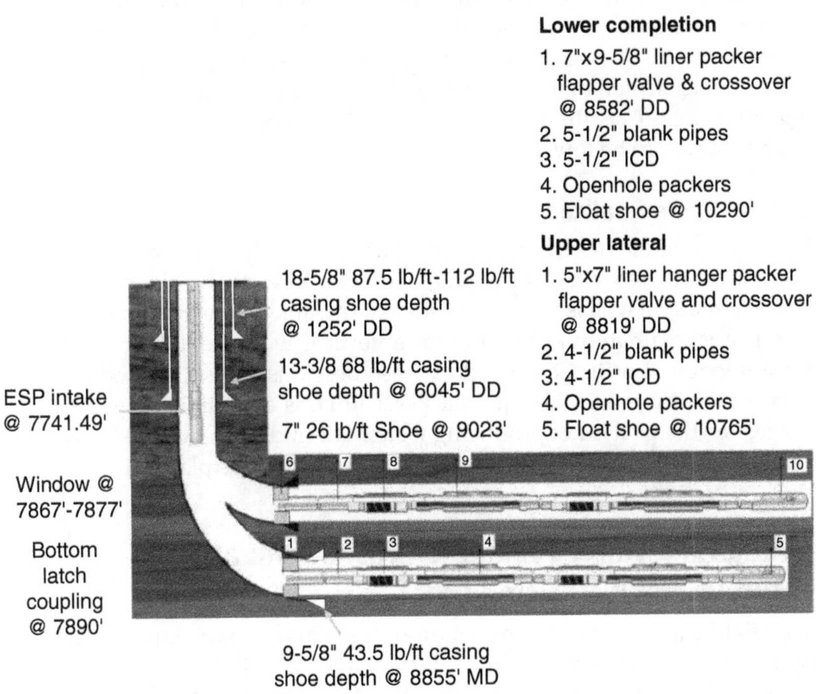

FIGURE 8.6 Well schematic of completion assembly (Chowdhuri et al., 2015).

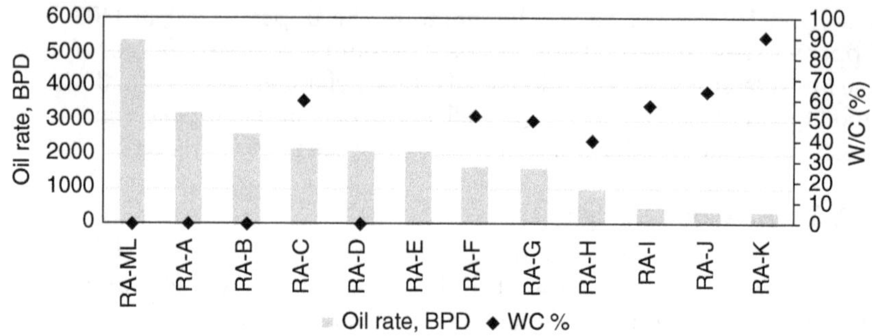

FIGURE 8.7 Production performance of the multilateral well (RA-ML) vs Other UB horizontal wells (Chowdhuri et al., 2015).

through a 48/64 choke size. The average production of previous horizontal wells is reported to be about 1500 BOPD. Also the best production from a single lateral horizontal well completed in the upper Burgan formation was approximately 3145 BOPD. Thus, the production of the multilateral well was found to be significantly higher than that of any other producing wells of the Upper Burgan formation in the Raudhatain structure. The well is producing 40% more than the best horizontal well and exhibits threefold more in production than the average horizontal production rate as shown in Fig. 8.7. The production rate of 5322 BOPD was exceptionally good considering the fact that the well was drilled in the eastern flank where the oil viscosity is high and production results from vertical wells drilled in the area have been reported to be low.

8.2.5 Value Realized

Use of a multilateral well with an ICD has efficiently accelerated production, improved sweep efficiency, and delayed water breakthrough. The multilateral well is located in the eastern flank where the expected rate is approximately 700 BOPD for a vertical well and 1500 BOPD for a horizontal well. Compared to these previous wells, the production result of the multilateral well (5322 BPD) is exceptionally good. Moreover, the well is not producing water. This indicates that the ICDs integrated to the upper and lower laterals functioned as planned.

8.3 Mitigating Water Production Risks from a Multilateral Well Using IWS

This field case describes completion designs and production results of the first maximum reservoir contact through multilateral well with IWS deployment in Ghawar field, Saudi Arabia. In the previous example, a passive ICD was employed to delay water breakthrough in a multilateral well. In this example, active inflow control valves are used to control lateral production to reduce or shut off water production. Comparing these two cases clarifies differences in water production control between ICD and IWS technology.

8.3.1 Ghawar Field Background

Ghawar field, discovered in 1948, is the largest oilfield in the world (Eyvazzadeh, 2003). Although Ghawar field is a single field, it is divided into six areas. From north to south, they are Fazran, Ain Dar, Shedgum, Uthmaniyah, Haradh, and Hawiyah. The Haradh area forms the southwest portion of the Ghawar oilfield, approximately 80 km onshore from the Arabian Gulf, in the Eastern Province of Saudi Arabia (Fig. 8.8). The producing formation belongs to the lower member of the Arab-D formation distinguished by a complex sequence of anhydrite and limestone with varying degrees of dolomitization. Fluid mechanism in this specific region is highly influenced by the presence of fracture corridors and strataform superpermeability. The initial developments used conventional vertical wells and the second developments employed long horizontal wells.

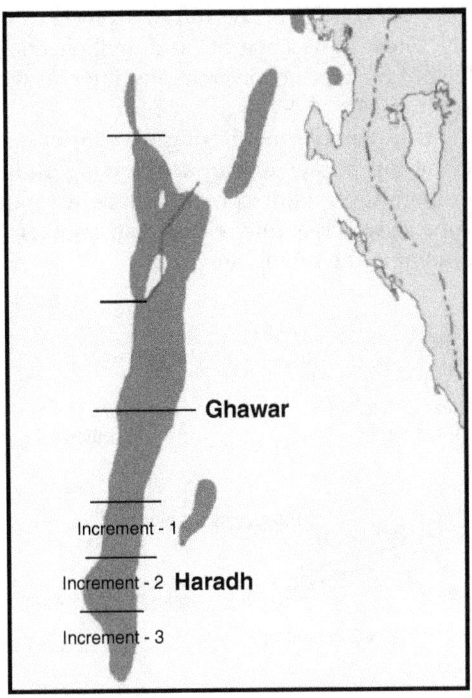

FIGURE 8.8 Ghawar field map (Al-Mubarak, 2008).

8.3.2 Completion Challenges

Geological complexities including the presence of fault and fracture systems, reservoir heterogeneities, and associated premature-water-breakthrough risks are the largest challenges in the Haradh development. The main objectives of the well designs and production are to sustain well productivity, improve sweep, enhance control of multiple laterals, manage water production, and minimize production interruptions.

8.3.3 Completion Design

Haradh-A12 well was drilled in April 2004 and was completed with a 7-in liner set horizontally into the Arab-D producing interval. A 6⅛-in horizontal openhole was then drilled out from the bottom of the 7-in liner. Because of heavy losses while drilling the main bore (L-0), a 4½-in liner was set to cover part of this openhole section. Two 6⅛-in horizontal sidetracks (L-1 and L-2) were then drilled from the 7-in liner and one of the lateral is split to two branches (L-1-1), completing the trilateral well. Figure 8.9 illustrates a well schematic of Haradh-A12 well. The well was completed and put on production from barefoot laterals. A year later, the well was worked over to install the IWS.

The IWS with three variable downhole-flow-control valves was designed to provide control of the inflow from each openhole section of the well. These valves operate as downhole chokes to restrict or completely shut off production from any interval with increasing water cut over time.

A permanent downhole monitoring system was selected to be placed above the top packer to monitor flowing and shut-in pressures and temperatures. Flow-control valves are equipped with 11 positions, one of which is fully closed and another is a full open position equivalent to the tubing flow area.

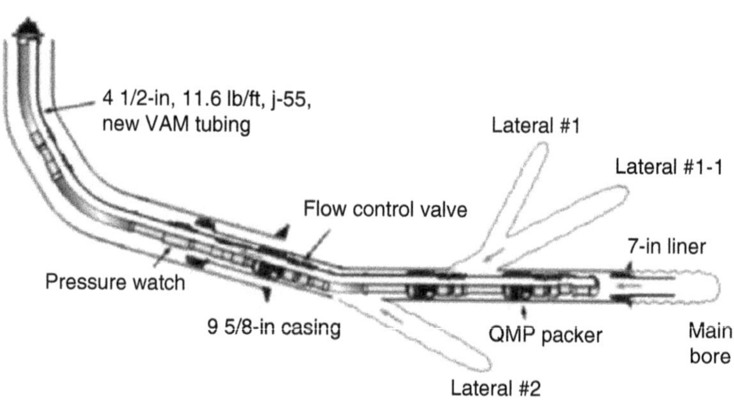

FIGURE 8.9 Haradh-A12 intelligent well completion schematic.

8.3.4 Completion and Production Results

The intelligent completion system was installed subsequently in early April 2005. During the equipment testing, installation, and subsequent flow testing of the well, each of the valves was actuated through more than 10 complete cycles ensuring the functionality of the downhole flow control valves.

Before the installation of the IWS, the well was put on production from barefoot laterals. The well was tested at a rate of 18,000 B/D of dry oil while the analysis of the transient test indicated a productivity index (PI) of 350 B/D/psi. The PI measured in this well is considered significantly higher than offset vertical or horizontal wells (17 or 31 B/D/psi).

Within 2 months of production, the well started producing water. The last test before the workover indicated a water cut of approximately 23% at an oil rate of 8000 B/D. During the testing of the well after the IWS installation, short-duration buildups were performed on each lateral. The productivity testing for each of the laterals was accomplished by testing an individual lateral while the other two laterals were closed. Buildup tests were conducted following the production-rate tests by shutting in the well by use of surface valves. This test was done to determine the PI of each lateral, which helped to decide the choke setting of each lateral when the production is commingled. The initial productivity indices of individual laterals are estimated to be 165, 60, and 80 B/D/psi, respectively. If it was not for the intelligent completion with downhole flow control valves, conducting individual-lateral pressure-transient test in a multilateral well would not be feasible and would require intensive intervention.

After 5 months of production, at an average oil rate of 8000 B/D, the water cut had increased to 30%. Upon that, a comprehensive rate test was performed on the well. The testing involves several downhole-choke-settings combinations with an objective to come up with optimized settings that honor the production strategy for the well and the region.

Test results indicated that L-0 was producing 100% water. The final configuration was adjusted to have the L-0 lateral closed while L-1 and L-2 laterals open at choke setting of 3 and 2, respectively. By use of the surface choke, the total withdrawal of the well was restricted to an oil rate of 5000 B/D and 0% water cut. Since then, the well has been producing at this rate with no water production for about 2 years (until the study was published).

8.3.5 Value Realized

The Ghawar field case shows that the IWS implemented in the Haradh-A12 well effectively reduced water production in the multilateral well. The effective water production reduction became possible with the capabilities of real-time surveillance in the IWS demonstrating a potential to reduce well interventions. Also the Ghawar field example shows that quality control of the system is a priority for successful implementation as illustrated by testing the completion functions in hole before setting the packers.

8.4 Acid Fracturing in a Deep, Hot Carbonate Reservoir

In general, acid fracturing is best applied in shallow, low-temperature carbonate reservoirs, in which the reservoir temperature is less than 200°F and the maximum effective stress on the fracture is less than 5000 psi. Low temperature limits the reaction rate between the acid and the formation, which allows the acid to penetrate deeper into the fracture before becoming spent. Because limestone reservoirs are ductile, a low effective stress on the fracture is required to maintain adequate fracture conductivity over the life of the well. Burgos et al. (2005) presented a field case from the Urdaneta West field, Venezuela, where acid fracturing is successfully applied to a reservoir at a high reservoir temperature of 280°F and reservoir depth greater than 15,000 ft.

8.4.1 Urdaneta West Field Background

The Urdaneta West block covers approximately 1150 km^2 and lies in the western part of the Maracaibo Basin. This basin is located in a triangular-shaped, tectonic block bounded by three major strike-slip faults (Fig. 8.10). A significant portion of the oil in-place volumes occurs in the tight and fractured carbonate of the Cretaceous Cogollo Group as a light (29° API mean) undersaturated, sour crude. The Cogollo Group is comprised of three formations, from bottom to top: the Apon, Lisure, and Maraca formations. It is on average 1000-ft thick consisting predominantly of tight carbonate.

The Maraca limestone formation is a 20- to 30-ft upper member of the Cogollo carbonate group at 15,000 to 16,000 ft deep. Reservoir volumes and productivity are attributed mainly by matrix development, in contrast with the rest of the Cogollo groups, where natural fractures play a major role in well production.

8.4.2 Initial Acid-Fracturing Stimulation Results and Stimulation Challenges

When stimulation and development strategy began focusing on the Maraca group, a number of matrix stimulations were performed, with most treatments done by coiled tubing (CT) at low pumping rates (~1 bpm) and volumes (60 to 80 gal/ft). The first few treatments used 15% hydrochloric acid (HCl), but the acid system was switched quickly to an organic acid blend (13% acetic and 9% formic) because of the high reactivity of the HCl and its tendency to form sludge after contacting the asphaltenic Maraca oil. These matrix treatments

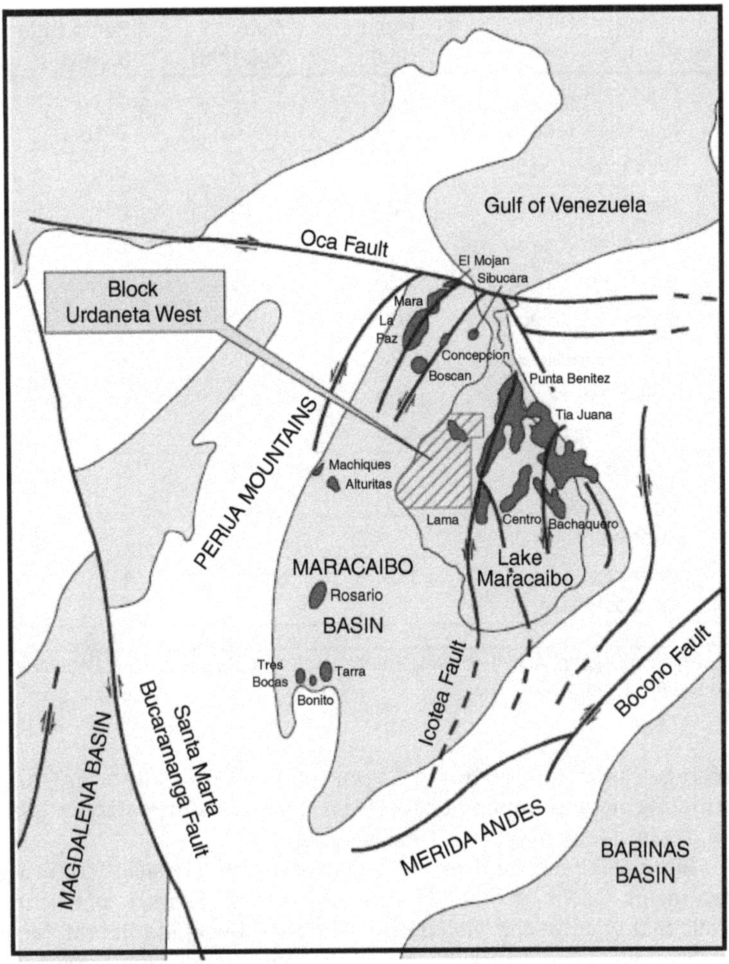

FIGURE 8.10 Location of Urdaneta West field within the Maracaibo Basin in northwestern Venezuela (Galarraga et al., 2005).

had modest results, and the gained productivity was lost after a few months of production. The PI obtained for four wells after matrix stimulations of the Maraca member using the organic acid blend ranged between 0.2 and 0.7 B/D/psi.

As an alternative way to effectively stimulate Maraca wells, acid fracturing was employed. The organic acid system (9% formic and 13% acetic acids) was selected rather than hydrochloric acid to minimize the potential for asphaltene flocculation from crude/acid contact, to have a more retarded acid, and provide easier and less costly corrosion inhibition at the downhole temperature of 280°F. Thirteen jobs were performed using the organic acid system at an

Stage	Fluid	Volume (bbl)	Cum Vol. (bbl)	Pump Rate (bpm)
1	Organic acid	5	5	2
2	Linear gel (cooldown)	145	145	2–18
3	Crosslinked pad	200	345	18
4	Gelled organic acid	220	565	18
5	Crosslinked pad	100	665	18
6	Gelled organic acid	220	885	18
7	Crosslinked pad	100	985	18
8	Gelled organic acid	220	1205	18
9	Linear gel	200	1405	18–20
10	Organic acid	170	1575	20–4
11	Organic acid	40	1615	4
12	Treat. water (overflush)	50	1665	4
13	Linear gel (displacement)	161	1826	4

TABLE 8.1 Typical Organic Acid-Fracturing Treatment Schedule in Maraca Limestone Formation (Galarraga et al., 2005)

average pump rate of 16 to 18 bpm followed by a closed fracture acidizing stage pumped at 2 to 4 bpm. Table 8.1 summarizes a typical organic acid-fracturing treatment schedule.

Figure 8.11 summarizes the results of the organic acid-fracturing treatments. Most of these treatments were performed in existing wells that were re-completed from a deeper depleted interval. Most of the wells did not produce after initial perforation due to perforation damage. After the acid-fracturing treatment, the well productivity dramatically improved. The post-stimulation skins are reported to be –3 to –4 after acid fracturing. Wells C and F showed relatively low production after acid fracturing but this was expected as the wells had permeabilities below 2 mD.

It was also confirmed that the gained productivity was sustainable over time in most of the wells even with the relatively high effective closure stresses of 4000 to 6000 psi on the etched fractures. The sustained fracture conductivity is thought to be caused by high rock strengths of the formation (Young's modulus in the 5 to 7×10^6 psi range). Although these acid-fracturing results are considered successful, further inflow performance analysis results indicated that effective fracture-half lengths are estimated to be only 50 to 100 ft suggesting that there is a potential for further production improvement with different acid-fracturing designs.

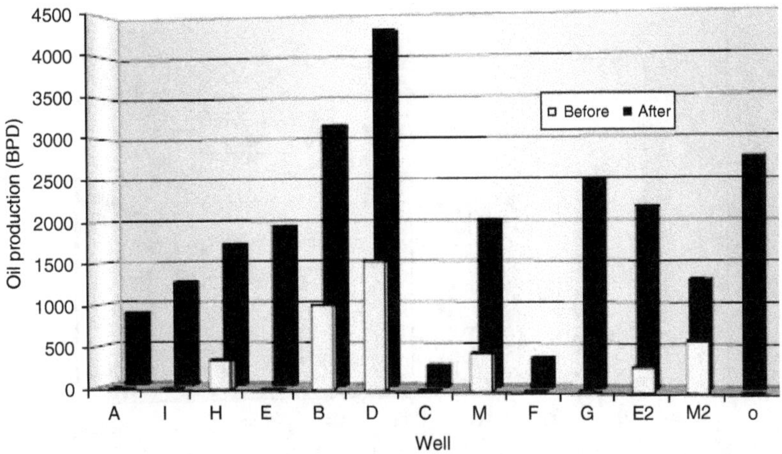

FIGURE 8.11 Production results of pre- and post-acid fractured wells using organic acids (Burgos et al., 2005).

8.4.3 New Stimulation Design

The new acid system employs 7% HCl–11% Formic system (7/11 system). A high carbonate dissolving capacity was obtained by adding 7% HCl improving the conductivity in the near wellbore part of the fracture. The formic acid starts spending only after most of the HCl is consumed penetrating deeper into the fracture. This results in an increased etched length. The concentration of formic acid was set at 11% to keep calcium formate from precipitating.

In the new acid-fracturing design, a surfactant-based system was chosen instead of a gelled acid or self-diverting crosslinked acid because the surfactant-based system maintains its viscosity across the spending process from an early stage during reaction and also has better clean up properties than gelled acid systems.

8.4.4 Stimulation and Production Results

The new formulation was tested in the field in new and previously fractured wells. After successful results on the first few wells, all the wells with at least 15-mD permeability were refractured with the new formulation and it became the standard for all wells. A total of 13 wells were treated in the third campaign, eight of which were refracturing executed to increase the PI and not because of significant productivity decline over time. The only well where the PI was not increased was well E. This well has the lowest permeability-thickness product of the group, and it is likely that the conductivity achieved with the first system was already close to the technical limit. The 35% average increase in productivity led to record production levels in the field and proved the merit of reformulating the

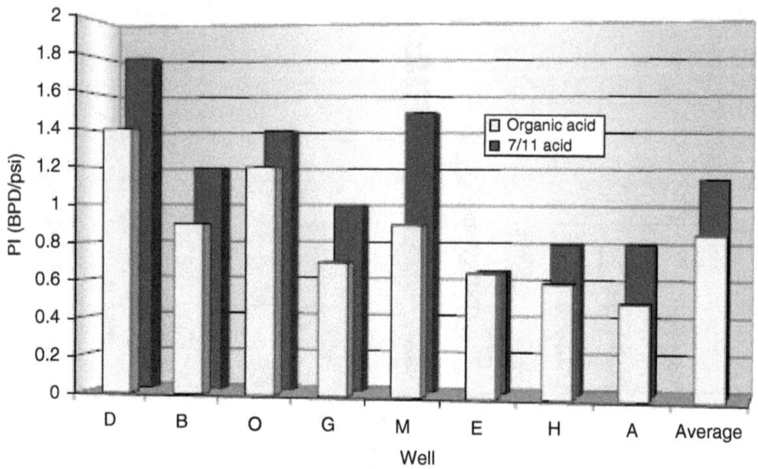

FIGURE 8.12 Productivity index history for wells re-fractured with the 7/11 formulation (Galarraga et al., 2005).

acid-fracturing system. Figure 8.12 summarizes production results before and after the re-fracturing campaign.

8.4.5 Value Realized

The acid-fracturing experience in the Maraca formation confirms that in many instances, significantly better field-tailored solutions can be obtained with proper laboratory testing and field trials of various alternatives. The equivalent pseudoradial skins achieved after acid fracturing the Maraca formation with a combination of HCl and formic acid are reported to be consistently between –5 and –7 and close to the technical limit. An optimum acid blend combines a high carbonate-dissolving capacity with sufficiently deep acid penetration. In the Maraca wells fractured with the HCl/formic acid system, it is reported that closed fracture acidizing-stage benefits were not evident following acid fracturing stages.

8.5 Horizontal Openhole Gravel Pack Completion in an Unconsolidated Formation with High Fines Content

This field case example is an application of horizontal openhole gravel pack completion for a shallow laminated gas pay presented by Hadfield et al. (2007). This example covers the selection process of downhole sand exclusion techniques in an unconsolidated formation with high fines content.

8.5.1 Shallow Clastics Field Background

The Shallow Clastics field consists primarily of two shallow gas-bearing reservoirs, H1 and H2, at approximately 2650 ft true vertical depth (TVD). These reservoirs are laterally extensive, covering an area of 200 km^2 with an estimated gas in place (GIP) in excess of 2 Tcf. The reservoirs are made up a sequence of highly laminated sand and shale deposits with significant sand-size variability and high fines content. Being highly unconsolidated, downhole sand exclusion is mandatory. The primary drive mechanism is a depletion drive based on the weak aquifers seen in existing fields in the area. The Shallow Clastics reservoirs overlay deeper Central Luconia carbonate gas reservoirs, which are already on production with further fields in development; therefore, a gas-processing and gathering system was already in place. Gas from all of these fields is produced to the Malaysian Liquefied Natural Gas (MLNG) plants at Bintulu, East Malaysia. Production from the Clastics field is intended to counteract decline from other fields and is critical to maintaining the security of the supply to MLNG.

Significant log data (Fig. 8.13) on the Shallow Clastics field were gathered from the appraisal and development wells of the deeper carbonate gas reservoirs; however, core data were limited to what could be generated from a single poor-quality core from E11-SC1. A dedicated Shallow Clastics appraisal/early-producer well (E11-SC2) had been drilled with a deviated wellbore through the H1/H2 targets, and a completion design consisting of a cased and perforated commingled completion inside 9⅝-in casing had been implemented. The sandface-completion design consisted of a large-OD expandable sand screen with a 150-μ-weave opening across the two zones. Upon completion, the reservoirs were cleaned up through a temporary facility to test productivity and evaluate integrity of the downhole sand exclusion installation. Fines production, possibly caused by a failure of the expandable screen, commenced during the cleanup and steadily increased to the extent that the well no longer could be produced.

8.5.2 Sand Control Challenges

The field-development plan for Shallow Clastics called for a phased multiwell development from four locations that would include the existing carbonate field, E11, and the future F13 drilling platforms. Given the existing infrastructure, it was logical to tie production from Shallow Clastics back to the processing system at E11. This plan, however, would require a dedicated booster compressor for the Shallow Clastics gas because of its lower reservoir pressure compared to that of the deeper carbonates.

The booster compressor and the existing infrastructure have a low tolerance to solids production and require a 98% level of reliability to maintain the security of the supply to MLNG. Further, high-rate

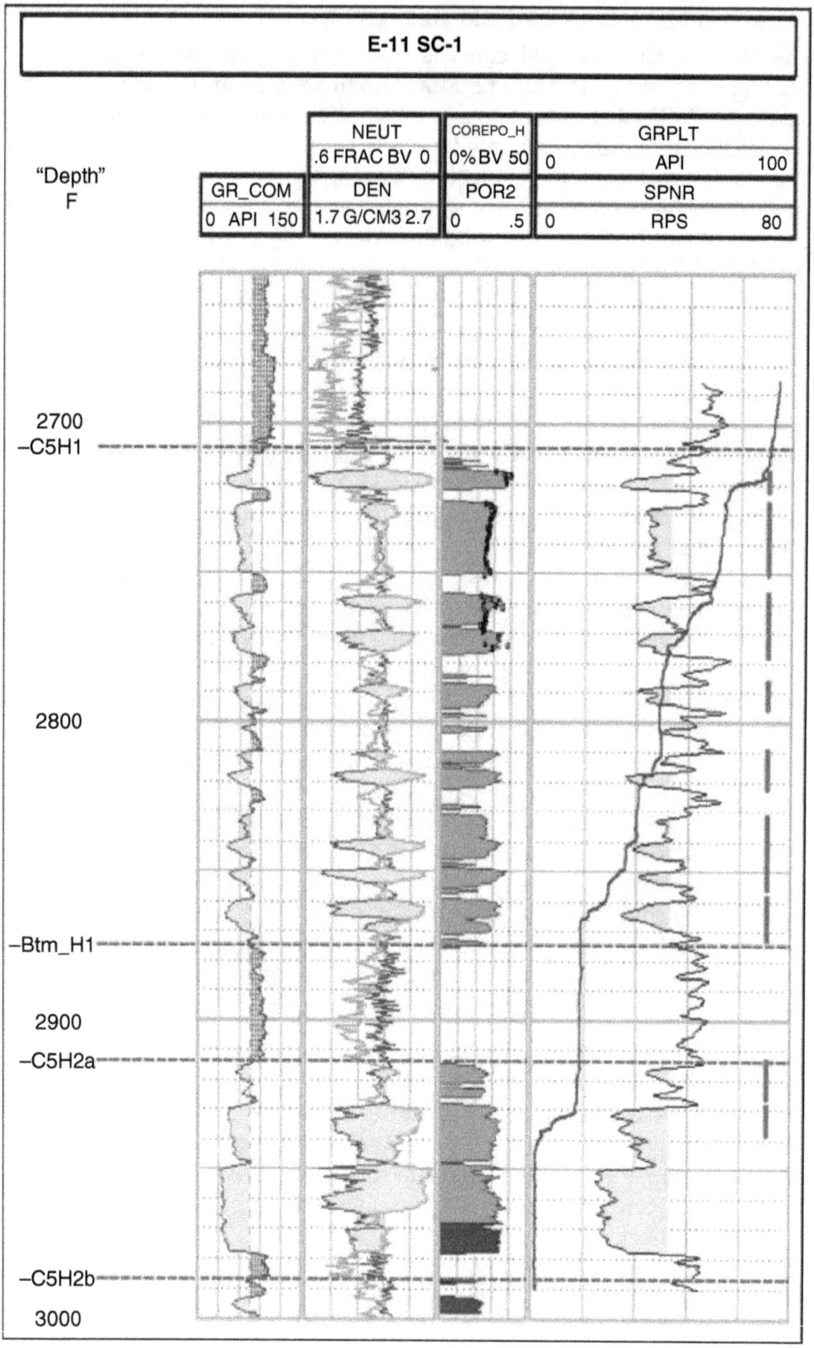

FIGURE 8.13 Log section through the Shallow Clastics Pay (Hadfield et al., 2007).

gas wells, by nature of the associated high-flow velocities, also have a low tolerance to solids production. Thus, effective downhole sand control is vital to the success of the development.

Before completing the E11-SC2 well, the development plan had consisted of highly deviated (70°) well with openhole, expandable sand screens penetrating the H1 and H2 reservoirs. However, in view of the results from the E11-SC2 well cleanup and test, which produced fines to surface, a re-evaluation of the field development plan and well concepts was initiated. The key challenges were to maintain planned well productivity (a minimum of 60 million scf/d) with minimum solids production and ensure effective drainage of the reservoir.

8.5.3 New Sand Control Completion Design

Analysis of the reservoir core sample indicated a smaller-than-expected sand-grain size (D50 80 to 220 μm). The uniformity coefficient (D40/D90), and indicator of formation-grain-size uniformity, varied between 1.5 and 34, indicating that the formation grain size is highly nonuniform (Fig. 8.14).

Defining fines are particles smaller than 44 μm, fines content in this formation can reach 34%, which is considered very high for attempting to effectively produce with minimum- to no-solids production. Fines movement in the formation is influenced mainly by flow velocity and associated drawdown. A means to minimize the production-flow velocity is by wellbore inflow area enlargement, either through fracturing or by drilling long, highly deviated, or horizontal openholes through the pay.

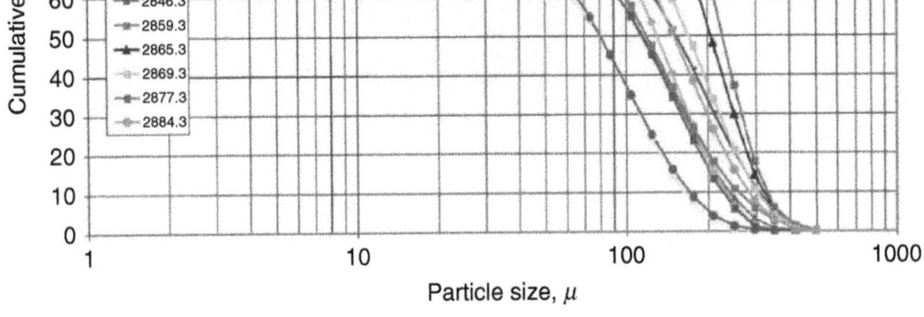

FIGURE 8.14 Laser particle analysis at various depths, H1 reservoir (Hadfield et al., 2007).

A full evaluation of sand control methods applicable to the field was performed including expandable screens, alpha/beta horizontal openhole gravel pack, alternate path horizontal openhole gravel pack and frac pack. Factors considered were installation record, suitability for highly nonuniform sand-size distribution, productivity, and life-cycle cost, limitations on well path and the effect on reservoir drainage. Based on the extensive selection evaluation, alternate-path horizontal openhole gravel pack is selected.

The final sand control completion design adopted comprised highly deviated (80° inclination), 8½-in boreholes through the pay completed with an openhole gravel pack. To minimize the risk of an incomplete gravel pack in the presence of shale throughout the pay, a multiple-path concentric-annuli packing system was incorporated into the design (Fig. 8.15).

The multiple-path concentric annuli openhole gravel pack system consisted of 4½ in, 12.6 lb/ft wire-wrapped screen and blank gravel pack assembly inside a 7-in, 17-lb/ft specially perforated shroud.

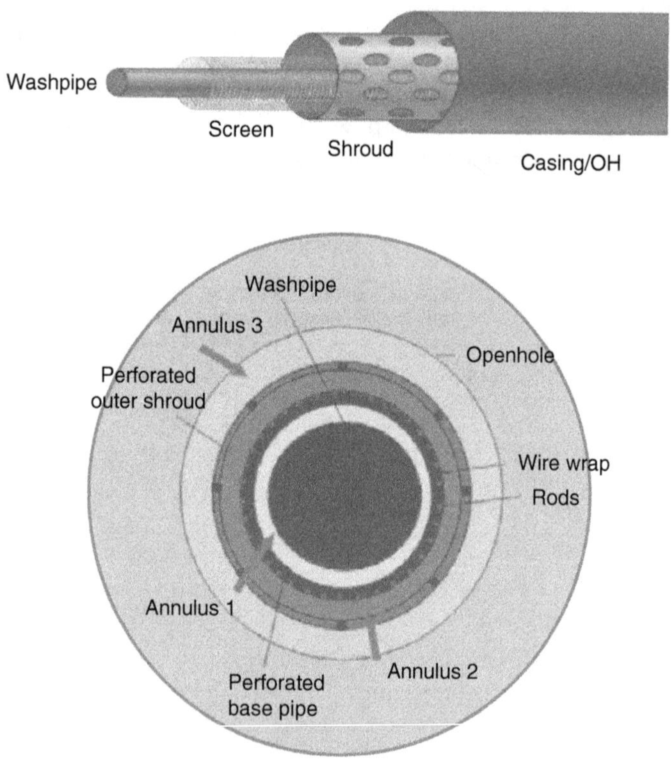

FIGURE 8.15 Concentric-annuli-packing system schematic (Hadfield et al., 2007).

Metallurgy selected for the gravel pack assembly was 13 Cr alloy to mitigate corrosion from the small amount of carbon dioxide (CO_2) expected in production. The perforated shroud, being expendable after the gravel pack was placed, was made of L-80 high-alloy steel.

Sand selection for gravel pack uses Saucier's criteria. For poorly sorted sands with a high level of fines, a gravel/sand D50 size ratio of 5 is recommended. Thus a 30/50-mesh gravel was selected. Gravel pack screen-slot size was selected so that the 30/50 gravel pack sand is retained. Therefore, an 8-gauge opening was recommended. The upper completion design consisted of a 7-in anchor latch that locates in the gravel pack packer, a disappearing plug, an autofill device, and a landing nipple below a production packer. A permanent pressure-gauge assembly, 7-in tubing, 7-in subsurface safety valve, and 7-in tubing completes the string above the production packer to the wellhead. All completion assembly components were 13 Cr steel. Figure 8.16 illustrates the final completion design.

During gravel packing, flow is split between the three annuli: with the gravel slurry transported along the outer two annuli, openhole/shroud, and shroud/screen. The sand-free carrier fluid is transported along the screen/wash pipe annulus. With no annular flow restriction, the outer two annuli will act as one annulus. When an annular blockage occurs in one annulus, the fluid flowing in that annulus is automatically rerouted to the other annulus, causing an increase in flow rate and, thus, an increase in pressure in those annuli. This increase in pressure in the open annuli will force the fluid back into the diverted annulus once the fluid is past the bridge and pressure tends to fall to the lowest point. The fluid flow in the respective annuli will automatically reapportion itself, and the alpha-beta packing continues to the toe.

8.5.4 Well Cleanup and Testing Results

Each of the wells was cleaned up and tested through a temporary test facility upon completion. The objectives were to confirm performance of the sand control design for solids production control and to measure well productivity. The test package consisted of a sand filter, a choke manifold, and inline erosion probes. Because of the dry nature of the Shallow Clastics gas, a separator was not used, and rate measurements were based on the choke equation. For the first well, a conservative bean-up approach was used to allow bridging of particles in the gravel pack to occur. Figure 8.17 shows the solids production as a function of choke size as the well was beaned up to a final gas-production rate of 60 million scf/d. Flush sand production after each bean change can clearly be seen tailing off as the particles were retained by the gravel pack. Final fines production at 60 million scf/d was less than 0.05 g/s, indicating that the rate limitation of 60 million scf/d imposed by erosional-velocity constraints could be reviewed further, depending on future production response.

Well Number	Open Hole Length (ft)	Depth MD (ft)	Depth TVD (ft)
E-11 SC-3	892	4966-5858	2780-2941
F13W SC-1	1104	4581-5685	2796-2931
F13W SC-2	831	4661-5492	2808-2948
F13W SC-3	1083	3632-4715	2791-2934

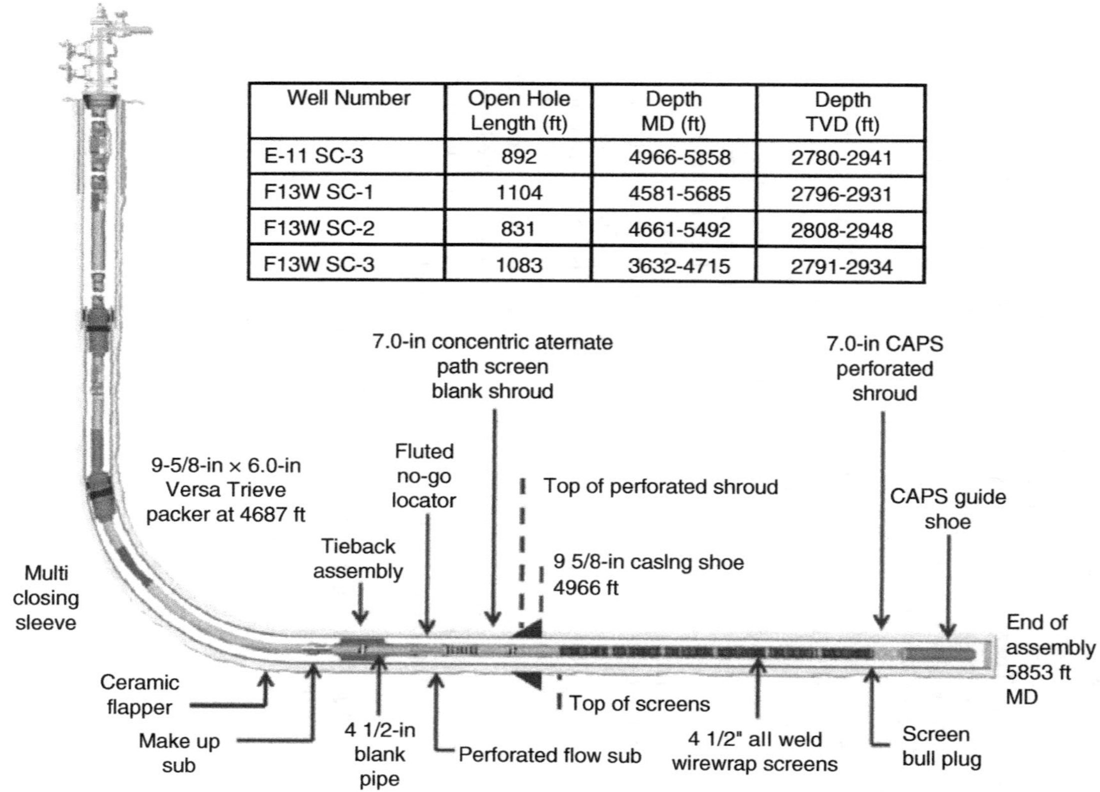

Figure 8.16 Final completion design (Hadfield et al., 2007).

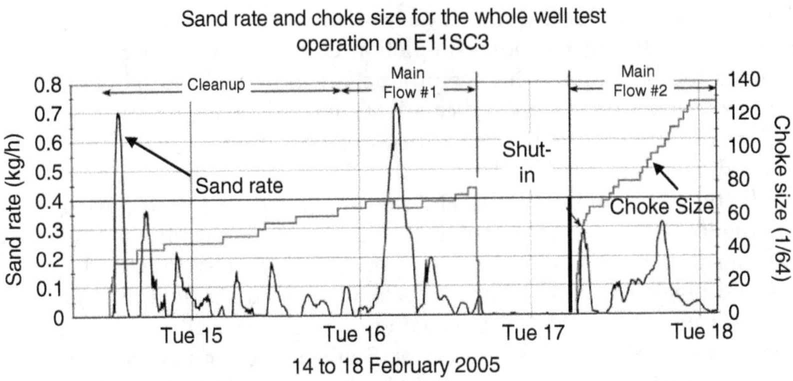

FIGURE 8.17 Sand rate measurements during post-gravel pack well cleanup and flow tests (Hadfield et al., 2007).

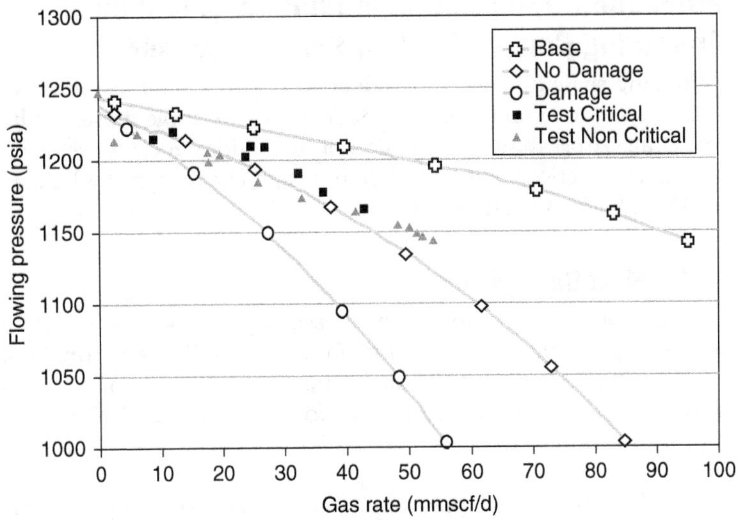

FIGURE 8.18 Inflow performance relationship and well test results of the new gravel pack system (Hadfield et al., 2007).

Figure 8.18 shows the test results at the perforations compared to predicted inflow performance relationships (IPRs). A model for expected well inflow performance was developed assuming a skin of +10 as the base case. Comparison of model predictions and well test results indicated better-than-expected inflow performance.

8.5.5 Value Realized

The fact that no sand production was observed during post-completion cleanup and testing proved the sand control concept for this environment. Fines production was minimal and stabilized to a low value during the post-completion cleanup and testing. This allowed the potential of higher production to be realized. Higher-than-expected production indicated minimal damage to the pay and good filter-cake removal. Post-completion cleanup and testing showed a low completion skin less than 5. There was minimal non-productive time, and no health and safety incidents occurred during completion execution. Time savings from using a concentric-annuli packing system, a single-trip gravel pack and treat system, and a delayed break filter-cake-removal system provided rig-time savings of about one day per well. Hole conditioning was well worth the extra time it took to circulate and swap fluids before running the concentric annuli pack system bottomhole assembly, which was run to the true depth without incident.

8.6 Production Optimization in Multistage Hydraulic Fracturing Completion in a Shale Reservoir

Optimizing shale oil and gas production is one of the most interesting topics in the industry today. The field case example presented by West et al. (2014) shows production optimization processes through drilling, completion, and stimulation design changes in the Bakken and Three Forks formations.

8.6.1 Field Background

The Williston basin portion of the Mississippian Lodgepole formation directly overlies the Mississippian/Devonian Bakken formation. The Bakken overlies the Devonian Three Forks formation, which is above the Devonian Birdbear or Nisku formation. Figure 8.19 shows a cross-section of the Bakken and Three Forks formations and includes boundary layers.

The Bakken is an Upper Devonian/Lower Mississippian formation that contains three members: the lower shale member, the middle sandstone member, and the upper shale member, the middle sandstone member, and the upper shale member. Both upper and lower Bakken members are organic rich marine shales of fairly consistent lithology formed under marine transgressional environments; they are petroleum source rocks and part of the continuous reservoir for hydrocarbons produced from the Bakken formation. The middle sandstone member varies in thickness, lithology, and petrophysical properties, and the local development of matrix porosity enhances oil production in both continuous and conventional Bakken reservoirs. The upper and lower Bakken shale members are also source rocks for oil produced from the overlying Lodgepole formation. The Lodgepole

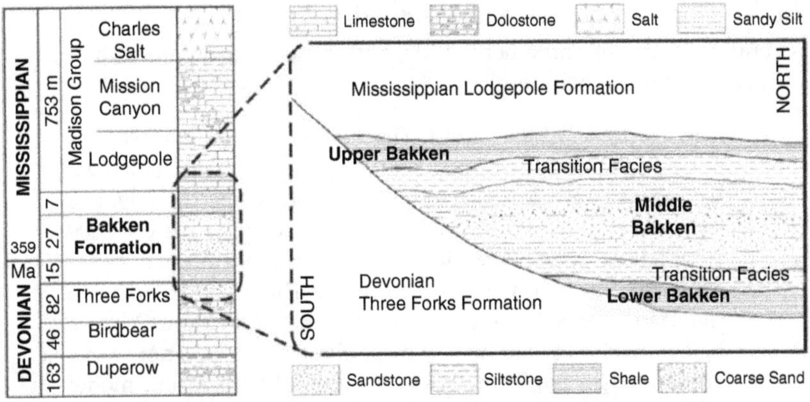

FIGURE 8.19 A cross section of the formations including the Lodgepole, Upper Bakken, Middle Bakken, Lower Bakken, and Three Forks (West et al., 2014).

formation, the lower portion of which is sometimes referred to as a "False Bakken," is an approximately 900-ft-thick Lower Mississippian limestone consisting of five major lithofacies representing a marine transgression. Below the Bakken lie the silty, shaley dolarentite Devonian Three Forks/Sanish formations.

8.6.2 Completion Challenges

Before 2009, in the South Antelope Field, Helis Oil and Gas Company wells landed six of nine wells in the Middle Bakken and, three of the nine wells in the Three Forks. The Middle Bakken wells' production responses were disappointing: Estimated Ultimate Recovery (EUR) is on the order of 300,000 Barrel of Oil Equivalent (BOE). The completion of these wells used various methods including multilaterals, openhole completions, short laterals with preperforated liners, single-stage, multiported subs with external-casing-packer isolation, multistage sliding sleeve with external-casing-packer isolation, and long stage intervals.

Wells with laterals treated in a single stage (preperforated liner with bio-balls and/or ported sub liners) had lower productivity than wells treated with multiple, isolated stages (sliding sleeves with external casing packers). Thus the potential for bypassing or not simulating zones was higher for completions with fewer stages. Further, many of the early completions were compromised by premature screenout and stages aborted as a result of excessive treating pressures. Thus, again, large sections of the lateral and reservoir were bypassed or understimulated. Wells in which the entire lateral was effectively stimulated and proppant was distributed into the far field had significantly higher production responses.

8.6.3 New Completion and Stimulation Design

Several changes are implemented to a new drilling, completion, and stimulation design including:

- *Alter horizontal landing target:* The vast majority of completions targeted the Middle Bakken, with the Three Forks considered a secondary target. In South Antelope, six of the nine considered a secondary target. In South Antelope, six of the nine pre-2009 completions were landed in the Middle Bakken interval. However, three of nine were landed in the Three Forks and were found to have better cumulative production than the six Middle Bakken wells. On the basis of these observations, the new wells are landed in the Three Forks. A hydraulic fracturing model indicated hydraulic fractures would start in the Three Forks and then grow vertically up into the Middle Bakken.
- *Increase wellbore lateral lengths:* In the new well design, the lateral length was doubled from approximately 4500 ft to approximately 9000 ft.
- *Improve formation contact per stage:* In addition, the stage length was reduced for each treatment. The objectives were to improve contact with the formation and increase the SRV. Because of wellbore-construction limits, stages in pre-2009 completions were generally approximately 1000 ft per stage. With advancements in materials and equipment, the stage length was reduced to 325 ft where each stage consists of three perforation clusters spaced approximately 110 ft apart. Although previous completion designs employed mainly sliding sleeves, the plug-and-perforation technique was selected because of more seismic activity observed from one of their microseismic survey study.
- *Refine pumpdown operations:* To reduce the risk of plugs stuck in the production liner, the fracture plug was changed from 3.6-in OD to a 3.4-in OD.
- *Implement critical-fracturing-mechanism diagnosis:* Diagnostic fracture injection tests (DFITs) and step-down rate tests (SRTs) were recommended to the completion program to help calibrate the fracture simulator and to design the most-effective treatments. The DFITs yield formation-modeling properties such as fracture-closure pressure, fracture gradient, fluid-leakoff coefficient, fluid efficiency, formation permeability, and reservoir pressure. The SRTs yield localized near-wellbore and perforation friction estimations that one can use to optimize individual treatments. In situations of high near-wellbore and perforation friction, more cleanup is pumped to reduce and minimize the risk of near-wellbore screenout. On-site, real-time

engineers make adjustments including changing pad volume, the gel loading, and the proppant concentration, and adding proppant slugs. Pre-2009 stimulation treatments had encountered premature screenout in more than 30% of the jobs. With the new change mentioned above, the screenout ratio becomes less than 2% in approximately 900 stages of the post-2009 completions.

- *Incorporate higher-strength proppant:* Proppant-pack modeling at the presumed closure pressure indicated that silica sand did not provide sufficient crush strength. The team recommended switching to at least 85% ceramic proppant to provide sufficient conductivity and crush resistance. Also to ensure high proppant concentration near the wellbore, increased proppant concentrations between 10 lbm/gal added (ppa) and 17 ppa (proppant spikes) was pumped at the end of the last sand stage (Fig. 8.20).
- *Refine flush procedure:* The key to a successful spike is the very abrupt transition from maximum proppant concentration to a clean flush. Blenders with tub bypass and conventional screw metering or blenders with proppant metering gates that open completely near the end of the last proppant concentration stage are required. It is extremely important that little or no sand be left in the wellbore after shutting down. If left, sand increases the risk of a subsequent gun and plug assembly becoming stuck.

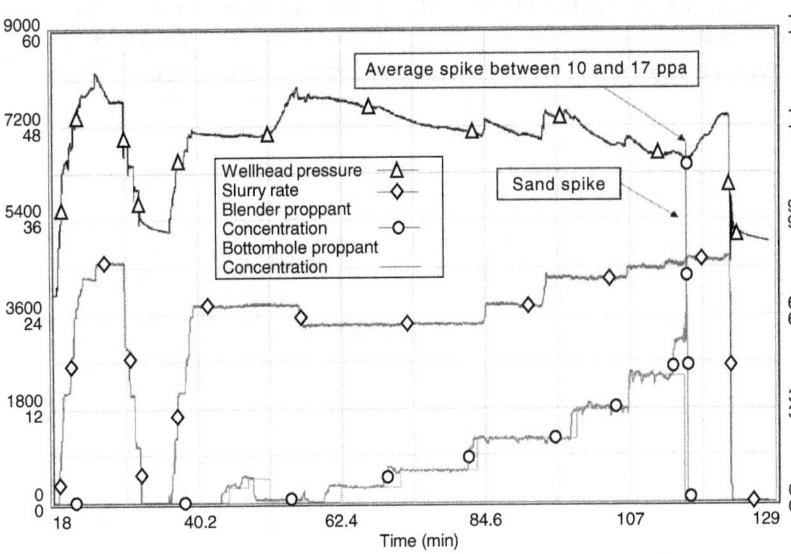

FIGURE 8.20 Treatment data showing proppant spike on the last sand-concentration stage and quick transition to flush (West et al., 2014).

- *Adjust treatment-fluid design:* Before 2009, a cross linked-fluid system of 40-lbm base gel loading was used as a primary fluid system. In the new fluid design, the gel loading was reduced to 25 lbm base gel which would provide sufficient viscosity and thermal stability to carry the proppant into the formation, while reducing formation damage.
- *Implement a controlled-flowback policy:* After the fracture treatments were completed, each well was backflowed conservatively. Restricting flowback rates, to maintain higher flowing pressure, prevents early breakout of gas within the formation by keeping the flowing bottomhole pressure above bubblepoint for a longer time.

8.6.4 Production Results

Compared with pre-2009 completions, wells completed after 2009 produced 50% to 75% more hydrocarbone. These percentages were calculated after normalizing for variations in lateral length and amount of sand pumped. The South Antelope post-2009 wells are some of the best producing wells in North Dakota. Figure 8.21 shows the annual increase in production between 2007 and 2011. The figure shows the industry improvement as completion methods were refined. Also included in the figure are the first six post-2009 wells for comparison. The figure illustrates the significant improvement from implementing the design changes in completion and stimulation discussed in the previous section.

Figure 8.22 shows the 30-, 90-, 180-, and 360-day cumulative production of offset Middle Bakken and Three Fork wells in a 30-Township area surrounding the South Antelope property compared with the post-2009

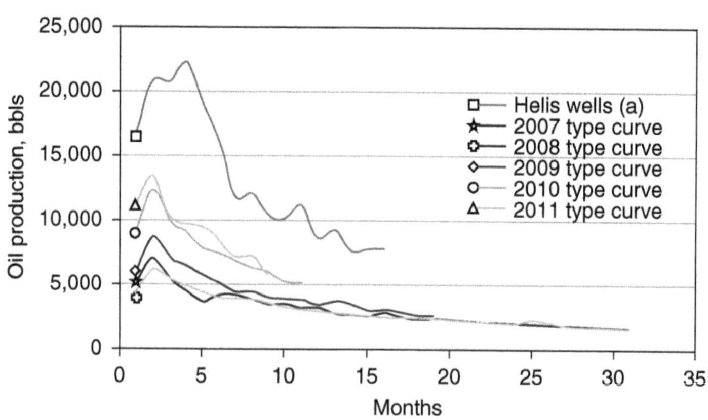

Figure 8.21 Typical production curves comparing first six case study wells with historical Bakken type curves from 2007 to 2011 (West et al., 2014).

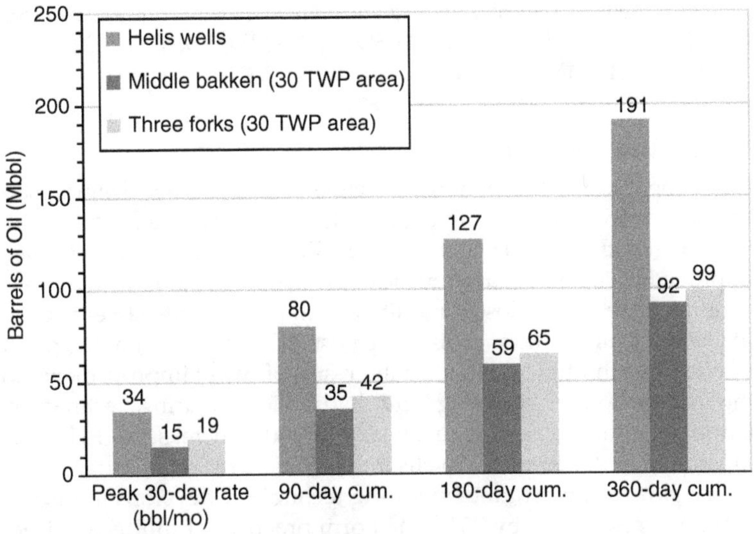

FIGURE 8.22 Post-2009 wells (labeled by Helis Wells) outperforming Bakken and Three Forks in 30-Township Area (West et al., 2014).

South Antelope wells. The findings show that the post-2009 South Antelope wells outperformed offset wells nearly twofold.

The same data are parsed by operator to help distinguish variations in completion effectiveness and reservoir rock properties. All comparison offset wells are post-2011 completions and illustrate the significant improvement achieved in this project. The cumulative production after 6 months for the South Antelope wells is 29% higher than the next-best performance and 300% better than the poorest performance (Fig. 8.23).

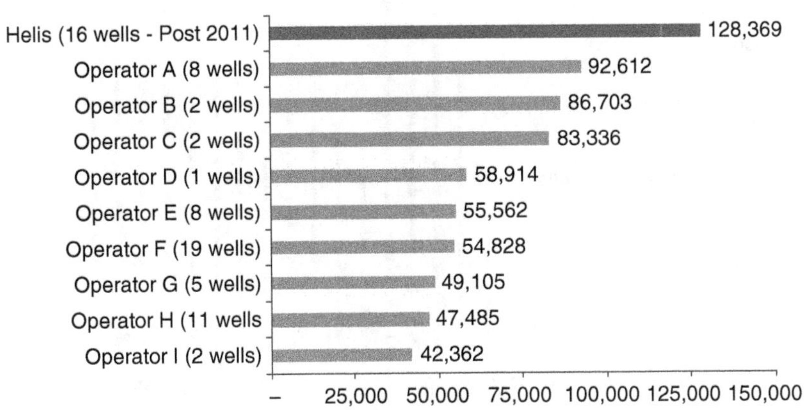

FIGURE 8.23 Post-2011, 6-month average cumulative production in bbl per well (West et al., 2014).

With the new completion and stimulation design implementation, EUR has improved from 300,000 BOE to 1,219,000 BOE (1,013,000 bbl of oil and 1241 MMscf of gas).

8.6.5 Value Realized

Adjusting the horizontal target, wellbore length, stage length, perforation spacing, and completion method improved the amount of SRV within the producing formation. Pumpdown procedures were improved, formation characteristics were assessed and modeled, pressure responses were closely monitored, and design adjustments were made in real time. The success of the post-2009 wells was not the result of one breakthrough, but rather the result of many improvements to the completions-engineering and hydraulic fracturing techniques. Ultimately, these changes resulted in a 400% increase in EURs, as illustrated in the post-2009 wells in Fig. 8.24.

As how fast hydraulic fracturing technology develops in unconventional reservoirs, by 2017, the norm practices of multistage fracturing in shale plays feature much higher volume, smaller proppant size (up to or even smaller than 100 mesh), smaller space length (on the order of 100 ft), and smaller cluster spacing (on the order of 15 to 20 ft). With these new trends in completion and stimulation, well performance improved dramatically in unconventional wells (Rassenfoss, 2017).

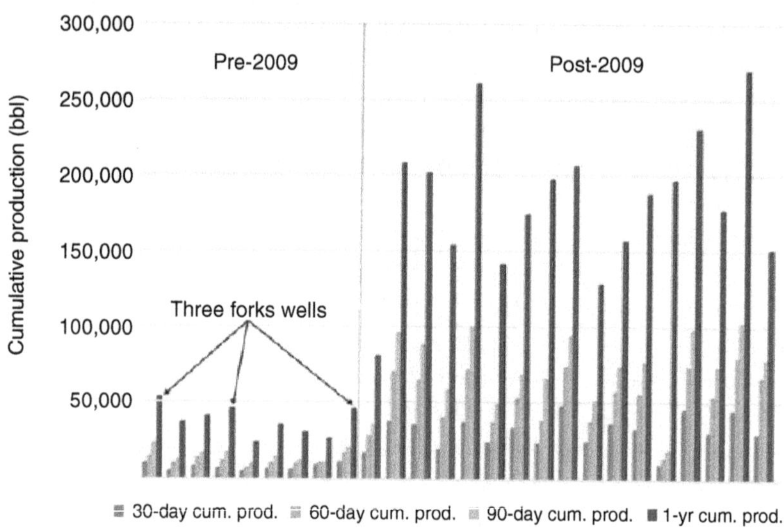

FIGURE 8.24 Cumulative production before and after completion changes (West et al., 2014).

References

Al-Mubarak, S. M., Pham, T. R., Shamrani, S. S., and Shafiq, M. (2008, November 1). "Case Study: The Use of Downhole Control Valves to Sustain Oil Production from the First Maximum Reservoir Contact, Multilateral, and Smart Completion Well in Ghawar Field," *Society of Petroleum Engineers*. doi:10.2118/120744-PA.

Burgos, G. A., Buijse, M. A., Fonseca, E., Milne, A., Brady, M. E., and Olvera, R. (2005, January 1). "Acid Fracturing in Lake Maracaibo: How Continuous Improvements Kept on Raising the Expectation Bar," *Society of Petroleum Engineers*. doi:10.2118/96531-MS.

Chowdhuri, S., Cameron, P., Gawwad, T. A., Madar, M. R., Sharma, S. S., AlMutairi, M. D., … Al-Ajmi, M. F. (2015, October 11). "Optimize Development Strategy of Upper Burgan Reservoir Through Multilateral Well with Inflow Control Device in Raudhatain Field, North Kuwait," *Society of Petroleum Engineers*. doi:10.2118/175227-MS.

Eyvazzadeh, R. Y., Cheshire, S. G., Nasser, R. H., and Kersey, D. G. (2003, January 1). "Optimizing Petrophysics: The Ghawar Field, Saudi Arabia," *Society of Petroleum Engineers*. doi:10.2118/81477-MS.

Galarraga, M. A., and Hansen, B. (2005, January 1). "Detailed 3D-Seismic Interpretation Using HFI Seismic Data, Fault Throw, and Stress Analysis for Fault Reactivation in the Cogollo Group, Lower Cretaceous, Urdaneta West Field, Maracaibo Basin," *Society of Petroleum Engineers*. doi:10.2118/95060-MS.

Hadfield, N. S., Terwogt, J. H., Van Kranenburg, A., Salahudin, S., and King, K. A. (2007, September 1). "Sandface Completion for a Shallow Laminated Gas Pay with High Fines Content," *Society of Petroleum Engineers*. doi:10.2118/111635-PA.

Kent, A. W., Burkhead, D. W., Burton, R. C., Furui, K., Actis, S. C., Bjornen, K., … Zhang, T. (2014, June 1). "Intelligent Completion Inside Uncemented Liner for Selective High-Rate Carbonate Matrix Acidizing," *Society of Petroleum Engineers*. doi:10.2118/166209-PA.

Rassenfoss, S. (2017, February 1). "Rebound to Test if Cost Cuts Will Last," *Society of Petroleum Engineers*. doi:10.2118/0217-0035-JPT.

West, D. R. M., Harkrider, J., Besler, M. R., Barham, M., & Mahrer, K. (2014, September 1). "Optimized Production in the Bakken Shale: South Antelope Case Study," *Society of Petroleum Engineers*. doi:10.2118/167168-PA.

Index

A

B

C

D

W

Y

Z

www.ingramcontent.com/pod-product-compliance
Lightning Source LLC
LaVergne TN
LVHW020418200725
816472LV00001B/12